POWDER METALLURGY STAINLESS STEELS

Processing, Microstructures, and Properties

ERHARD KLAR
PRASAN K. SAMAL

**The Materials
Information Society**

ASM International®
Materials Park, Ohio 44073-0002
www.asminternational.org

First printing, June 2007

Comments, criticisms, and suggestions are invited, and should be forwarded to ASM International.

Prepared under the direction of the ASM International Technical Book Committee (2006–2007), James C. Foley, Chair.

ASM International staff who worked on this project include Scott Henry, Senior Manager of Product and Service Development; Steven R. Lampman, Technical Editor; Diane Grubbs, Editorial Assistant; Bonnie Sanders, Manager of Production; Madrid Tramble, Senior Production Coordinator; Patti Conti, Production Coordinator; and Kathryn Muldoon, Production Assistant

Library of Congress Cataloging-in-Publication Data
Klar, Erhard.
Powder metallurgy stainless steels / Erhard Klar and Prasan K. Samal.
p. cm.
Includes bibliographical references and index.
ISBN-13: 978-0-87170-848-9
ISBN-10: 0-87170-848-5
1. Powder metallurgy. 2. Steel, Stainless—Metallurgy—Industrial applications. I. Samal, Prasan K. II. ASM International. III. Title.

TN697.I7K57 2007
672.37—dc22   2007060659
SAN: 204-7586

ASM International®
Materials Park, OH 44073-0002
www.asminternational.org

Printed in the United States of America

Contents

Preface

The treatment of sintered stainless steels in this book addresses the need to more clearly understand the many factors that affect the corrosion resistance of powder metallurgy (PM) stainless steels. For over a half-century, PM technology has been an effective method of net shape processing to produce structural parts for the automotive and other industries. Conflicting literature on the factors that influence the corrosion resistance of PM stainless steels has led to some widespread misconceptions, which generally attribute poor corrosion resistance to just the presence of pores (whereby the presence of pores increases the effective surface area of a sintered part and thereby increases the corrosion rate in the passive region). A crevice-sensitive density region does exist in which a neutral chloride environment can give rise to crevice corrosion. Nonetheless, many or most cases of underperformance cited can be traced to inappropriate sintering practices that result in poor metallurgical soundness.

Recent progress in the understanding of corrosion and corrosion-resistance properties of sintered stainless steels has led to renewed interest in their application in the automotive sector for the benefit of net-shape processing and more efficient materials utilization. To obtain good corrosion-resistance properties, sintered stainless steels require careful processing, starting with powder selection, avoidance of contamination, efficient delubrication, and through to controlled sintering and cooling. There are several distinct, process-related corrosion issues with sintered stainless steels that the PM industry had to cope with over the years:

- Contamination with less noble constituents, causing galvanic corrosion
- Crevice-corrosion-prone density range in neutral saline environments
- Excessive carbon content (from various sources), causing sensitization and intergranular corrosion
- Excessive nitrogen content, due to sintering in H_2–N_2 mixtures containing large amounts of nitrogen (i.e., dissociated ammonia), in combination with slow cooling rates produces sensitization and intergranular corrosion
- Inadequate cooling after sintering, which, in the presence of excessive carbon, can cause sensitization and intergranular corrosion
- Excessive dewpoints and/or inadequate cooling causes reoxidation during cooling and susceptibility to pitting
- Surface chromium losses due to sintering in a vacuum furnace can impair the corrosion resistance of sintered stainless steels
- Pitting corrosion due to incomplete reduction of original residual oxides.

Solutions to all of these problems are at various stages of implementation in the industry. Because of this, and the fact that corrosion resistance is usually the prime property when stainless steel is selected as a material, the subject of PM stainless steel processing, and, more specifically, of optimal processing, from powder to final part pervades the entire book.

For stainless steel parts manufacturers, this book serves as a guide to making parts that possess improved corrosion-resistance properties, thereby opening new market opportunities. Although some of the aforementioned problems can also be present in wrought and cast stainless steels, some are specific to PM, and all have special PM processing-related characteristics. The general approach is first to present the phenomenological aspects of a subject, including problem areas. This is followed by a description of its underlying principles and then a discussion and illustration of available solutions. The perspectives taken are often those of a powder producer, reflecting the authors' affiliation. Significant portions of the data are from Professor Maahn and coworkers of the Technical University of Denmark, who, in a three-year effort (1990 to 1993) in cooperation with industry, made important contributions to this subject.

The structure of the book more or less follows the sequence of the production process. After a brief historical background, the chapters include metallurgical background and alloy compositions, powder manufacture and properties of powders, compaction and shaping, sintering and corrosion, optimal sintering and surface modification, with concluding chapters on mechanical and magnetic properties, corrosion-resistance testing and properties, secondary operations, and applications. Emphasis is concentrated on the press-and-sinter technology of PM, although some consideration is given to metal injection molding, powder extrusion, and hot isostatic pressing. The discussion of optimal sintering in Chapter 6, "Alloying Elements, Optimal Sintering, and Surface Modification in PM Stainless Steels," although based largely on press-and-sinter technology, is relevant, with appropriate restrictions, to other modes of PM shaping and consolidating.

Introductory books on PM and corrosion science provide a useful basis for this text, because the reader is assumed to possess a basic knowledge of metallurgy, powder metallurgy, and corrosion science. Suggested references are *Powder Metallurgy Science* by R.M. German (Ref 1); *Powder Metal Technologies and Applications*, Volume 7, *ASM Handbook*, 1998 (Ref 2); *Corrosion Engineering* by M.G. Fontana (Ref 3); *Corrosion and Corrosion Control* by H.H. Uhlig and R.W. Revie (Ref 4); and *Corrosion: Fundamentals, Testing, and Protection*, Volume 13A, *ASM Handbook*, 2003 (Ref 5). Standards on metal corrosion are found in Volume 3.02 of the *Annual Book of ASTM Standards*. Nevertheless, the authors have attempted to keep the text simple and to facilitate understanding through the use of numerous pictures, illustrations, and references. A brief glossary of definitions of powder metallurgy and corrosion terms is shown in an Appendix, with more complete versions available in ASTM standard B-243 (Ref 6) and in the aforementioned *ASM Handbook* on corrosion.

In this context, it is hoped that this work provides a contribution to the more effective processing of sintered stainless steels to achieve improved corrosion resistance and successful applications in more demanding environments. Evidence for this is presented, and the authors believe that it will be only a matter of time until the versatility of the PM process closes any gaps that still exist with wrought or cast forms. As the industry implements the solutions to the previously mentioned problems, knowledge from wrought and cast stainless steel technology can be used and applied more effectively to PM stainless steels. This then should develop into a more comprehensive use and representation of PM stainless steels within the overall field of metals technology.

Erhard Klar
Prasan K. Samal

References

1. R.M. German, *Powder Metallurgy Science*, Metal Powder Industries Federation, 1994
2. *Powder Metal Technologies and Applications,* Vol 7, *ASM Handbook,* ASM International, 1998
3. M.G. Fontana, *Corrosion Engineering*, McGraw-Hill, Inc., 1986
4. H.H. Uhlig and R.W. Revie, *Corrosion and Corrosion Control*, John Wiley & Sons, 1985
5. *Corrosion: Fundamentals, Testing, and Protection,* Vol 13A, *ASM Handbook*, ASM International, 2003
6. "Standard Terminology of Powder Metallurgy Terms," B-243, ASTM International

Acknowledgments

Some of the work reported in this book was performed in the research laboratories of SCM Metal Products/OMG Americas (currently North American Hoganas) during the authors' affiliation with those companies. We would like to thank all those involved in this work as well as for general support and permission to publish the data. We also are grateful to Professor Randall M. German, Dr. Chaman Lall, and Professor Alan Lawley for reviewing the manuscript and providing valuable suggestions to improve the book. The support and the encouragement of the Publications Managers of ASM International, namely Scott Henry and Steven Lampman, are greatly appreciated. We also thank a number of our friends in the PM industry who have provided assistance in the forms of technical information and review of selected chapters of the manuscript. They include James P. Adams, Professor Paul Beiss, David F. Berry, Peter dePoutiloff, Dr. Olle Grinder, Jack A. Hamill, Jr., Dr. Kishor M. Kulkarni, Suresh O. Shah, Howard I. Sanderow, and Maryann Wright. The authors express their gratitude to Metal Powder Industries Federation (MPIF) for permission to use various copyrighted figures and data. The authors would also like to acknowledge the significant efforts and contributions of our present and former colleagues. Most prominent among them are Harry Ambs, Ingrid Hauer, J. Patrick Hughes, Richard Ijeoma, Owe Mars, Samir Nasser, George Novak, Mary Pao, David Ro, Ronald Solomon, Mark Svilar, and Joseph Terrell.

About the Authors

Erhard Klar studied at the University of Tuebingen and the Technical University of Berlin where he received his Ph.D. in physical chemistry. This was followed by postdoctoral studies at the University of Pittsburgh. Dr. Klar's work on the powder metallurgy of stainless steels and other materials was conducted at the Metals Group of SCM Corporation, where he was the Director of Research. Dr. Klar is now retired.

Prasan K. Samal received his B. Tech. in Metallurgy from the Indian Institute of Technology-Madras, and his M.S. and Ph.D. in Materials Engineering from the University of Maryland-College Park and Case Western Reserve University, respectively. He started his career with Kennecott Copper Corporation, then joined the Metals Group of SCM Corporation, which later became a part of OMG Americas, and subsequently acquired by North American Hoganas. Dr. Samal holds ten U.S. patents and has published more than forty technical papers.

CHAPTER 1

Introduction

POWDER METALLURGY (PM) COMPO-NENTS produced from corrosion-resistant alloys are a growing area of PM application, of which stainless steel PM alloys span a variety of industries, including aerospace, automotive, chemical processing, medical, and recreational. Recent progress has also led to the understanding that proper processing and sintering of PM stainless steel are critically important factors in achieving corrosion resistance for increasingly demanding applications. In fact, many (if not most) cases of underperforming PM stainless steel parts in terms of corrosion resistance can be traced to metallurgical defects due to improper processing. Therefore, improved understanding of the PM stainless steel processing factors can have important results in terms of corrosion resistance and the extended use of PM technology for its well-known economic value in terms of net shape processing and more efficient material utilization in a large number of applications for the automotive and other industries.

1.1 Historical Background

In North America, laboratory and small-scale exploration of PM stainless steels in the 1930s and 1940s led to their commercial production in the late 1940s. Initially, stainless steel compositions were simply copied from known wrought stainless steels, and mixtures of elemental powders of iron, chromium, and nickel were pressed and sintered in dry hydrogen (Ref 1, 2). High sintering temperatures and uneconomically long sintering times were required to achieve full homogenization of the microstructure. The so-called sensitization method made use of sensitized stainless steel sheet, that is, stainless steel with grain boundaries depleted of chromium.

When leached in acid, such sheet disintegrated into a fine powder consisting essentially of the individual grains of the stainless steel sheet (Ref 3–5). This powder actually had several promising properties and was commercially produced in the 1940s. Its compacting properties, however, were marginal.

In the late 1940s, Vanadium Alloys Steel Company began to use water atomization for making alloyed stainless steel powder. In spite of the initially high oxygen contents of these powders, they had adequate green strength and could be sintered in reasonable times. Subtle but critical modifications to wrought stainless steel compositions, as well as improvements in the atomization process itself, led to much improved powders. Even though the corrosion resistance of sintered parts was still low, it was sufficient for applications requiring only moderate corrosion resistance. The first such large-volume application was the rear-view mirror bracket in passenger cars. One of the reasons for using stainless steel in this application is the requirement that the material must match the coefficient of thermal expansion of the windshield glass.

In the early years of commercial use of sintered stainless steels, emphasis was placed on improving the compacting properties, compressibility, and green strength of stainless steel powders. Compressibility was even more important for stainless steel powders than it was for iron powders, because of the higher hardness of the former and therefore the high compacting pressures required to obtain useful green densities.

Stainless steel powder shipments (Fig. 1.1) illustrate the evolution and growth of sintered stainless steels in North America.

A low growth rate of approximately 5% in the 1970s and 1980s was due in large part to the relatively low corrosion properties of sintered

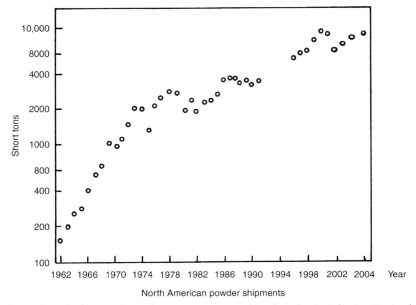

Fig. 1.1 Stainless steel powder shipments for North America. Source: Metal Powder Industries Federation. Reprinted with permission from MPIF, Metal Powder Industries Federation, Princeton, NJ

stainless steels in those years. Contributing to this lower growth were generalized conclusions and statements in the literature about the corrosion resistance of sintered stainless steels that were half-truths at best. In technical seminars presented to powder metallurgists, one of the authors summarized such statements (Table 1.1) to illustrate the problem areas.

These statements and their contradictions capture the insecure and sometimes chaotic state of the industry in those years regarding the corrosion resistance of sintered stainless steels. All of the statements in Table 1.1 reflect problem areas that, at least in part and to a varying degree, are still with us. They are dealt with in detail, including solutions, particularly in the chapters on sintering and corrosion-resistance testing and evaluation.

Table 1.1   Half-truths about the corrosion resistance of sintered stainless steels

Sintering only in hydrogen gives good corrosion resistance.
Vacuum sintering gives the best corrosion resistance.
Vacuum sintering gives poor corrosion resistance because of chromium losses.
Sintering in dissociated ammonia gives poor corrosion resistance because of the formation of chromium nitrides.
Good corrosion resistance cannot be obtained at low (1150 °C, or 2100 °F) sintering temperatures because of the lack of reduction of chromium oxides.
It is possible to obtain good corrosion resistance of parts sintered in laboratory furnaces but not in industrial furnaces.
Sintered stainless steel will always have inferior corrosion resistance because of the presence of pores that give rise to crevice corrosion.

By the mid-to-late 1980s, both stainless steel powder manufacture and the sintering processes for stainless steel parts had been improved sufficiently to qualify for the second large-volume application: antilock brake system sensor rings in cars. This application made increased demands on both corrosion resistance and magnetic properties of (ferritic) stainless steels. As is shown in the chapter on sintering, sintering conditions conducive to good magnetic characteristics are also conducive to good corrosion-resistance characteristics.

With continuing progress, and attendant with an increase in the compound growth rate of stainless steel powders to over 15%, the third automotive application followed in the 1990s: coupling flanges and sensor bosses in car exhaust systems. Requirements, in addition to general and hot exhaust gas corrosion resistance, included elevated-temperature oxidation resistance, resistance to thermal cycling, leak resistance, and weldability, as well as improved room- and elevated-temperature mechanical properties. In some respects, the PM parts were even superior to their wrought counterparts (Chapter 11, "Applications").

Even though sintered stainless steels benefit from the same economic advantages as carbon steel structural parts, namely energy efficient, environmentally acceptable, and nearly scrap-free mass production, it is clear from the aforementioned that the gradual improvement of corrosion properties over a period of 20 to 30 years had been the dominant growth characteristic for this industry. This also emerges from state-of-the-art reviews on the powder metallurgy of stainless steels by Ambs et al. (1977) (Ref 6), Dyke et al. (1983) (Ref 7), and Klar (1987) (Ref 8).

Industrial growth of metal injection molding (MIM) commenced in the 1980s with some aerospace applications. The technology is presently still in its rapid growth phase. Over half of all injection-molded metal parts are stainless steel parts; of these, 316L is the most widely used, followed by ferritic and precipitation-hardened stainless steels. In this regard, injection molding parallels the early history of conventional compacted and sintered stainless steels.

The market for MIM parts was estimated to be $150 million for North America and $360 million globally in 2005 (Ref 9). The MIM stainless steel tonnage was estimated worldwide to be 1450 metric tons in 2005 (Ref 10). Relatively low capital investment cost, in comparison to conventional press-and-sinter technology, as well as further reductions of the cost of MIM-grade powders due to economies of scale and improvements in atomizing technologies are likely to further drive the impressive growth of this technology.

1.2 Present State and Scope

As pointed out in the preface, the PM industry is in the midst of applying and implementing the fundamental requirements for optimizing the corrosion-resistance properties of sintered stainless steels for new applications. At present, several types of corrosion (i.e., stress corrosion, corrosion fatigue, erosion and cavitation corrosion, elevated-temperature oxidation) have been investigated only sporadically for sintered stainless steel PM parts, and the corrosion properties of existing commercial parts are sometimes still not as uniform and consistent as those of wrought stainless steels. Then, there also are cases where sintered, that is, porous stainless steels, exhibit corrosion-resistance properties superior to those of their wrought counterparts. Furthermore, many uses of stainless steels do not require the "full" corrosion resistance of an alloy, although the more severe corrosive applications require that the stainless steel be in the

condition of its best corrosion resistance. For instance, 316L, sintered for 30 min at 1120 °C (2048 °F) in a 90%H_2-10%N_2 atmosphere, has been found to possess critical potentials equal to those of wrought 304 stainless steel (Ref 11). In many applications, the pitting and crevice corrosion behavior, as defined by the critical potentials of a material, is believed to describe the corrosion performance of that material.

In spite of certain limitations of sintered stainless steels on account of their porosity, optimal sintering, as described in the following chapters, will go a long way to produce sintered stainless steel parts that can satisfy many applications. Optimally sintered 317L and SS-100 (20Cr18Ni5Mo), for instance, have shown corrosion resistances in long-term immersion tests in 5% NaCl approaching or equaling those of wrought 316L (Chapter 6, "Alloying Elements, Optimal Sintering, and Surface Modification in PM Stainless Steels"). Exploiting possibilities unique to PM, such as certain kinds of surface modification (Chapter 6) or liquid-phase sintering (Chapter 5, "Sintering and Corrosion Resistance"), further extends the uses of sintered stainless steels. In neutral chloride exposure testing, optimally sintered tin-copper surface-modified stainless steels exhibit corrosion-resistance improvements of an order of magnitude over their unmodified equivalents, in addition to machinability improvements that equal or exceed (the free machinability grade) 303L. Boron-assisted liquid-phase sintering of 316L and of a higher-alloyed austenitic stainless steel (23Cr18Ni3.5Mo0.25B) has demonstrated that corrosion characteristics similar to wrought 316L are possible.

With the implementation of recent insights regarding the control of corrosion-resistance properties, that is, with "optimal" sintering, the authors believe that both sintered, that is, porous stainless steels, as well as nearly fully dense stainless steels will find many new uses. Optimal sintering (Chapter 6) is a recurring theme of this book. For practical reasons, it is defined here as control of processing and sintering that eliminates and avoids all metallurgical defects—with the exception of some residual oxides (i.e., oxides originating in the water atomization process and that, under conditions of commercial sintering, remain partially unreduced)—as well as a crevice-sensitive density region for certain alloys (Chapter 5) exposed to a neutral saline environment. Without such control, progress will be slow and the full potential of sintered stainless steels will be realized later rather than sooner. The role of residual oxygen or oxides in sintered stainless steels was one of the first observed but last addressed. It is still not entirely clear how much and/or in which circumstances residual oxides can be tolerated in regard to corrosion resistance.

The opportunities for surface modification of the porous surfaces of sintered stainless steels, quite amenable to PM processing, have not yet been fully exploited and should, possibly together with the improved control of residual oxides, lead to the elimination of crevice corrosion in neutral saline solutions for the crevice-sensitive density range.

Promising results with transient and persistent liquid-phase sintering of stainless steels will open up the entire density range to applications demanding excellent corrosion resistances, thus more broadly justifying the primary purpose of the use of stainless steels.

More PM-specific corrosion standards and more corrosion data will support these opportunities. In comparison to five families of stainless steels and 158 standard and nonstandard wrought stainless steels listed in the ASM *Metals Handbook Desk Edition*, 2nd ed., 1998, the 2007 MPIF standard 35 lists only 14 sintered stainless steel compositions comprising three families of stainless steels, and three compositions for metal injection molding.

REFERENCES

1. D. Shaw, W.V. Knopp, and B.A. Gruber, *Prec. Met. Mold.*, Vol 11, 1953, p 42–45, 73–76
2. F. Eisenkolb, *Stahl Eisen*, Vol 78, 1958, p 241–248
3. A. Adler, *Mater. Method.*, Vol 41, 1955, p 118–120
4. B. Sugarman, in *Symposium on Powder Metallurgy*, 1954, p 175
5. D.A. Oliver, in *Symposium on Powder Metallurgy*, 1954, p 180
6. H.D. Ambs and A. Stosuy, Chapter 29, The Powder Metallurgy of Stainless Steel, *Handbook of Stainless Steel*, D. Peckner and I.M. Bernstein, Ed., McGraw-Hill, 1977
7. D.L. Dyke and H.D. Ambs, Chapter 5, Stainless Steel Powder Metallurgy, *Powder Metallurgy—Applications, Advantages, and Limitations*, E. Klar, Ed., American Society for Metals, 1983
8. E. Klar, Corrosion of Powder Metallurgy Materials, *Corrosion*, Vol 13, *Metals Handbook*, 9th ed., ASM International, 1987, p 823–845
9. Estimates by American Metal Injection Molding Association
10. Mark Schulz, BASF, private communication
11. T. Mathiesen, "Corrosion Properties of Sintered Stainless Steel," Ph.D. thesis, Institute of Metals, Technical University of Denmark, 1993

CHAPTER 2

Metallurgy and Alloy Compositions

2.1 Introduction

STAINLESS STEELS, as a class of ferrous alloys, are mainly distinguished by their superior resistance to corrosion. They are also recognized for their excellent resistance to oxidation and creep at elevated temperatures. Primarily, corrosion resistance of stainless steels stems from their ability to form an adherent, chromium-rich passive film on the surface. The fact that this characteristic is displayed by alloys of iron that contain a minimum of approximately 10.5% Cr serves to define stainless steels as alloys of iron containing at least this amount of chromium. Figure 2.1 illustrates this effect (Ref 1). In practice, however, some iron-chromium alloys containing as low as 9% Cr are also considered stainless steels. Other alloying elements that are highly essential in specific grades of stainless steel include nickel, molybdenum, silicon, carbon, manganese, sulfur, titanium, and niobium.

To a large extent, compositions of powder metallurgy (PM) stainless steels have been derived from some of the popular grades of wrought stainless steels. As a result, the characteristics of most PM stainless steels parallel those of their wrought counterparts. Nevertheless, the compositional ranges of PM stainless steels, particularly those based on water-atomized powders, often differ by small but important amounts from those of their wrought counterparts (Chapter 3, "Manufacture and Characteristics of Stainless Steel Powders"). This is particularly true with regard to the carbon, silicon, and manganese contents.

A vast majority of wrought stainless steels have a maximum permissible carbon content of 0.08% or higher. Only a few selected wrought stainless steels are available in a low-carbon version with a maximum carbon content of 0.03%, and these are designated as "L" grades. By contrast, almost all PM stainless steels, with the exception of the martensitic grades, are specified to be "L" grades or the low-carbon versions of the alloys. The need for the low-carbon requirement is twofold. Low carbon content renders the stainless steel powder soft and ductile, thus making it easier to compact. Secondly, low carbon content minimizes the potential for chromium carbide formation or sensitization during cooling from the sintering temperature (Chapters 3 and 5). The latter reason is also the basis for selecting "L" grades of wrought stainless steels for applications requiring welding. Stainless steel components sintered in a nitrogen-bearing atmosphere will contain large amounts (typically several thousand parts per million) of nitrogen and, as a result, these do not qualify as "L"-grade materials. Similarly, parts sintered under conditions of inadequate delubrication may contain greater than 0.03% C and hence would not meet the "L"-grade criterion.

Other PM preferences within the standard ranges of composition are also related to the

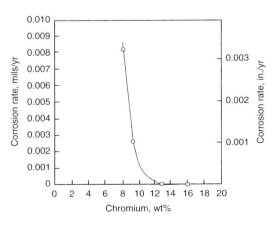

Fig. 2.1 Corrosion rates of iron-chromium alloys in intermittent water spray, at room temperature. Source: Ref 1

effects of respective constituents on powder production and powder compaction. The importance of adhering to relatively narrow ranges in PM for silicon, manganese, and phosphorus is treated in Chapter 3. Similarly, additives that are reactive in nature (such as titanium, zirconium, and aluminum) are excluded from the compacting grades of PM alloys, in order to avoid formation of stable, unreducible surface oxides during water atomization of the powder.

2.2 Identification and Specifications

The American Iron and Steel Institute (AISI) numbering system is the oldest and most popular identification system for all steels in the United States. This system specifies the composition of an alloy based on its ladle analysis. However, it does not specify other requirements and properties. It uses a three-digit numbering system, with the prefix "type," for identification of the steel. Some numbers may take on a one- or two-letter suffix to indicate modifications to the composition (e.g., type 303Se, for selenium-containing 303). Proprietary and other nonstandard alloys often bear a trade name or unique identifying number. The AISI designations and compositions serve as the primary standards for most industries.

The Society of Automotive Engineers International (SAE) uses a five-digit numbering system, which is in compliance with the compositional limits set forth by AISI standards. The last three digits of the SAE numbering system match with the AISI designation of the alloy.

The Unified Numbering System (UNS) is a five-digit identification system that is designed to catalog similar alloys specified by various standards organizations and nations. Each of these five-digit UNS numbers is a designation assigned to the chemical composition of an alloy, without requiring the composition to be a specification. Each five-digit designation is preceded by a letter code that indicates the broad class to which the alloy belongs (e.g., "S" for stainless steels). For alloys that have an AISI designation, the first three digits usually correspond to the alloy's AISI designation. When the last two digits are "00," the number designates a basic AISI grade. A modification of the basic alloy is designated by a number not ending with zeros. The system also assigns UNS numbers to alloys that are primarily recognized by their trade names.

The systems of designation described previously contain only a portion of the information necessary to properly describe a steel product for procurement purposes.

The American Society for Testing and Materials (ASTM) International calls for performance requirements in addition to composition. ASTM standards often specify the minimum, as well as some typical values for various mechanical and physical properties. ASTM International also provides standards for test methods.

The PM industry, under the auspices of Metal Powder Industries Federation (MPIF), has been developing standards covering standard grades of PM materials that specify compositions as well as some sintered properties. These are presented in a publication called *MPIF Standard 35—Materials Standards for PM Structural Parts* (three other volumes of MPIF standard 35 are also available that cover PM self-lubricating bearings, powder-forged steel parts, and metal injection molded parts). The MPIF also specifies standard practices for testing PM materials, powders and sintered products, which are covered in the *MPIF Standard Test Methods for Metal Powders and Powder Metallurgy Products*. Chemical composition ranges of MPIF standards closely follow those specified by AISI. Each grade of alloy is divided into three or four classes of material, with each class representing a specific set of sintering conditions, selected from popular commercial practices. Within each designated class of material, further classification is made based on the sintered density, leading to a series of material codes. Thus, a material code identifies the alloy composition, the sintering conditions employed, and an approximate sintered density. This makes the standard useful for procurement purposes. As for the performance, the standard specifies only the minimum values of yield strength and tensile elongation while providing typical values for some of the other mechanical and physical properties. The sintering conditions and sintered densities listed are meant to serve as guidelines, with the idea that a parts producer has the option to make necessary adjustments to the process (including selection of green density and sintering time) in order to meet the specified minimum values for yield strength and tensile elongation. This system of materials designation gives the parts producer sufficient flexibility in processing. Table 2.1 lists the various material codes, along with their sintering conditions. Mechanical properties data listed in MPIF standard 35 are held as important benchmark properties for PM part design and use (Appendixes 1 and 2).

Table 2.1 Material designations in accordance with Metal Powder Industries Federation (MPIF) standard 35

Base alloy	MPIF material designation code(a)	Sintering atmosphere	Sintering temperature °C	°F	N$_2$(b) (typical), %
303 304 316	SS-303N1-XX SS-304N1-XX SS-316N1-XX	Dissociated ammonia	1149	2100	0.20–0.60
303 304 316	SS-303N2-XX SS-304N2-XX SS-316N2-XX	Dissociated ammonia	1288	2350	0.20–0.6
304 316	SS-304H-XX SS-316H-XX	100% hydrogen	1149	2100	<0.03
303 304 316	SS-303L-XX SS-304L-XX SS-316L-XX	Vacuum	1288	2350	0.03
410	SS-410-HT-XX(c)	Dissociated ammonia	1149	2100	0.20–0.60
430 434	SS-430N2-XX SS-434N2-XX	Dissociated ammonia	1288	2350	0.20–0.60
410 430 434	SS-410L-XX SS-430L-XX SS-434L-XX	Vacuum	1288	2350	<0.06

(a) "XX" refers to minimum yield strength. (b) Data shown are for information only; these are not part of the standard. (c) SS-410HT-XX is processed by adding up to 0.25% graphite to a 410L powder. After sintering, the material is tempered at 177 °C (350 °F).

The Powder Metallurgy Parts Manufacturers Association (PMPA) Standards Committee of MPIF is the main body responsible for developing standards for PM steels, including stainless steels. Standards in use today were developed mostly between 1992 and 1997, under the guidance of the PMPA Standards Committee, using funds from MPIF and the U.S. Navy. Sample preparation and testing were carried out at several parts fabricators and the laboratories of Concurrent Technologies Corp. (Ref 2). Much of the data generated were also published by Sanderow and Prucher (Ref 3) at various technical conferences. The committee continues to add newer materials, processes, and additional properties to these standards.

In addition to the standard grades of PM stainless steel, a number of nonstandard grades (custom and proprietary alloys) are in widespread use in the PM industry. In contrast to wrought, such alloys make up a significant portion of the total number of alloys. This is partly due to the fact that the PM process is highly flexible and amenable to the development of custom alloys via sintering of mixtures of metal powders. Also, compared to the wrought steel industry, the PM industry employs much smaller melting furnaces, which makes it more convenient to produce custom alloys.

2.3 Basic Metallurgical Principles

A basic knowledge of the metallurgy of stainless steels is essential for understanding the classification system used for wrought as well as PM stainless steels. Stainless steels are commonly grouped into five families, four of which are based on microstructure. These are known as ferritic, austenitic, martensitic, and duplex. Duplex is a hybrid of austenitic and ferritic structures. The fifth family, known as the precipitation-hardening family, is distinguished by its unique strengthening mechanism. This system of classification has been in use ever since the discovery of the first stainless steels in the early 1900s and is based on the fact that both the metallurgy and physical properties of an alloy are strongly influenced by its crystal structure. The crystal structure of an alloy, in turn, is determined by its chemistry and thermal history.

Structurally, pure iron exists at room temperature in a body-centered cubic (ferritic) structure. As it is heated above 910 °C (1670 °F), it undergoes transformation into a face-centered cubic (fcc) (austenitic) structure, known as the gamma phase (γ). Upon further heating through 1400 °C (2552 °F), it undergoes transformation back to the ferritic structure. The lower-temperature version of the ferritic phase is called alpha ferrite (α), and the higher-temperature version of the ferritic phase is known as delta ferrite (δ). (Both alpha and delta ferrites are physically indistinguishable from each other; the nomenclature serves to identify the condition under which they are formed.) When pure iron is alloyed with increasing amounts of chromium, the temperatures of transformation from ferrite (α) to austenite and from austenite to ferrite (δ) both decrease gradually until approximately 7% Cr (Fig. 2.2, Ref 4). Further addition of chromium to the alloy increases the temperature of transformation of ferrite (α) to austenite (γ) while still lowering the temperature of transformation from austenite (γ) to ferrite (δ). This tends to restrict the temperature range in which austenite is stable. As the chromium content is increased beyond approximately 13%, the alloy remains ferritic at all temperatures (Fig. 2.3). Because the addition of chromium leads to an increase in the stabilization of the ferrite phase (i.e., reduction of the austenitic region in the phase diagram), chromium is called a ferrite-forming element or a ferritizer. Other elements, often present in stainless steels, that produce a similar effect are molybdenum, silicon, niobium, titanium, tantalum, and aluminum. It may be noted that with the exception of aluminum, all of these ferrite-forming elements have a body-centered cubic structure at room temperature.

Alloying of either iron or an iron-chromium alloy with the fcc metal nickel produces a much different effect. Nickel promotes transformation of ferrite to austenite. Nickel addition results in the expansion of the γ-phase region as well as that of the α + γ region located below it. Alloying with nickel makes it possible to have high-chromium-containing Fe-Cr-Ni alloys in the austenitic form over wide ranges of temperatures, including room temperature. The large substitutional nickel atoms diffuse very slowly in the ferrous matrix, and hence, the phases present at room temperature are not predictable from the equilibrium diagrams of the Fe-Ni or Fe-Cr-Ni system. Typically, the actual amount of austenite present in most alloys is higher than what is indicated by the equilibrium diagram.

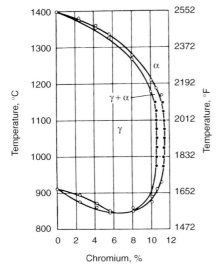

Fig. 2.2 Iron-chromium partial phase diagram showing the gamma loop for a 0.004% C- and 0.002% N-containing alloy. Source: Ref 4

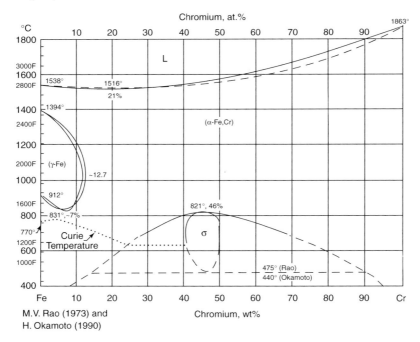

Fig. 2.3 Binary iron-chromium equilibrium phase diagram. Source: Ref 5, 6

All austenitic stainless steels contain at least 16% Cr; wrought alloys containing 16 to 19% Cr and at least 9% Ni are predominantly austenitic at room temperature. Occasionally, a small amount (typically less than 15%) of δ ferrite may be present in an austenitic stainless steel at room temperature. In addition to the relative amounts of iron, chromium, and nickel present, the microstructure of the alloy is influenced by the presence of some minor elements. Austenitizers, other than nickel, are manganese, carbon, and nitrogen. The total effect of all ferrite-forming elements can be expressed as the chromium equivalent of the alloy, and that of all austenite-forming elements as the nickel equivalent of the alloy.

The combined effect of all austenitizing and ferritizing elements can help determine which phase or phases are expected to be present in the alloy at room temperature. Such prediction is possible with the help of a diagram originally developed by Schaeffler (Ref 7). Figure 2.4 shows the Schaeffler diagram along with a set of commonly accepted equations for chromium equivalence and nickel equivalence. This diagram was originally developed for estimation of relative amounts of ferrite and austenite present in the microstructures of stainless steel welds; subsequently, a number of modifications of the diagram were proposed, much of which were

determined empirically (section 10.2 in Chapter 10, "Secondary Operations").

The strong austenitizing effect of manganese has been exploited in wrought metallurgy to create austenitic alloys where nickel is partially substituted by manganese (200-series stainless steels); however, this is not an option in PM, for reasons discussed earlier. Similar to nickel, carbon, and nitrogen tend to expand the $\alpha + \gamma$ region of the iron-chromium phase diagram. However, unlike nickel, these elements are effective in stabilizing the austenite phase mainly at high temperatures, as shown in Fig. 2.5 (a and b) (Ref 4). Because of their relatively small atomic size, atoms of these alloying elements can diffuse through the alloy matrix rapidly and are able to locate themselves at interstitial sites (hence called interstitials). Even very small concentrations of carbon and/or nitrogen can lead to stabilization of austenite at elevated temperatures in alloys containing as much as 30% Cr. The solubility limit of carbon is fairly high in the austenitic Fe-Cr and Fe-Cr-Ni matrices at elevated temperatures. However, at temperatures below 371 °C (700 °F) the solubility limit of carbon decreases rapidly, to below 0.03% at room temperature. When an austenitic iron-chromium alloy containing carbon (or nitrogen) in excess of its solubility limit is cooled rapidly

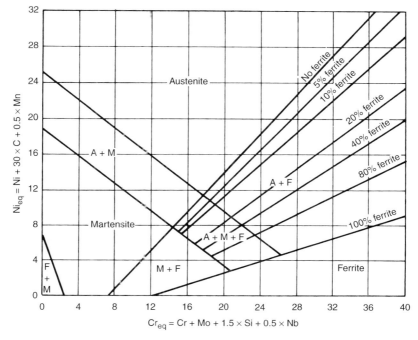

Fig. 2.4 Schaeffler diagram for determining phases formed upon solidification, based on chemistry

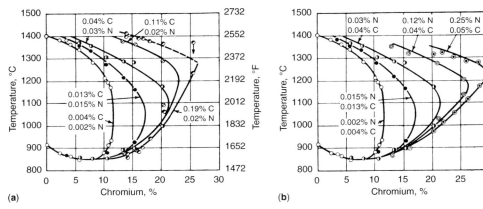

Fig 2.5 Effects of (a) carbon and (b) nitrogen addition on the (α + γ)/α boundary of the iron-chromium phasediagram. Source: Ref 4

to room temperature, the atoms of the interstitial elements remain trapped in the matrix, resulting in distortion of the matrix. The austenite phase existing at the elevated temperature transforms via lattice shear into a nonequilibrium phase called martensite. The distorted matrix takes the form of a body-centered tetragonal structure. Martensite has higher hardness and lower ductility compared to either austenite or ferrite.

Duplex stainless steels possess a microstructure that comprises both austenitic and ferritic grains at room temperature. Typically, these alloys contain approximately 65 to 70% Fe, 18 to 25% Cr, and 3 to 6% Ni, with minor additions of molybdenum, copper, nitrogen, silicon, titanium, and tungsten. When cooled to room temperature, these alloys tend to form as 100% ferrite (their compositions always fall in a fully ferritic region of the Schaeffler diagram). However, by providing sufficient time for atomic diffusion at an elevated temperature, a mixed microstructure of austenite and ferrite is produced. The relative amount of the two phases is nearly the same in most commercial alloys. Because the iron content of most duplex alloys is approximately 70%, a pseudobinary phase diagram of Fe-Cr-Ni, with the iron content fixed at 70% (Fig. 2.6) (Ref 8), may be used to explain the phase structure in duplex stainless steels. In this case, because the α/(α + γ) and (α + γ)/γ phase boundaries are not vertical, the relative amounts of α and γ vary with temperature. The α phase is more stable at high temperatures, and it is also the equilibrium phase at room temperature. However, due to the sluggish diffusion of nickel, γ would typically remain as the predominant phase. Hot working and annealing in the temperature range of 1010 to 1120 °C (1850 to 2050 °F) promotes precipitation and growth of

austenite in the ferritic matrix; a duplex microstructure is developed. Austenite nucleates at ferrite grain boundaries and along preferred crystallographic directions within the ferrite matrix, with the austenite-stabilizing elements (copper, nickel, carbon, and nitrogen) enriching the austenite phase and the ferrite-stabilizing elements (chromium, molybdenum, silicon, and tungsten) enriching the ferrite phase. Enhancement of strength and ductility is aided by high nitrogen content (in solution) as well as by the fine grain size of the dual-phase microstructure.

The precipitation-hardening (PH) family of stainless steels is a relatively new family of alloys. These are designed to offer remarkably high strength and toughness via the formation of submicroscopic precipitates in the matrix. Precipitation-forming elements include copper, molybdenum, aluminum, niobium, and titanium. When solution annealed and cooled to room temperature, the matrix is supersaturated with the precipitate-forming elements. Upon aging (i.e., isothermal hold for several hours), second-phase precipitates nucleate uniformly throughout the matrix. Aging treatment is designed to keep the size of the precipitates at a submicroscopic level so that the strength of the alloy is maximized. These alloys offer a unique advantage in wrought metallurgy, because they permit fabrication of components from an alloy in its relatively ductile, solution-annealed condition. The components are then strengthened using a low-temperature aging treatment.

The PH alloys are further classified as martensitic, semiaustenitic, and austenitic, based on their martensite start and finish (M_s and M_f) temperatures. For example, the popular martensitic precipitation-hardening alloy 17-4 PH, with an M_f temperature just above room temperature, transforms to a fully martensitic matrix upon air cooling from the solution-annealing temperature. It essentially has a low-carbon martensitic matrix (lower strength and higher ductility in the solution-annealed condition compared to martensitic 410). Hardening is accomplished by aging at 482 to 648 °C (900 to 1200 °F) for 1 to 4h. The precipitate formed is a copper-rich fcc phase. In the case of a semiaustenitic FH alloy, such as 17-7 PH, the M_s temperature is lower than room temperature. In the solution-annealed condition, the matrix is fully austenitic, exhibiting good ductility. After forming, a conditioning heat treatment in the range of 730 to 955 °C (1346 to 1750 °F), is provided, which raises both the M_s and M_f temperatures by precipitating out carbon and some alloying elements from the matrix. The temperature of the conditioning treatment determines the proximity of M_f to room temperature, thus influencing the relative amounts of martensite and austenite in the matrix. Transformation may be aided by refrigeration or cold work. Finally, aging is carried out to accomplish precipitation hardening. The austenitic PH stainless steels, such as A-286 and 17-10P, have their M_s temperatures well below room temperature, thus preventing any transformation to martensite. Strengthening is accomplished by the precipitation of intermetallic compounds in an austenitic matrix.

In order to realize the full benefit of their high strengths, it is essential that the PH materials are processed to their full or near-full density. Hence, metal injection molding (MIM) is the most suitable PM process for producing PH stainless steel components.

2.4 Characteristics and Chemical Compositions of Wrought and PM Stainless Steels

Three out of the five families of stainless steels, namely austenitic, ferritic, and martensitic, are eminently suitable for manufacture via conventional PM. Selected alloys from all five families can be processed via MIM. With the martensitic grades considered useful mainly for high-wear-resistance-type applications, alloys from the ferritic and austenitic families represent the bulk of PM stainless steel grades. Selection of an alloy for a given application is dependent on a number of factors, corrosion resistance usually being the most important. Other criteria that are frequently taken into account for alloy selection are mechanical properties, resistance

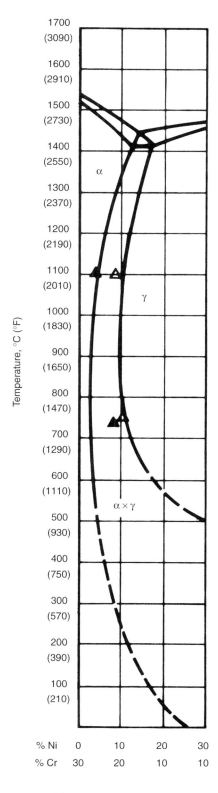

Fig. 2.6 Pseudobinary phase diagram of Fe-Cr-Ni, with iron content fixed at 70%. Source: Ref 8

to oxidation and creep at elevated temperatures, fabricability, thermal and magnetic properties, as well as cost.

The following section covers the key characteristics of each of the five families of stainless steels, with specific reference to PM, along with the chemical compositions of popular grades of PM stainless steels.

2.4.1 Ferritic Grades

Ferritic stainless steels are essentially alloys of iron and chromium having a ferritic structure at room temperature. Compared to austenitic grades, ferritic stainless steels are less corrosion resistant and their elevated temperature strength is lower. However, they are the optimal choice in many applications because of their lower cost. Compared to austenitic grades, they have a lower rate of work hardening and somewhat better machinability (section 10.1 in Chapter 10, "Secondary Operations"). They offer good formability and ductility, which can be useful in a sizing or repressing operation. Ferritic grades are selected in some applications because of their magnetic behavior. The addition of niobium is essential to impart weldability. In PM processing, ferritic alloys undergo greater rates of shrinkage during sintering compared to austenitic alloys, thus resulting in higher sintered densities.

Compared to austenitic grades, the ferritic stainless steels have a relatively lower coefficient of thermal expansion and a higher thermal conductivity. These characteristics make them more resistant to thermal fatigue as well as to oxide spalling in applications involving thermal cycling in air (Ref 9). This has been observed in both PM and wrought ferritic stainless steels.

Compositions of PM Ferritic Alloys. Wrought ferritic stainless steels are broadly divided into three classes based on their chromium content, namely, low, medium, and high chromium. The PM ferritic grades represent only the low- (10 to 14%) and medium- (15 to 19%) chromium classes (high-chromium ferritic alloys in a PM version would suffer from low compacting properties). The standard PM ferritic grades are 409L, 409LE, 410L, 430L, and 434L. Table 2.2 lists the compositions of these alloys as specified by MPIF standard 35 (material standards for PM structural parts), along with typical compositions of some nonstandard grades. Grades 409L and 409LE contain a small amount of niobium (columbium), which serves to stabilize the alloy against sensitization and renders the alloy weldable. Niobium combines with carbon present in the matrix, thus preventing formation of chromium carbide. Sintering in a nitrogen-bearing atmosphere is entirely unacceptable, because it will lead to the formation of excessive amounts of niobium and chromium nitrides. The PM versions of 409L and 409LE stainless steels came into use in the late 1990s with the introduction of PM stainless steel exhaust flanges and HEGO bosses. In wrought ferritic stainless steels, the stabilizer is most often titanium, although it can be a combination of titanium and niobium, or only niobium. In stainless steels with low concentrations of niobium, the form of niobium carbide is NbC. Theoretically, for all carbon to be converted to NbC, the amount of niobium required is 8 times the carbon content; this translates to a minimum niobium content of 0.24% for an "L"-grade stainless steel. In the case of most PM stainless steels, however, the specifications call for niobium content in the range of 0.4 to 0.8%. The excess is intended for tying up any residual carbon that may arise from a marginal delubrication/sintering practice. Grade 409LE, which contains a higher amount of chromium, is preferred by some exhaust system manufacturers because of the proven success of the PM 410L grade of stainless steel as sensor rings in antilock brake sensor (ABS) systems. The PM 410L stainless steel exhibits a fully ferritic

Table 2.2 Compositions of powder metallurgy ferritic stainless steels

Grade	Standard	Non standard	Cr	Ni	Mn	Si	S	C	P	Mo	N	Nb
409L	X		10.50–11.75		0–1.0	0–1.0	0–0.30	0–0.03	0–0.04		0–0.03	0.4–0.8
409LE	X		11.50–13.75	0–0.5	0–1.0	0–1.0	0–0.30	0–0.03	0–0.04		0–0.03	0.4–0.8
410L	X		11.50–13.50		0–1.0	0–1.0	0–0.30	0–0.03	0–0.04		0–0.03	
430L	X		16.00–18.00		0–1.0	0–1.0	0–0.30	0–0.03	0–0.04		0–0.03	
430LN2	X		16.00–18.00		0–1.0	0–1.0	0–0.30	0–0.08	0–0.04		0–0.06	
434L	X		16.00–18.00		0–1.0	0–1.0	0–0.30	0–0.03	0–0.04	0.75–1.25	0–0.03	
434LN2	X		16.00–18.00		0–1.0	0–1.0	0–0.30	0–0.08	0–0.04	0.75–1.25	0–0.06	
434LNb		X	16.00–18.00		0–1.0	0–1.0	0–0.30	0–0.03	0–0.04	0.75–1.25	0–0.03	0.4–0.8
434L-Modified		X	17.00–19.00		0–1.0	0–1.0	0–0.30	0–0.03	0–0.04	1.75–2.25	0–0.03	
444L		X	17.50–19.50	0.8 typical	0–1.0	0–1.0	0–0.30	0–0.03	0–0.04	1.75–2.50	0–0.03	0.4–0.8

structure, provided that the carbon plus nitrogen content is held below 0.03%. In order to keep the structure fully ferritic, the low-chromium grades should be more stringently restricted in their interstitial content when compared to the medium-chromium grades (15 to 19% Cr) (Ref 10).

The medium-chromium ferritic grades, such as 430L and 434L, offer significantly higher corrosion resistance compared to the low-chromium grades, such as 410L and 409L. The corrosion resistance of the low-chromium ferritic materials is still adequate for most atmospheric conditions, providing structural integrity over long periods of exposure, with or without the degradation of surface appearance. The presence of a small amount of molybdenum in 434L enhances its resistance to crevice and pitting corrosion. Two common examples of nonstandard ferritic alloys are listed in Table 2.2. The niobium-stabilized 434LNb (similar to AISI 436L) is intended for applications requiring welding. A high-chromium, high-molybdenum version of 434L, known as 434L-Modified, is selected by some ABS sensor system manufacturers for use as sensor rings because of its superior corrosion resistance. Studies based on wrought stainless steels have shown that the pitting resistance of a stainless steel strongly correlates with its composition by an empirical equation (Ref 11):

$$PREN = \%Cr + 3.3\% \, Mo + 16\% \, N \qquad (Eq \ 2.1)$$

where PREN is the pitting resistance equivalence number.

The multiplier for nitrogen (in solution) has been variously quoted between 12.8 and 30, but the (%Cr + 3.3% Mo) portion of the equation is held in good agreement. In one study involving three grades of PM 400-series stainless steels, the validity of the PREN equation was confirmed (Ref 12). In this study, optimally sintered (Chapter 6, "Alloying Elements, Optimal Sintering, and Surface Modification in PM Stainless Steels") 410L, 434L, and 434L- modified samples, in the form of ABS sensor rings, were subjected to a 1000 h salt spray test. Removal of surface rust by sand blasting revealed the formation of corrosion pits in all samples. Two types of pits had formed: single pits and pit clusters. The latter were comprised of two to five pits in close proximity to each other. Pit counts of the samples (both as total number of pits and the number of pit clusters) are plotted in Fig. 2.7 against the PREN numbers of the alloys. In addition to showing a direct correlation between the PREN number and pit count, these data show the strong beneficial effect of molybdenum in combating pitting corrosion.

Potential Problems with Embrittlement. Several embrittlement mechanisms have been identified and well documented for wrought stainless steels. These occur only in the medium- and high-chromium ferritic stainless steels and are the

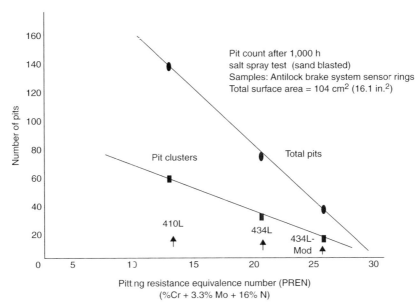

Fig. 2.7 Relationship between number of corrosion pits formed and pitting resistance equivalence number (PREN) for three powder metallurgy 400-series stainless steels. ABS, antilock brake sensor

results of specific structural changes in the alloy. Although these phenomena can occur in PM stainless steels that have similar chemistry and thermal history, the probability of their occurrence is much smaller because of the low interstitial contents and minimal residual stresses in PM materials. Only some of the PM ferritic stainless steels from the medium-chromium class may possibly be prone to such behavior. Described as follows are three embrittlement phenomena as observed in wrought ferritic stainless steels:

- *Sigma-phase embrittlement:* The sigma phase (σ) is a hard, brittle phase (essentially an intermetallic precipitate of iron-chromium) that forms in medium- and high-chromium ferritic alloys upon long-term exposure to the critical temperature range of 500 to 800 °C (932 to 1472 °F). This is an equilibrium phase in the iron-chromium phase diagram (Fig. 2.3). At chromium contents of less than 20%, this phase is difficult to form. However, the presence of molybdenum, silicon, manganese, and nickel can shift this limit to lower levels. Wrought ferritic stainless steel containing 18% Cr and 2% Mo is reported to suffer from sigma-phase formation when exposed to the critical temperature range for several thousand hours (Ref 13). Cold work enhances the rate of sigma-phase formation. Formation of sigma phase leads to significant loss of ductility and toughness, with a small increase in the hardness. Among the PM grades of stainless steels, only the 434L-Modified (18Cr-2Mo) may be prone to sigma-phase embrittlement. This phenomenon needs be taken into consideration if a medium-chromium ferritic alloy is to be exposed to the critical temperature range in service. Sigma phase usually manifests itself as a continuous network in the microstructure. Because it has a significantly lower corrosion resistance compared to the ferrite matrix, its presence can be detected by etching in a metallographic examination. Sigma phase can be eliminated (dissolved) by heat treating above 850 °C (1562 °F) for approximately an hour, followed by air cooling.
- *475 °C (885 °F) embrittlement:* Iron-chromium alloys containing 15 to 70% Cr may exhibit a pronounced increase in hardness, accompanied by severe loss of ductility and corrosion resistance, if exposed to the temperature range of 400 to 540 °C (752 to 1004 °F) for

significantly shorter time periods than is required for sigma-phase formation (the peak hardness usually occurs at 475 °C, or 885 °F, and hence the name). In fact, it can occur during slow cooling from an elevated temperature as well as during elevated-temperature service. For alloys containing 18% Cr, the onset of embrittlement is fast enough to require rapid cooling from the annealing temperature in order to ensure optimal ductility. Alloys containing greater than 16% Cr should not be used at 375 to 540 °C (707 to 1004 °F) for extended periods of time or cycled from room temperature through this critical range. This embrittlement phenomenon is believed to be due to the formation of a submicroscopic, coherent precipitate that is induced by the presence of a solubility gap below approximately 550 °C (1022 °F) in a chromium range where sigma phase forms at higher temperatures. Cold work intensifies the rate of 475 °C (885 °F) embrittlement, especially for the higher-chromium alloys. Reheating the alloy to above 550 °C (1022 °F) for a few minutes completely removes 475 °C (885 °F) embrittlement (Ref 14).
- *High-temperature embrittlement:* Medium- and high-chromium ferritic alloys, containing moderate amounts of carbon and/or nitrogen, develop this type of brittleness if cooled slowly from above 950 °C (1742 °F). The mechanism is similar to that of sensitization, and it also leads to severe intergranular corrosion. Work on two wrought ferritic stainless steels containing 18 and 25% Cr, respectively, has shown that the maximum amount of carbon plus nitrogen tolerable for good room-temperature toughness is 0.055% for the 18% Cr alloy and 0.035% for the 25% Cr-containing alloy (Ref 13). Also, there is an equivalency in the effect by carbon and nitrogen. Because PM stainless steels must have very low levels of interstitials (<0.03% total C + N) in order to avoid sensitization, no additional effort is necessary to combat high-temperature embrittlement.

2.4.2 Austenitic Grades

Austenitic stainless steels offer superior corrosion resistance compared to both ferritic and martensitic grades. Austenitic grades are also the preferred grades for applications requiring exposure to elevated temperatures. While ferritic and martensitic stainless steels show noticeable reduction in oxidation resistance, leading to scaling, at temperatures above 700 °C (1292 °F), the austenitic stainless steels exhibit satisfactory resistance to oxidation at temperatures as high as 900 °C (1652 °F). Austenitic stainless steels also exhibit a superior resistance to creep when compared to ferritic grades. Austenitic stainless steels tend to work harden rapidly and also are difficult to machine. These alloys can tolerate slightly higher levels of interstitials, as compared to the ferritic alloys, and hence, the "L" versions of austenitic grades are most often weldable without the use of a stabilizer. Austenitic alloys are nonmagnetic. This behavior results from the fact that the addition of nickel to the ferritic iron-chromium not only forces the γ/α-phase boundary to lower temperatures but also causes the magnetic (Curie) transformation boundary (dashed line in Fig. 2.3 separating the ferromagnetic from the paramagnetic region) to below room temperature. Cold working or cooling to a subzero temperature can transform an austenitic stainless steel to a ferromagnetic martensitic structure. In wrought metallurgical processing, it is not uncommon to alloy either nitrogen or manganese as partial substitutes for nickel. In PM, it is difficult to keep large amounts of nitrogen in solution during cooling from the sintering temperature; manganese causes excessive oxidation during water atomization (Chapter 3, "Manufacture and Characteristics of Stainless Steel Powders"). Nitrides of chromium, which form easily during cooling, not only affect corrosion resistance but also can drastically lower the chromium equivalence of the alloy matrix. It is not uncommon to find heavily nitrided austenitic stainless steels in which chromium depletion is so severe that the composition of the alloy matrix falls in the martensitic regime of the Schaeffler diagram, making the alloy weakly magnetic.

Sensitization is a potential problem with wrought austenitic stainless steels and high interstitials containing PM austenitic stainless steels, which can result in loss of corrosion resistance and ductility. Sigma-phase embrittlement is a potential problem with high-chromium- and -molybdenum-containing austenitic alloys also, due to the fact that the small amount of ferrite phase present in these alloys can undergo sigma-phase transformation in the same manner as it occurs in ferritic stainless steels, that is, under conditions of slow cooling and annealing.

This will also have a severe adverse effect on corrosion resistance.

Compositions of PM Austenitic Alloys. The compositions of the standard grades and some of the more common custom grades of PM austenitic stainless steels are listed in Table 2.3. The low-carbon modifications of the three most popular wrought alloys, namely 303L, 304L, and 316L, make up the standard grades for PM processing. Type 304L is known as the general-purpose austenitic stainless steel and also the most economical austenitic material. Its composition is derived from an earlier established wrought grade known as 18-8 stainless steel. Type 303L has a composition similar to that of 304L, except for its high sulfur content. Sulfur combines with the manganese present in the alloy to form manganese sulfide, which enhances machinability. Austenitic stainless steels are difficult to machine because they tend to gall and smear on the cutting tool. Hence, for applications requiring machining, type 303L is often selected. A disadvantage may be experienced in terms of the lower overall corrosion resistance of 303L compared to 304L (section 10.1 in Chapter 10, "Secondary Operations"). Type 316L contains a small amount of molybdenum for enhanced corrosion resistance. Molybdenum is especially effective in increasing resistance to crevice and pitting forms of corrosion. Crevice corrosion plays a significant role in PM stainless steels due to the presence of porosity in the material. Hence, it is not surprising to note that PM parts made of 316L alloy are significantly superior in corrosion resistance to those made from 304L. Type 316L also contains a slightly higher amount of nickel to counter the ferritizing effect of molybdenum.

Within the broad range provided in AISI specifications, the nickel content of the "L" version of a wrought stainless steel is typically kept approximately 2% higher than that of its standard-grade counterpart, in order to compensate for the loss of austenitizing potential from carbon (Ref 15). At the same time, because the solubility of carbon in an austenitic stainless steel decreases with increasing nickel content, it is preferable that the maximum carbon content of the austenitic "L" grades be limited to 0.02%, rather than 0.03%, in order to ensure freedom from sensitization under slow cooling conditions (section 3.1.3 in Chapter 3, "Manufacture and Characteristics of Stainless Steel Powders"). The preference to keep the nickel content of PM austenitic stainless steels closer to the upper end

Table 2.3 Compositions of powder metallurgy austenitic stainless steels

Grade	Standard	Non standard	Fe	Cr	Ni	Mn	Si	S	C	P	Mo	N	Sn	Cu	Other
303L	X		bal	17.0–19.0	8.0–13.0	2.0(a), 0.2(b)	1.0(a), 0.8(b)	0.15–0.30	0.03(a)	0.2(a), 0.1(b)		0.0–0.03			
303N1, N2	X		bal	17.0–19.0	8.0–13.0	2.0(a), 0.2(b)	1.0(a), 0.8(b)	0.15–0.30	0.03(a)	0.2(a), 0.1(b)		0.2–0.6			
304L	X		bal	18.0–20.0	8.0–12.0	2.0(a), 0.12(b)	1.0(a), 0.8(b)	0.03(a)–0.01(b)	0.03(a)	0.04(a), 0.01(b)		0.0–0.03			
304N1, N2	X		bal	18.0–20.0	8.0–12.0	2.0(a), 0.12(b)	1.0(a), 0.8(b)	0.03(a)–0.01(b)	0.03(a)	0.04(a), 0.01(b)		0.2–0.6			
316L	X		bal	16.0–18.0	10.0–14.0	2.0(a), 0.12(b)	1.0(a), 0.8(b)	0.03(a)–0.01(b)	0.03(a)	0.04(a), 0.01(b)	2.0–3.0	0.0–0.03			
316N1, N2	X		bal	16.0–18.0	10.0–14.0	2.0(a), 0.12(b)	1.0(a), 0.8(b)	0.03(a)–0.01(b)	0.03(a)	0.04(a), 0.01(b)	2.0–3.0	0.2–0.6			
303LSC		X	bal	17.0–19.0	8.0–13.0	2.0(a), 0.2(b)	1.0(a), 0.8(b)	0.03(a)–0.01(b)	0.03(a)	0.2(a), 0.1(b)		0.0–0.03	1.0(b)	2.0(b)	
303L–Ultra		X	bal	17.0–19.0	8.0–13.0	2.0(a), 0.2(b)	1.0(a), 0.8(b)	0.03(a)–0.01(b)	0.03(a)	0.2(a), 0.1(b)		0.0–0.03	1.5(b)	0.8(b)	
304LSC		X	bal	18.0–20.0	8.0–12.0	2.0(a), 0.12(b)	1.0(a), 0.8(b)	0.03(a)–0.01(b)	0.03(a)	0.04(a), 0.01(b)		0.0–0.03	1.0(b)	2.0(b)	
304L–Ultra		X	bal	18.0–20.0	8.0–12.0	2.0(a), 0.12(b)	1.0(a), 0.8(b)	0.03(a)–0.01(b)	0.03(a)	0.04(a), 0.01(b)		0.0–0.03	1.5(b)	0.8(b)	
316LSC		X	bal	16.0–18.0	10.0–14.0	0.2(a), 0.12(b)	1.0(a), 0.8(b)	0.03(a)–0.01(b)	0.03(a)	0.04(a), 0.01(b)	2.0–3.0	0.2–0.6	1.0(a)	2.0(b)	
316L–Ultra		X	bal	16.0–18.0	10.0–14.0	0.2(a), 0.12(b)	1.0(a), 0.8(b)	0.03(a)–0.01(b)	0.03(a)	0.04(a), 0.01(b)	2.0–3.0	0.2–0.6	1.5(b)	0.8(b)	
317L		X	bal	19.5(b)	14.8(b)	0.0–1.0	1.0(a), 0.8(b)	0.03(a)–0.01(b)	0.03(a)	0.04(a), 0.01(b)	3.5(b)	0.0–0.03			
SS-100		X	bal	20.0(b)	18.0(b)	0.0–1.0	1.0(a), 0.8(b)	0.03(a)–0.01(b)	0.03(a)	0.04(a), 0.01(b)	6.0(b)	0.0–0.03			
316-B	ASTM B853		bal	23(b)	18(b)	1.0(a)	1.0(a)	0.05(a)	0.05(a)	0.2(a)	3.5(b)	0.1(a)			0.35 B(b)

(a) Maximum. (b) Typical. LSC and Ultra are trade names.

of the AISI specification limits stems from this and from the beneficial effect of nickel on the compressibility of the powder. Takeda and Tamura (Ref 16) found the porosity of compacts made from 18% Cr, 4 to 14% Ni, and balance iron alloy powders, to decrease with increasing nickel content, up to approximately 12% Ni (Fig. 2.8).

High chromium contents have a detrimental effect on compressibility, which explains why high-chromium stainless steels are not widely used in press-and-sinter PM technology. In Fig. 2.9, pressure-density curves for three austenitic stainless steels illustrate that the influence of chromium (17% for 316L to 20% for SS-100)

dominates over that of nickel (13.5% for 316L to 18% for SS-100). The combined effect of chromium and nickel on the compressibility of chrome-nickel steels, from work by Kato and Kusaka (Ref 18), is shown in Fig. 2.10.

Regression analyses performed on the properties of a series of 316L stainless steel powders also indicate that compressibility increases with increasing nickel content. Furthermore, it decreases with increasing chromium content (Ref 17) and with increasing contents of oxygen and nitrogen (Ref 19). Thus, because of the great importance of compressibility (low porosity), most commercial 304L and 316L powders intended for compaction at room temperature

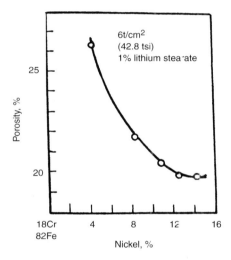

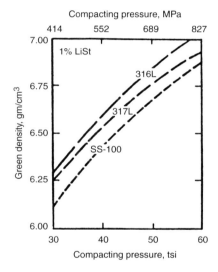

Fig. 2.8 Influence of nickel content on compressibility of 316L stainless steel powder. (Martensite formation is a significant contributor to the loss of compressibility in samples containing 8% and less nickel.) Source: Ref 17

Fig. 2.9 Pressure-density curves of three austenitic stainless steels. Unpublished data

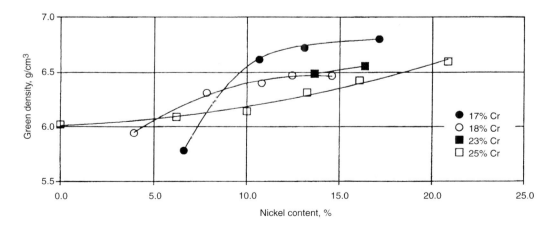

Fig. 2.10 Effect of chromium and nickel on compressibility of chrome-nickel steels. Source: Ref 19. Reprinted with permission from MPIF, Metal Powder Industries Federation, Princeton, NJ

tend to have nickel concentrations approaching their upper AISI specification limit and their chromium content approaching their lower AISI specification limit.

Nonstandard Austenitic Alloys. A significant number of nonstandard austenitic alloys are in use in the PM industry. Almost all of these are intended to offer enhanced corrosion resistance over the standard alloys:

- *Tin-modified grades:* Addition of 1 to 2% Sn to the austenitic alloy is found to enhance its corrosion resistance significantly. Tin addition is often made in conjunction with a small addition of copper. This technique is found to be most effective when the additives are prealloyed in the powder. Tin-modified versions of the standard alloys 303L, 304L, and 316L are widely used in the PM industry. Tin-modified alloys are also more forgiving to marginal sintering atmospheres. Additions of tin and/or copper also offer benefits in terms of improved machinability (Ref 20, 21).

- *Higher-alloy grades:* The PM alloys containing higher amounts of chromium, nickel, and molybdenum than in 316L are specified for more demanding applications. Table 2.3 lists two such alloys. Alloy 317L is derived from the standard 317L wrought alloy. SS-100, developed by Reen (Ref 22), contains significantly higher amounts of nickel and molybdenum. Molybdenum enhances resistance to crevice and pitting corrosion, but because it promotes ferrite formation, it becomes essential to increase the nickel content in order to stabilize the austenite phase. Optimally sintered PM SS-100 and its variations, containing as low as 5.5% Mo, are found to exhibit corrosion resistance that is equivalent to the corrosion resistance of wrought 316L (Ref 23). These alloys make good candidates for use in marine and food-processing applications, despite their higher cost. The significantly improved crevice and pitting resistance of the alloy is attributed to its high PREN value. Based on Eq 2.1, alloy SS-100 has a PREN of 55, compared to 37 for 316L. The PM version of a high-chromium (24 to 26%), high-nickel (19 to 22%), and high-silicon (1.5 to 2.5%) alloy, known as 310B, offers excellent elevated-temperature oxidation resistance and hence is being considered for use as particulate filters for diesel engines.

- *Boronized grades:* Boron promotes the formation of a liquid phase during sintering in stainless steels as well as in many other PM steels. Typically, a boron addition in the range of 0.15 to 0.25% is found to be sufficient for achieving full or near-full sintered densities. Prealloying of boron is generally preferred over elemental addition of either pure boron or boron compounds, such as CrB, NiB, and FeB. Several researchers have determined that addition of boron to a standard grade of PM stainless steel, such as 316L, is effective in producing near-full sintered densities and excellent corrosion resistance (Ref 24, 25). However, Reen (Ref 26) has suggested that for successful boronization, a PM alloy should be further enriched in chromium in order to compensate for chromium that becomes tied up in secondary phases. Commercial success of boronization is limited, due to the fact that the technique requires stringent control of boron levels, sintering temperature, and the dewpoint of the sintering atmosphere.

2.4.3 Martensitic Grades

Martensitic grades of PM stainless steels exhibit high strength and wear resistance, combined with a fair resistance to corrosion. They are magnetic. Both corrosion resistance and magnetic characteristics are somewhat inferior to those of the standard PM ferritic stainless steels. The ductilities of PM martensitic stainless steels are quite low, especially when sintered densities are significantly below full theoretical density.

Standard PM Martensitic Alloys. Grades 410 and 420 can be readily produced by blending carbon (in the form of a fine graphite powder) with 410L prior to compacting and sintering. Typical carbon addition levels are 0.15% for grade 410 and 0.30% for grade 420 (Table 2.4). During sintering, carbon goes into solution in the alloy matrix, stabilizing the austenite phase at the sintering temperature. During cooling from the sintering temperature, austenite is

Table 2.4 Nominal compositions (wt%) of Powder Metallurgy martensitic stainless steels

Grade	Standard	Fe	Cr	C	Ni	Si	Mn	Other	Hardness
410	Yes	bal	12.5	0.15	0.0	0.9	0.35		21 HRC
420	No	bal	12.5	0.30	0.0	0.9	0.35		25 HRC
440C	No	bal	17.0	1.0	0.0	0.9	0.35	0.75 Mo, 0.1 B	55 HRC
409LNi	No	bal	12.0	0.02	1.2	0.9	0.35	0.5 Nb	89 HRB

converted to martensite. Cooling rates normally achieved with industrial belt or pusher furnaces are sufficient for the formation of martensite in these chromium-containing materials; water quenching or rapid cooling is not essential. Svilar and Ambs (Ref 27) have observed only marginal improvement in mechanical properties by reaustenitizing at 1010 °C (1850 °F), followed by oil quenching.

Alloying of 410L with nitrogen instead of carbon (or in combination) will also produce a martensitic material (although technically not a 410 or 420). Although this can be readily achieved by sintering 410L in a nitrogen-bearing atmosphere, the control of the amount of nitrogen absorbed is rather difficult. Also, significant amounts of chromium nitrides could form in the material, which will not only lead to inconsistent mechanical properties but also degrade the alloy corrosion resistance and magnetic behavior.

Tempering of as-sintered martensitic alloys is often recommended, because it enhances mechanical strength, toughness, and ductility. These materials are suitable for the manufacture of wear-resistant bushings and blades for blenders and choppers.

Nonstandard PM Martensitic Grades. The addition of a small amount of nickel (1 to 3%) to a low-carbon, low-chromium 400-series stainless steel can increase its yield and tensile strength via formation of body-centered cubic (bcc) martensite in the alloy. This phase is often called α' to differentiate it from the equilibrium bcc ferrite (α). Ideally, the composition of such an alloy falls in the ferritic + martensitic regime of the Schaeffler diagram (Ref 23). Table 2.4 lists the composition of nickel-modified 409L (409LNi), which is being used for the manufacture of automotive exhaust flanges. Mechanical properties of this alloy are covered in Chapter 7, "Mechanical Properties." Nickel modification of 434L is also found to be beneficial in terms of increased strength and brazeability.

A PM version of the popular wrought alloy type 440C is also being produced commercially. A small amount of boron is used in order to help achieve full theoretical density via liquid-phase sintering. Stringent control of the process parameters (sintering and heat treatment) is essential, not only to achieve full theoretical density but also to optimize hardness and toughness. Despite these restrictions, the PM process is still considered attractive because of its near-net shape capability.

2.4.4 Duplex and Precipitation-Hardening Grades

Duplex stainless steels combine some of the positive and negative attributes of both micro-constituents. Although some duplex alloys can exhibit superior corrosion resistance compared to 304L or 316L, due to their molybdenum and nitrogen contents, the characteristics of these alloys most commonly fall in between those of the austenitic and ferritic families. The strongest attributes of these alloys are their high strength and excellent resistance to chloride stress-corrosion cracking when compared to austenitic alloys. They can be optimal choices in specific applications. Duplex stainless steels have not been processed via conventional PM routes. They have, however, been made via the MIM route.

The preferred process route for alloys of these two families is MIM. In the case of the high-strength alloys, such as 17-4 PH, achievement of full or near-full density is essential in order to realize the full benefit of their superior mechanical properties. Nevertheless, the feasibility of using the conventional PM process route to produce 17-4 PH with sintered densities greater than 7.3 g/cm^3 has been demonstrated by Reinshagen and Witsberger (Ref 29). In their study, these relatively high sintered densities were achieved by selecting finer-than-conventional, prealloyed powders, combined with high-sintering-temperature (1260 °C, or 2300 °F) sintering. Sintered and heat treated 17-4 PH materials produced in their study showed yield strengths greater than 690 MPa (100 ksi), with tensile elongations of 7%. The chemical composition of 17-4 PH alloy is listed in Table 2.5.

2.5 MIM Grades

Processing by MIM typically produces near-full-density components, primarily from gas-atomized fine powders. Due to their low porosity and low oxygen contents, MIM stainless steels exhibit significantly better corrosion resistance and mechanical properties compared to the conventionally processed PM stainless steels. Their corrosion resistance is generally equal to those of their wrought counterparts. Currently, 316L is the most widely used MIM grade of stainless steel. It is the material of choice when corrosion resistance is the primary requirement. For applications requiring high strength, MIM technology

Table 2.5 Chemical compositions of metal injection molding (MIM) grades of stainless steels

Grade	Standard/ Nonstandard	Fe	C	Cr	Ni	Mo	Cu	Si	Mn	Other	Source/Ref
316L	MPIF	bal	0.03(a)	16.0–18.0	10.0–14.0	2.0–3.0		1.0(a)	2.0(a)		
316L	EPMA	bal	0.03(a)	16.0–18.5	10.0–14.0	2.0–3.0		1.0(a)	1.0(a) 1.0(a)		EPMA MIM standards
17-4 PH	MPIF	bal	0.07(a)	15.5–17.5	3.0–5.0		3.0–5.0	1.0(a)	1.0(a)	Nb + Ta: 0.15–0.45	
17-4 PH	EPMA	bal	0.07(a)	15.0–17.5	3.0–5.0		3.0–5.0	1.0(a)	1.0(a)	Nb + Ta: 0.15–0.45; others: 1.0(a)	EPMA MIM standards
430L	MPIF	bal	0.05(a)	16.0–18.0				1.0(a)	1.5(a)		
440C	JIS	bal	0.95–1.20	16.0–18.0							Ref 30
310S	JIS	bal	0.26–0.4	24.0–26.0	19.0–22.0				1.5(a)		Ref 30
420J	JIS	bal		12.0–14.0							Ref 30
904L	DIN 91.4539	bal	0.03(a)	21.6(a)	24.8(a)	4.47(a)	1.5(a)	0.02(a)			Ref 31
PANACEA	Proprietary	bal	0.095(a)	17.1(a)	0.03(a)	3.4(a)		0.6(a)	12.3(a)	N = 0.73(b)	Ref 31
312 (duplex)	Nonstandard	bal	0.03(a)	24.0–26.0	5.5–6.5	1.2–2.0		1.0(a)	2.0(a)		

(a) Maximum. (b) Typical.

offers a number of 400-series martensitic alloys and the precipitation-hardening 17-4 PH alloy. The 400-series martensitic alloys processed via MIM do exhibit significantly higher toughness and ductility compared to their conventionally processed PM counterparts.

Table 2.5 lists a number of MIM-grade stainless steels that are currently in commercial use. The composition of standard MIM-grade 316L is identical to that of the PM-grade 316L. The compositions of other standard MIM grades vary slightly from the compositions of standard PM grades.

A fairly large number of nonstandard grades of stainless steels are being processed via MIM, some of which are proprietary alloys. Compared to conventional PM technology, MIM has greater flexibility with alloy compositions, because it is not limited by the restrictions imposed by the water-atomization process or by the effect of alloying additions on powder compressibility. Material cost is also a lesser concern. Taking advantage of these factors, for example, Wohlfromm et al. (Ref 31) have demonstrated the feasibility of producing a nickel-free, non-magnetizing grade of stainless steel for use in the manufacture of cases for wristwatches (eliminating nickel allergy). Substitution of nickel by manganese and nitrogen helps retain its austenitic structure. It far exceeds MIM 316L in crevice corrosion resistance, and in the solution-annealed condition, it exhibits a yield strength of 690 MPa (100 ksi), along with a tensile elongation of 35%. Table 2.5 lists its composition. This proprietary alloy was developed by ETH (Zurich) under the trade name Catamold PANACEA (protection against nickel allergy, corrosion, erosion, and

abrasion). High nitrogen content (0.8 to 1.0%) is achieved via sintering in a nitrogen-rich atmosphere. Similarly, 17-4 PH alloys containing higher chromium (up to 19%), molybdenum (up to 6%) or silicon (up to 3%) have also been processed via MIM. Properties of a number of standard and nonstandard grades are reported by Achitika (Ref 30). Fukuda et al. (Ref 32) have demonstrated the benefits of increasing silicon content of MIM 410L, up to 3%, in terms of its soft magnetic properties.

REFERENCES

1. C.W. Kovach and J.D. Redmond, Austenitic Stainless Steels, *Practical Handbook of Stainless Steel and Nickel Alloys*, S. Lamb, Ed., ASM International, 1999, p 160
2. L.F. Pease III and W.G. West, *Fundamentals of Powder Metallurgy*, MPIF, Princeton, NJ, 2002, p 242
3. H.I. Sanderow and T. Prucher, Mechanical Properties of PM Stainless Steels: Effect of Composition, Density and Sintering Conditions, *Advances in Powder Metallurgy and Particulate Materials*, ed. M. Phillips, J. Porter, Vol 7, Part 10, Metal Powder Industries Federation, Princeton, NJ, 1995, p 10-13 to 10-28
4. D. Peckner and I.M. Bernstein, *Handbook of Stainless Steels*, McGraw-Hill Publications, New York, NY, 1977
5. M.V. Rao, *Metallography, Structures and Phase Diagrams*, Vol 8, *Metals Handbook*, 8th ed., Amercan Society for Metals, 1973, p 291

6. H. Okamato, *Binary Alloy Phase Diagrams*, 2nd ed., T.B. Massalski, Ed., ASM International, 1990

7. A.J. Schaeffler, Constitution Diagram for Stainless Steel Weld Metal, *Met. Prog.*, Vol 56, Nov 1949, p 680

8. R.A. Lula, *Stainless Steels*, American Society for Metals, 1986, p 73

9. J.H. Reinshagen and R.P. Mason. The Basics of 400-Series PM Stainless Steels, *Advances in Powder Metallurgy and Particulate Materials*, ed. R. McKotch, R. Webb, Vol 9, MPIF, Princeton, NJ, 1997, p 9-3 to 9-17

10. A. Sabata and W.J. Schumacher, Martensitic and Ferritic Stainless Steels, *Practical Handbook of Stainless Steels*, S. Lamb, Ed., ASM International, 1999, p 132

11. A.J. Sedriks, *Corrosion of Stainless Steels*, 2nd ed., sponsored by The Electrochemical Society, Princeton, NJ, and John Wiley & Sons, New York, 1996, p 111

12. S.O. Shah, J.R. McMillen, P.K. Samal, S.A. Nasser, and E. Klar, "On the Real Life Performance of Sintered Stainless Steel ABS Sensor Rings," Paper 970423, SAE Congress and Exposition (Detroit, MI), 1997

13 J. Demo, Structure and Constitution of Wrought Ferritic Stainless Steels. *Handbook of Stainless Steels*, D. Peckner and I.M. Bernstein, Ed., McGraw-Hill Book Co., 1977, p 5–28

14. J.R. Davis, Ed., *Stainless Steels*. ASM Specialty Handbook, ASM International, 1994, p 51

15. C.J. Novak, Structure and Constitution of Wrought Austenitic Stainless Steels, *Handbook of Stainless Steels*, D. Peckner and I.M. Bernstein, Ed., McGraw-Hill Book Co., 1977, p 4–10

16. T. Takeda and K. Tamura, Pressing and Sintering of Chrome-Nickel Austenitic Stainless Steel Powders, trans. H. Brucher, *J. Jpn. Soc. Powder Powder Metall.*, Vol 17 (No. 2), 1970, p 70–76

17. D.J. McMahon and O.W. Reen, The Prediction of Processing Properties of Metal Powders, *Modern Developments in Powder Metallurgy*, Vol 8, Metal Powder Industries Federation, Princeton. NJ, 1974, p 41–60

18. T. Kato and K. Kusaka, On the Recent Development in Production Technology of Alloy Powders, *Mater. Trans.*, *JIM*, Vol 31 (No. 5) 1990, p 363–374

19. E. Klar and W.M. Shafer, On Green Strength and Compressibility in Powder Metal Compaction, *Modern Developments in Powder Metallurgy*, Vol 9, Metal Powder Industries Federation, Princeton, NJ, 1976, p 91–113

20. K. Kusaka, T. Kato, and T. Hisada, Influence of S, Cu, and Sn Additions on the Properties of AISI 304L Type Sintered Stainless Steel, *Modern Developments in Powder Metallurgy*, ed. E. Aqua, C. Whitman, MPIF, Princeton, NJ, 1984, p 247–259

21. P.K. Samal, O. Mars, and I. Hauer, Means to Improve Machinability of Sintered Stainless Steel, compiled by C. Ruas and T.A. Tomlin, *Advances in Powder Metallurgy and Particulate Materials*, Vol 7, MPIF, Princeton, NJ, 2005, p 66–78

22. O.W. Reen, U.S. Patent 3,980,444, 1976

23. O.W. Reen and G.O. Hughes, Evaluating Stainless Steel Powder Metal Parts, Part II: Corrosion Resistance, *Precis. Met.*, Aug 1977, p 53–54

24. P.K. Samal and J.B. Terrell, Corrosion Resistance of Boron-Containing 316L, Part 7, *Advances in Powder Metallurgy and Particulate Materials*, ed. H. Ferguson, P. Whychell, Jr., MPIF, Princeton, NJ, 2000, p 7-17 to 7-31

25. J. Kazior, I. Cristofolini, A. Molinari, and A. Tiziani, Sintered Stainless Steel Alloyed with Boron, *Proceedings of PM World Congress* (Paris, France), EPMA, 1984, p 2097–2100

26. O.W. Reen, U.S. Patent 4,014,680, 1977

27. M. Svilar and H.D. Ambs, PM Martensitic Stainless Steels: Processing and Properties, *Advances in Powder Metallurgy and Particulate Materials*, ed. E. Andreotti, P. McGeehan, Vol 2, MPIF, Princeton, NJ, 1990, p 259–272

28. P.K. Samal, J.B. Terrell, and S.O. Shah, Mechanical Properties Improvements of PM 400-Series Stainless Steels via Nickel Addition, *Advances in Powder Metallurgy and Particulate Materials*, ed. C. Rose, M. Thibodeau Vol 3, MPIF, Princeton, NJ, 1999, p 9-3 to 9-14

29. J.H. Reinshagen and J.C. Witsberger, Properties of Precipitation Hardening Stainless Steel Produced by Conventional Powder Metallurgy, *Advances in Powder Metallurgy and Particulate Materials*,

ed. C. Lall, A. Neupaver, Vol 7, MPIF, Princeton, NJ, 1994, p 7-313 to 7-339

30. M. Achitika, Development of MIM Components for Automobile and Power Tools, *Ninth Case Studies on New Product Development*, JPMA Session I, PM World Congress (Kyoto, Japan), Nov 16, 2000, p 25–34

31. H. Wohlfromm, M. Bloemacher, D. Weinand, P. Uggowitzer, and O. Spiedel, Novel Stainless Steel for Metal Injection Molding, *Proc. of the 1998 Power Metallurgy World Congress and Exhibition*, Vol 3 (Granada Spain), EPMA, p 3–8

32. M. Fukuda, Y. Soda, and Y. Yoshida, Improvements in the Soft Magnetic Properties of the Ferritic Stainless Steels, *Proc. of PM World Congress 2000* (Kyoto, Japan), p 503–505

CHAPTER 3

Manufacture and Characteristics of Stainless Steel Powders

THE GOAL in stainless powder manufacture is to control fundamental powder properties, both chemical and physical, such as composition, particle size, particle size distribution, and particle shape, so as to produce powders that are suited for their intended uses and to conduct the powder production process in a cost-effective way. Intended uses of powders are most often discussed in terms of specific requirements of their engineering properties, such as apparent density, flowability, green strength, compressibility, and so on. In most cases, the dependence of these engineering properties on fundamental properties is known only qualitatively, which accounts for the still large empirical content in powder metallurgy (PM) processing. This also holds true for corrosion resistance.

For stainless steel powders that can be cold pressed in a die, the so-called compacting-grade powders, water atomization is the preferred powder production process, because water atomization renders the powder particles irregular (Fig. 3.1).

Stainless steel powders used for consolidation through extrusion or hot pressing are usually made by gas atomization, and these PM techniques typically produce fully dense parts. Gas-atomized stainless steel powders have a spherical particle shape (Fig. 3.2) and superior packing densities, properties desired for these modes of consolidation.

Although the cooling rates in water atomization are higher than those in gas atomization, both are sufficient to produce powders free of macrosegregation and which, after processing, yield homogenous microstructures. These advantages, when combined with consolidation to full density, can in some alloy systems produce

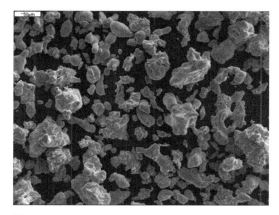

(a)

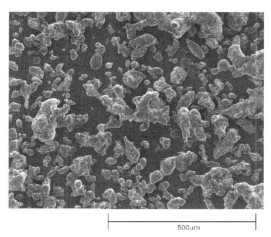

500μm

(b)

Fig. 3.1 Examples of water-atomized stainless steel powder. SEM of (a) water-atomized 409L powder, (b) water-atomized 316 stainless powder of high apparent density (slightly more rounded edges); original magnified 100 times

10 μm

Fig. 3.2 SEM of gas-atomized 316L

properties superior to those attainable with wrought alloys, for example, extended alloy solubility and improved formability, fatigue, and impact strength.

Stainless steel powders used for injection molding are generally made by gas, water, hybrid gas-water, or centrifugal atomization. The emphasis in this case is to make a powder with a high yield of fines (<20 μm) and with a particle shape that is nearly spherical so as to combine the requirements of a high fill density and of shape retention during debinding (section 4.2 in Chapter 4, "Compacting and Shaping").

Some typical atomization parameters for both kinds of atomization are listed in Table 3.1.

With increasing market size, some induction furnaces are expected to be replaced by larger-sized electric arc furnaces that, in combination with argon oxygen decarburization, will permit tighter control of composition as well as the use of less expensive raw materials.

The choice of raw materials and melting of gas-atomized stainless steels follow, more or less, those used for wrought and cast stainless steels. However, because water-atomized stainless steels require good compacting properties, their raw materials selection, melting, and atomization differ in several important aspects.

3.1 Water Atomization of Stainless Steel Powders

3.1.1 Brief Process Description

Water atomization of metals is described in detail in Ref 1. In the following, process details that differ from general water atomization of metals are highlighted as they relate to basic and engineering properties specific to stainless steel powders. Also, much finer powders are required for metal powder injection molding than for conventional compaction.

Compacting-Grade Powders. Figure 3.3 shows a schematic of a water-atomizing system. It typically consists of a power unit, a melting furnace, an atomization tank, and a collection vessel. For stainless steels, high-frequency induction furnaces are common, because they permit, through induction stirring, efficient alloying of the various constituents of a steel. Open-air melting is also common.

Table 3.1 Typical industrial process parameters for water and gas atomization of stainless steels

Parameter	Water atomization	Gas atomization
Type of furnace	Induction	Induction
Furnace capacity	Up to 4500 kg (10,000 lb)	Up to 5000 kg (11,000 lb)
Melting	Open air	Open air
Atomizing medium	Water	Argon (nitrogen)
Water pressure	11–18 MPa (1500–2500 psi)	...
Gas pressure	...	0.76–2.6 MPa (110–380 psi)
Water flow rate (single-orifice nozzle)	200–400 L/min (53–106 gal/min)	...
Gas flow rate (single-orifice nozzle)	...	20–40 m³/min (800–1600 scfm)
Metal flow rate	45–90 kg/min (100–200 lb/min)	45–90 kg/min (100–200 lb/min)

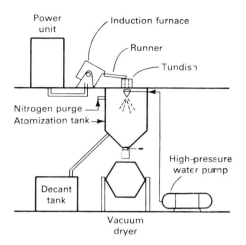

Fig. 3.3 Schematic of a water-atomization system. Source: Ref 2

After the melt charge reaches its pouring temperature, typically 50 to 75 °C (100 to 150 °F) above the melting point of a particular alloy, the induction furnace is tilted to allow liquid metal, via a runner, to flow into a tundish, from which the metal emerges as a well-defined stream within the atomization tank in which it is disintegrated into a powder. Oxidation of the powder decreases significantly with decreasing pouring or atomizing temperature (Ref 3). However, a suitably low pouring temperature must take into account the increasing likelihood of freezing of the liquid metal within the tundish nozzle as the pouring temperature approaches the melting point of the metal. The atomization tank is made of stainless steel. In the interest of low-oxygen-content powders, it is partly filled with water and purged with nitrogen; thus, air leakage into the tank is avoided. Powder producers use proprietary atomizing heads that deliver high-pressure water jets for efficient atomization and maximum yield of useable powder.

High-Pressure Water Atomization. Attempts to produce fine powders by water atomization go back to the 1970s. Water pressures were several times those used in conventional atomization for the production of compacting-grade powders. Special wear-resistant nozzles were used. With water pressures ranging from 60 to 150 MPa (8700 to 21,750 psi), Tanaka et al. (Ref 4) were able to produce steel powders with particle sizes as fine as 5 μm.

Okamoto et al. (Ref 5) describe a high-pressure water-atomization system that allowed them, with a water pressure of 70 MPa (10,150 psi) and

in a protective atmosphere of nitrogen, to produce SUS 430 stainless steel powders with mean particle sizes between 12 and 15 μm and with oxygen contents between 0.2 and 0.3%. Fine-particle-size stainless steel powders produced with this system are available from Kobe Steel and include SUS 304L, SUS 316L, SUS 430, SUS 410L, and SUS 630 (16Cr4Ni4 Cu00.3Nb).

Kikukawa et al. (Ref 6) used a swirl jet, that is, a waterjet that forms a spiral cone, to render the particle shape of high-pressure (83 MPa, or 12,040 psi) water-atomized powders more spherical. For metal powder injection molding (section 4.2 in Chapter 4, "Compacting and Shaping"), a nearly spherical particle shape is considered optimal.

3.1.2 Physical Powder Characteristics

Particle Size and Particle Size Distribution. Particle size is controlled by water pressure; higher pressures produce finer powders and vice versa (Fig. 3.4). Water pressures in the vicinity of 13.8 MPa (2000 psi) disintegrate the stainless steel metal stream into a predominantly −100 mesh (−150 μm) powder, used for most structural parts. So-called V-jet nozzles are widely used for directing the high-pressure water onto the liquid metal. Through different arrangement of such nozzles, various waterjet configurations are possible (Ref 7), that in turn give rise to a variety of particle size distributions. Because atomized powders, when atomized at constant pressure and constant flow rates of metal and atomizing medium, have particle size distributions that form straight lines when plotted on log-normal paper (Fig. 3.5), it is easy to

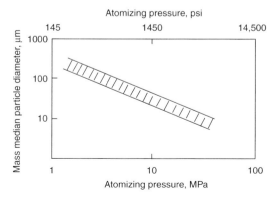

Fig. 3.4 Typical particle size-pressure relationship of water-atomized stainless steels. Source: Ref 34

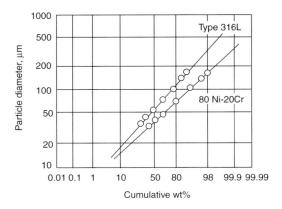

Fig. 3.5 Log-normal plots of cumulative undersized particle size distributions of water-atomized (80Ni-20Cr and type 316L) metal powders. Source: Ref 2

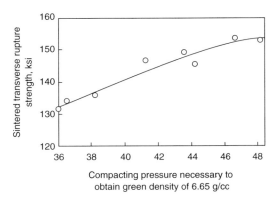

Fig. 3.6 Correlation between compressibility and sintered transverse rupture strength of 316L powders of varying apparent densities. Source: Ref 34

manipulate particle size data, because only two numbers, d_m, the mean mass diameter, and Φ_g, the geometric standard deviation (a measure for the breadth or width of a distribution), define the entire particle size distribution. The geometric standard deviation, Φ_g, is obtained as the ratio of the particle size diameters taken at 84.1 and 50.0% of the cumulative undersized weight plot, respectively (Ref 8).

A narrow particle size distribution assures a high yield of useable powder, because the amount of fines, –325 mesh (–45 μm), should not exceed approximately 35 to 55% for compacting-grade stainless steel powders. Larger amounts of fines impair the powder flowability. Several variables, including the design of the atomizing head, which determines the interaction of the liquid metal with the high-pressure water, affect the width of a particle size distribution. With a good water-atomization system, it is possible to produce powders with a standard deviation, Φ_g, of approximately 2.0. This corresponds to a powder where two-thirds of its particle size distribution (by volume) have an upper-to-lower size ratio of 4 to 1.

More details on atomization systems and atomization process parameters can be found in *Atomization of Melts* by Yule and Dunkley (Ref 1).

Particle Shape. The particle shape of a water-atomized powder has a major influence on the apparent density, flow properties, green strength, and compressibility of the powder; it also affects sintered properties, including dimensional change and mechanical properties. The fundamental property of importance in this case is probably the particle coordination number, because irregularly shaped particles form

more contact points with neighboring particles during pressing. This requires higher compacting pressures to achieve a certain green density, and it generates superior mechanical properties after sintering to the same density. Figure 3.6 illustrates such a relationship for water-atomized 316L powders of apparent densities from 2.8 to 3.4 g/cm^3.

Although statistical methods for the determination of particle shape have been developed, the apparent density, or better yet, the tap density, of a narrow size fraction of a powder is a convenient measure of particle shape that can be quickly and easily determined and correlated with other properties.

Notwithstanding the fact that water-atomized powders in general have irregular particle shape, quite a broad range of particle shapes can be produced by water atomization. Some shape control is possible by modifying chemical composition, but normal control is through the atomizing head design. The appropriate head design is typically arrived at empirically. The patent literature provides a number of specific designs (Ref 9, 10). Improved atomization efficiency is attributed to the so-called two-step atomization technique that allows the production of powders that are finer than those possible with either gas, or water atomization alone. In this technique, the liquid metal is first preatomized with gas, followed by water atomization. Independent control of the two closely linked types of atomization allow for wide control of particle shape (Ref 1).

Physical powder characteristics that increase the apparent density of a powder, for example, a more spherical particle shape or a particle size distribution that provides improved packing,

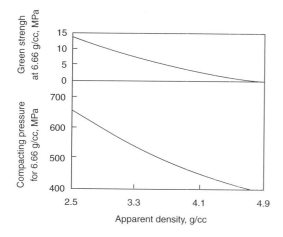

Fig. 3.7 Effect of apparent density on green strength and compressibility of 316L stainless steel powders. Source: Ref 34

improve compressibility at the expense of green strength. This relationship is illustrated in Fig. 3.7 which is based on regression analysis data of 316L powders.

Because for most structural applications, green strength values of approximately 5.5 to 8.3 MPa (800 to 1200 psi) permit pressing and safe handling of green parts, the powder producer tends to keep the apparent densities of stainless steel powders in the range of 2.8 to 3.0 g/cm³, which can provide for this minimum green strength without detracting excessively from the more valued compressibility property. Strictly speaking, the ideal green strength for a particular part is the one that is just sufficient to accomplish safe pressing and handling. This approach will deliver maximum compressibility for that application.

With the advent of warm compaction, which drastically increases green strength (Ref 11), it may be possible to use spherical or nearly spherical stainless steel powders for compaction. Besides the benefit in compacted density, spherically shaped powders with their higher apparent densities permit faster compaction due to faster flow rates and lower die fill heights. Furthermore, the more uniform density distribution arising from warm compaction, as well as the potential to use low-oxygen-content gas-atomized powders, should benefit the sintering process as well as the corrosion properties of the sintered material, because the high sintering temperatures presently used in the sintering of stainless steels (Chapter 5,

"Sintering and Corrosion Resistance") serve mainly to reduce the surface oxides of the water-atomized stainless steel powder.

3.1.3   Chemical Powder Characteristics

The control of chemical powder characteristics is of equal importance to that of the physical characteristics of a powder. This is mainly due to the intrinsic effects of some constituents on compacting properties, and the preferred oxidation of silicon and manganese during water atomization.

Standardized wrought and cast stainless steel compositions typically have composition ranges for their major and minor constituents (Chapter 2, "Metallurgy and Alloy Compositions") and maximum limits for a number of trace elements. Manufacture within these limits has only moderate effects on the properties of a steel. For compacting-grade stainless steel powders, however, some constituents even within these ranges exert a profound influence on the compacting properties of a powder and indirectly, on the corrosion-resistance properties of the sintered parts.

For good compressibility, a powder should have good deformability, as indicated by low hardness, and/or low yield strength. In comparison to plain iron powder, stainless steel powders require substantially higher compacting pressures because of their high alloy content, which increases their hardness and work-hardening rates. With these properties in mind, powder manufacturers try to optimize the concentrations of several constituents in water-atomized stainless steel. The effects of nickel and chromium on compressibility have already been discussed in Chapter 2, "Metallurgy and Alloy Compositions."

The negative effects on compressibility are particularly strong for the two interstitials, carbon and nitrogen (Ref 12). Daido Steel achieved very low interstitial contents in ferritic grades of stainless steel powders by using argon oxygen decarburized melt stock combined with additional refining (Ref 13). Annealing at 830 °C (1526 °F) in hydrogen further increases compressibility while preventing any caking or sintering of the powder. Annealing softens the powder by stress relieving the rapidly solidified (quenched) powder.

The importance of a low carbon content is also reflected in the fact that the commercial austenitic stainless steel powder grades 304L and 316L are of the low-carbon variety, that is, the "L" designation that specifies a maximum carbon content of 0.03%. A maximum carbon content of 0.03% is meant to prevent carbide precipitation during cooling, in accordance with the solubility of carbon in austenitic stainless steels. However, it is known from wrought austenitic stainless steels that nickel contents above 10% decrease carbon solubility and therefore increase susceptibility to intergranular corrosion. Figure 3.8 shows the effect of carbon and nickel on the intergranular corrosion penetration of 18 wt% Cr-base stainless steels.

For water-atomized grades of 304L and 316L that have nickel contents close to their upper ranges, that is, 13 and 14%, respectively, it is advisable to keep their carbon contents to ≤0.02%. While it is possible to achieve such low carbon contents with an induction furnace and typical commercial sintering, it is safer to forgo the benefits of high nickel content and seek compressibility improvements through other means. The potential for intergranular corrosion represents an ever-present challenge because of the slow cooling rates employed in most industrial sintering operations.

The beneficial effects of nickel and the detrimental effects of carbon and nitrogen on compressibility act in the same direction for the property of green strength. Low levels of interstitials are also beneficial to magnetic and corrosion-resistance properties, particularly for the ferritic stainless steels. In wrought and cast

stainless steel technology, very low levels of interstitials are achievable through argon oxygen degassing, usually in combination with electric arc melting. In PM, based on the use of induction furnaces, low levels of carbon and nitrogen are feasible through control of raw materials in the melt charge and by minimizing nitrogen pickup from the air during melting. Control of oxygen is more complex and is discussed in the following section.

Effects of Silicon and Manganese. Silicon is probably the most critical of all constituents. This is because most of the oxidation taking place during water atomization of a stainless steel is silicon dioxide, SiO_2, which, in the typical industrial sintering process, is reduced only in part and which therefore, depending on sintering conditions, gives rise to variable amounts of residual oxides (second-phase oxides) in a sintered stainless steel part. This metallurgical defect, together with excessive and variable amounts of carbon and nitrogen (giving rise to second-phase carbides and nitrides), accounts for variable dynamic mechanical properties such as impact and fatigue strength as well as for variable corrosion-resistance properties. Silicon dioxide also can form on the surfaces of a part during cooling after sintering and impair corrosion resistance. The higher the density of a sintered part, the greater the negative effect on mechanical properties, including scatter or variability of properties, due to the declining effect of porosity.

In early experiments, before the role of silicon had been appreciated and atomizing tank atmospheres were not always kept inert, stainless steel powders often had oxygen contents of 3000 to 5000 ppm, were dark colored, and had inferior compacting properties. When it was discovered that higher silicon contents, typically around 0.8 to 1.0%, produced lower oxygen contents and lower apparent densities, the stage was set, in the 1950s, to make water atomization the accepted commercial process for the production of stainless steel powders. Later, it was discovered that low manganese levels, lower than those found in most wrought stainless steels of comparable composition, were able to further reduce oxidation during atomization.

Manganese, a benign and often useful constituent in many wrought stainless steels, appears to increase oxidation during water atomization. For high-manganese-content stainless steels, oxidation can be so severe that substantial quantities of hydrogen are formed,

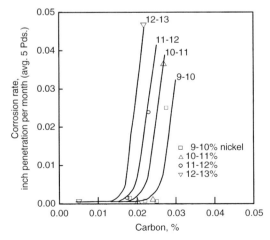

Fig. 3.8 Effect of carbon and nickel content on intergranular corrosion penetration rate of 18 wt% Cr-base stainless steels. Alloys sensitized for 100 h at 550 °C (1022 °F). Immersion in boiling 65% nitric acid. Pds., periods (48 h) of exposure. Source: Ref 14

which, for safety reasons, must be taken into account in the design of an atomization system.

Already in 1966, Takeda and Tamura (Ref 15) suggested that, to achieve a low-oxygen-content powder, the manganese content of the liquid steel should not exceed 0.15%. In 1976, Dautzenberg and Gesell (Ref 3) reported the opposite effects of silicon and manganese with regard to oxidation during water atomization. Apparently, the opposite behavior of these two elements is also present in the solid-state oxidation of 20/25-Cr/Ni steels and thereby finds its explanation.

Dunkley (Ref 16) shows the combined effect of silicon and manganese on the oxide contents of water-atomized 304L as a function of their arithmetic differences (Fig. 3.9).

With typical silicon and manganese contents of 0.8 to 0.9 and 0.1%, respectively, it is possible to obtain oxygen contents as low as 1500 to 1700 ppm. Kato and Kusaka (Ref 17) show oxygen contents as low as 0.10%, for silicon = 0.7%, and manganese = 0.07%.

Kusaka et al. (Ref 18) report the beneficial effect of copper on compressibility and corrosion resistance in austenitic stainless steels. Commercial tin-containing stainless steel powders, designated as LSC and Ultra (Chapter 6, "Alloying Elements, Optimal Sintering, and Surface Modification in PM Stainless Steels"), also make use of copper additions. They also studied the effect of silicon on apparent density.

A high-silicon version of 316L, 316B, or 316L-Si contains over 2% Si. The larger amount of silicon causes the liquid stainless steel to form a more irregularly shaped powder, with a lower apparent density of less than 2.0 g/cm^3 and a higher green strength. The coarser mesh fractions of this powder are used for making stainless steel filters possessing large pore sizes and large porosities.

Nyborg et al. (Ref 19) found the thicknesses of oxide layers for gas- and water-atomized stainless steel powders to be constant or to increase with particle size. The different behavior is attributed to mass transfer of oxygen being the rate-determining step in gas atomization, and to exposure to water vapor in water atomization, which renders the cooling rate of individual particles as the critical parameter.

Fine powder particles develop a thinner oxide layer due to their faster rate of cooling, but they possess a relatively high oxygen content because of their large specific surface area. Coarse powders, on the other hand, develop a thick oxide layer that, despite their low specific surface areas, leads to higher oxygen contents. Thus, because of the opposite effects of cooling rate and surface area with respect to particle size, a plot of oxygen content versus particle size exhibits a minimum that, for the conditions employed, lies within the subsieve particle size range ($1/d \sim 0.06$ or ~ 17 μm).

The oxygen contents of water-atomized stainless steel powders, and also of sintered stainless steel parts made from such powders, compared with 200 ppm for wrought and cast stainless steels are an order of magnitude higher. This difference, together with inherent porosity, are the two most significant differences between wrought and sintered PM stainless steels. As is seen later, reduction and minimization of oxides will markedly improve the corrosion resistance as well as the dynamic mechanical properties of sintered stainless steels.

Both Metal Powder Industries Federation (MPIF) and ASTM International, the two prominent organizations for PM material standards in the United States, use the composition ranges of the American Iron and Steel Institute (AISI) for wrought stainless steels, with few exceptions. For example, MPIF standard 35 of 2007 and ASTM standard B 783 of 2004 show the chemistry ranges for silicon and manganese, not to mention other constituents, to be identical to their wrought counterparts. Such statements are misleading and can give rise to misunderstandings between powder producers and stainless steel part users, because significant departures from the PM-optimized chemistries

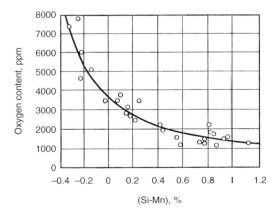

Fig. 3.9 Effect of silicon and manganese on the oxygen content of 304L powders

will result in substandard powders similar in quality to those manufactured in the 1950s.

Increasing the density of stainless steels through warm compaction is discussed in Chapter 4, "Compacting and Shaping."

3.1.4   Raw Materials and Melting

Melting of stainless steels for the purpose of water atomization differs in several respects from the practice in wrought and cast stainless steel. Because little refining takes place during melting in an induction furnace, the melt charge must be protected from excessive oxidation during melting. Furthermore, without oxygen injection, no carbon removal takes place, and much of the carbon introduced with raw materials remains in the charge. This mandates the use of low-carbon iron and low-carbon ferroalloys. When chromium-bearing constituents enter the melt charge and the temperature is high enough, the chromium will readily absorb nitrogen from the air. This imposes the use of low-nitrogen ferroalloys and a melt practice that minimizes nitrogen pickup. The latter can be accomplished by shrouding the melt surface with inert gas, as described in the SPALL process. Manganese must be kept low, which calls for using stainless steel scrap only sparingly, if at all. Silicon must also be controlled carefully. Silicon, in addition to its effect on oxygen content of the powder and particle shape, serves not only as a deoxidizer but, together with phosphorus, significantly affects the viscosity of the melt.

During melting, a viscous, semisolid slag forms that protects the melt from excessive oxidation. A low manganese content, essential for controlling oxidation during atomization, also keeps the slag from becoming too fluid. Just prior to pouring, the slag is removed and the liquid metal is poured, via a runner, into a preheated tundish. Care must be taken, especially at the end of the pour, to prevent slag from entering the tundish nozzle and becoming part of the atomized powder.

3.1.5   Atomization

As mentioned earlier, a major difference between water and gas atomization is the substantial oxidation taking place with the former. The liquid stainless steel, after its deoxidation with ferrosilicon, has a low oxygen content of less than 200 ppm. However, when it comes into contact with water, like most other metals, it reacts in accordance with:

$$xMe + yH_2O \rightarrow Me_xO_y + yH_2$$

The metal becomes partly oxidized, even though the atomizing tank has been purged with an inert gas to minimize oxidation of the powder. Reinshagen and Neupaver (Ref 20) reported an empirical correlation of a number of atomization parameters with the amount of oxidation, particle size, particle shape, and powder dispersion for a stainless steel powder.

Until 1980, it was widely believed that mainly chromium oxides formed during atomization of stainless steels, due to the large concentration of this constituent in stainless steels, its affinity to oxygen, and the fact that powders and parts occasionally had a greenish tint. However, surface analytical techniques, particularly Auger and electron spectroscopy for chemical analysis (ESCA), have shown that the surfaces of unsintered 316L were greatly depleted of chromium and iron and instead enriched by silicon and oxygen (Fig. 3.10). Surface analysis by ESCA has shown these oxides to be silicon oxides. Larsen and Thorsen (Ref 22) confirmed these findings and reported that the SiO_2-oxide film remained continuous during low-temperature (1120 °C, or 2048 °F) sintering in dissociated ammonia, whereas at higher temperature (1250 °C, or 2282 °F), it formed discrete particles.

In essence, the critical role of silicon during water atomization and sintering is similar to what happens at a somewhat higher temperature in conventional stainless steel technology. During refining (oxygen blowing), silicon becomes oxidized and enters the oxide slag phase of the slag/steel equilibrium, whereas chromium predominantly remains unoxidized in the steel phase. Any chromium that has become oxidized and has entered the slag phase is recovered by subsequent deoxidation with silicon. The main difference between wrought and PM stainless steel is that, in the former, the undesirable silicon dioxide is removed via the slag, whereas in PM it remains in the product unless it is removed or reduced during sintering. In some high-silicon steels, the preferred oxidation of silicon during water atomization can also be demonstrated visually. The powder particles, when dissolved in aqua regia, leave behind hollow, whitish particles, presumably consisting of silicon dioxide.

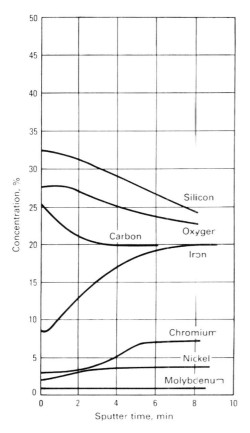

Fig. 3.10 Auger composition depth profile of a type 316L stainless steel green part. Source: Ref 21. Reprinted with permission from MPIF, Metal Powder Industries Federation, Princeton, NJ

Even plain iron powders made by water atomization, with silicon contents of approximately 0.1%, have surfaces that are highly enriched in SiO_2 (Ref 23). The low concentration, however, tends to mask the negative effects of that oxide layer.

Silicon, with its greater affinity for oxygen than chromium and its high mobility in the liquid state, migrates to the surface of the liquid particle during water atomization, in preference of other constituents and despite its small concentration in the alloy. The silicon dioxide-enriched surface is protective and limits total surface oxidation to approximately 1000 to 2500 ppm oxygen in stainless steel powders. With small amounts of silicon and/or with high levels of manganese, surface oxidation can rise to over 5000 ppm. This explains why a certain critical amount of silicon is required to minimize oxidation and why it can mask, within limits, the negative effect of manganese, although it is

better to keep the manganese concentration as low as possible.

At low temperatures, the surface oxides formed on stainless steels are those of iron, manganese, and chromium. It is the presence of chromium-rich hydrated oxides that provide the so-called passive layer in stainless steels and account for their excellent corrosion properties. Most other surface oxides interfere with the formation and stability of a passive layer. Silicon dioxide appears to be such an oxide, and its formation in sintered stainless steels must therefore be minimized.

The phenomenon of preferred silicon oxidation has major implications because of the high affinity of silicon to oxygen and the difficult reduction, even in a matrix of stainless steel, of SiO_2 during sintering. When left in the powder and sintered part, oxygen/oxides will impair the compacting properties of the powder and the mechanical and corrosion properties of a sintered part. Perhaps some of the rare earth metals, alone or in combination with silicon and other deoxidants, would be superior to silicon, in that they generate a low level of oxidation during atomization, produce an irregular particle shape for good compactibility, and their oxides are more reducible than SiO_2. Stainless steel parts could then be sintered at "ordinary" low temperatures of approximately 1149 °C (2100 °F) without impairing their corrosion properties. Furthermore, a superior deoxidant may also slow the fast cooling rates necessary after sintering and aid in preventing the formation of silicon dioxide on the surface of a part (section 5.2.1 in Chapter 5, "Sintering and Corrosion Resistance").

In this connection, it is pertinent to mention that compacting-grade stainless steel powders with very low-oxygen-content surfaces can be made by thermal agglomeration of fine, that is, predominantly –325 mesh (–45 μm), low-oxygen-content gas-atomized spherical powder into –100 mesh (–150 μm) agglomerates (Ref 24). The agglomerates are strong enough to withstand deformation during compaction. A less expensive approach would be to agglomerate a low-oxygen-content spherical stainless steel powder through the use of an organic binder, similar to powder injection molding (Ref 25). The binder should decompose or volatilize without a carbon residue during the heat-up phase of the sintering process. If properly formulated, such a powder would have maximum compacting properties and could be sintered at "normal" temperatures for maximum corrosion resistance. Or, using warm compaction (Chapter 4, "Compacting and Shaping"), because of the greatly improved green strength, lower oxygen content, and more spherically shaped powders, perhaps even inert gas-atomized powders (with appropriate binders) could be used to bring the oxygen content, prior to sintering, to a level approaching that of wrought stainless steels.

Despite the negative effects of silicon dioxide, experience has taught that it is preferable to cope with a small amount of this oxide rather than the alternative of having much larger amounts of other (manganese, iron, chromium) oxides with their detrimental effects on compressibility. How to cope with silicon dioxide reduction in the sintering process is dealt with in Chapter 5, "Sintering and Corrosion Resistance".

Therefore, a major goal in the water atomization of stainless steels is to produce powders with oxygen contents as low as possible yet sufficient to render the particle shape irregular for adequate green strength. In 1968, Dautzenberg (Ref 26) viewed the oxygen content of stainless steel powders and sintered parts as a major quality criterion on the basis of the detrimental effect of oxygen content ($\geq$0.2%) on tensile strength and elongation of sintered 18/10 Cr/Ni steel. As is seen later, this statement can be extended to include lower oxygen contents below 0.2% as well as corrosion-resistance properties.

As is clear from the previous paragraph, producing "good" water-atomized stainless steel powders requires control of overall chemical composition and powder surfaces to a much greater extent than is required in ordinary PM steel powders.

In this context, the authors stress some changes in process and quality control that have occurred in recent years as a result of ever-increasing demands for quality and consistency of PM powders and parts. When a powder or parts producer changes or modifies a production process, powder or parts may not always perform as intended, even if the specified properties are met. This problem arises from the fact that the typical PM specification refers to only a few major or critical properties, assuming that other, nonspecified properties will fall into place. Unfortunately, this is not always the case and can cause costly mistakes and delays. Today (2007), the competent and capable manufacturer is aware of this and places great emphasis on rigorous process control. If he changes or modifies his process, he informs his customer beforehand to minimize any potential problems.

3.2 Gas Atomization of Stainless Steel Powders

In gas atomization, melting is often conducted under a protective atmosphere or under vacuum, in order to protect reactive constituents from becoming oxidized. Such systems are designed for dry collection of the atomized powder, and the atomization tanks are then quite tall, usually from 6 to 10 m (20 to 33 ft), to ensure solidification of the powder particles before they reach the bottom of the atomizing tank. Figure 3.11 shows a schematic of such a system.

Horizontal gas atomization using horizontal tanks is used for the same purpose. Anval Nyby Powders of Sweden uses such a system for making stainless steel powders for extrusion into pipes. The horizontal design is said to be less expensive than a vertical design. Melting is performed in open air. Nevertheless, because of the absence of oxidation during atomization, the powder possesses practically the same chemical composition as the melt, and oxygen contents are quite low, typically less than 200 ppm. Although low-oxygen, inert-gas-atomized powders also show surface enrichment of its high-oxygen-affinity constituents, such oxide layers are only a few atomic layers thick, and any negative effects on interparticle bonding can be minimized or eliminated by including shearing elements in the consolidation process. Also, much coarser powders than those used in conventional cold pressing and sintering decrease surface-related problems and are preferred for extrusion and hot pressing.

The atomizing heads used in gas atomization are often of the so-called confined or close-coupled design (Fig. 3.12).

Coupling, that is, minimizing the distance between liquid metal and high-pressure gas at the tip of the tundish, maximizes the energy transfer from the gas to the liquid metal and results in more efficient atomization. It also can produce a narrower particle size distribution, an important economical goal in many uses. Fundamentally speaking, the thinner the metal stream exposed to the high-pressure gas stream, the more uniform the gas-metal interaction and

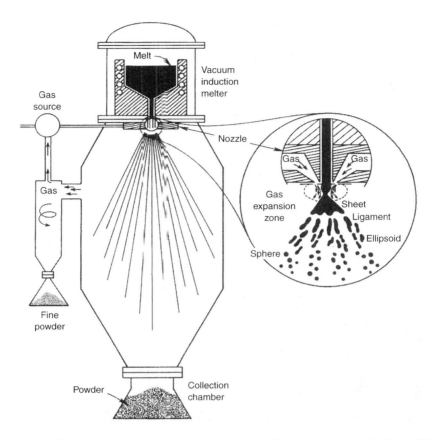

Fig. 3.11 Schematic of inert gas atomization system with expanded view of the gas expansion nozzle. Source: Ref 27. Reprinted with permission from MPIF, Metal Powder Industries Federation, Princeton, NJ

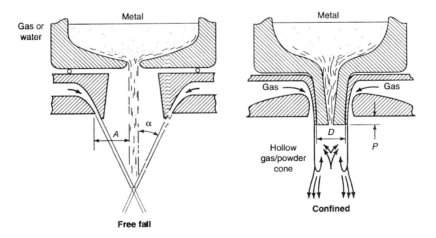

Fig. 3.12 Two fluid-atomization designs. Source: Ref 1

the narrower the particle size distribution of the atomized powder. Concentric coupled nozzles prefilm the liquid metal into hollow cylinders of liquid metal. The standard deviation of a powder distribution (section 3.1.2 in this chapter) increases with increasing metal flow rate; that is, the size distribution becomes broader. Several of the atomizing parameters must be carefully balanced to prevent freezing of the metal at the exit of the tundish nozzle. As in water atomization, particle size in gas atomization is controlled by the pressure of the atomizing medium.

In general, conventional gas atomization employing concentric nozzles produces only small yields of the fine particle sizes preferred for powder injection molding. The reason for this lies in the insufficient and nonuniform energy transfer from gas to liquid metal when the metal flow rate is increased through the use of a larger-diameter tundish nozzle. The so-called WIDEFLOW melt atomization technique (Ref 28) represents a solution to this problem and is expected to provide tonnage quantities of injection molding-grade powders at reduced cost. This technique uses a small linear slot that permits a high gas energy concentration onto the liquid metal over the length of the slot, thereby enabling fine particle sizes and good production rates at the same time.

A recently developed hybrid gas atomization/centrifugal atomization technique uses gas atomization to form a thin and stable liquid metal film on a rotating disk (Ref 29). This permits more efficient melt breakup at the rim of the disk. For a low-melting tin alloy, the authors reported a fine mean particle size of approximately 10 μm and a narrow particle size distribution (Φ_g 1.3 to 1.7). Particle shape was spherical. For more information on gas atomization, see Ref 1.

3.3 Drying, Screening, Annealing, and Lubricating

After water atomization, stainless steel powders are dried and screened. After passing a series of quality-control tests, typically for chemical analysis with inclusion of oxygen and nitrogen, screen analysis, flow rate, apparent density, green strength, compressibility, sintered strength, elongation, and dimensional change during sintering, the powder is ready for shipping. The addition of custom-blended lubricants to a powder by the powder manufacturer has become a common option.

Drying and Screening. After water atomization, the fine powder particles suspended in the powder-water slurry in the atomizing tank are allowed to settle so that excess water can be decanted. Removal of the remaining water is accomplished by centrifuging, filtrating,

vacuum or heat treating, or combinations of these. The dried powder is then screened, either to remove oversize particles, nominally +100 mesh (+149 μm) for compacting-grade powders, or to generate various screen cuts for stainless steel filter applications.

Annealing. Austenitic stainless steel powders are used in their as-atomized condition. Although such powders often are weakly magnetic as a result of the fast quenching action during atomization, which retains a ferritic phase in the inhomogeneous and dendritic microstructures of the particles, they are nevertheless relatively soft and possess good compacting properties. During sintering, these powders become homogenized and fully austenitic.

Due to its low carbon content of <0.03%, 410L has a ferritic microstructure and fairly good compacting properties in the as-atomized condition. However, because of the presence of several hundred parts per million of carbon and nitrogen, the powder green strength and compressibility can be improved by annealing. With very low levels of carbon and nitrogen, <~150 to 200 ppm, annealing of 410L is less effective, because such powder is already quite soft. Annealing transforms the powder into a lightly sintered cake. Light milling nearly re-establishes the original particle size distribution, flow rate, and apparent density.

Lubricating. The two most widely used lubricants for stainless steel powders are lithium stearate and Acrawax C (ethylene bis stearamide). Although the basic function of a lubricant is to reduce die wall friction and tool wear in the compaction process (for this purpose, 0.75 to 1.0% additions are most widely used), lubricants have several other functions and consequences.

Figures 3.13 and 3.14 show the effect of three different lubricants on compressibility and green strength of 316L.

Parts producers take advantage of these differences in that they select lithium stearate when high sintered density matters, and stearic acid when superior green strength is required, or they use combinations of lubricants (Chapter 4, "Compacting and Shaping"). The major cause for these differences lies largely in the effect of a lubricant on particle or powder packing characteristics. Most grades of lithium stearate produce better packing, that is, a higher apparent density, which in turn accounts for improved compressibility and reduced green strength. Acrawax, a synthetic wax, is a popular lubricant,

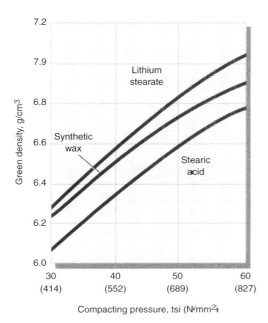

Fig. 3.13 Effect of lubricant on green density of 316L. Source: Ref 34

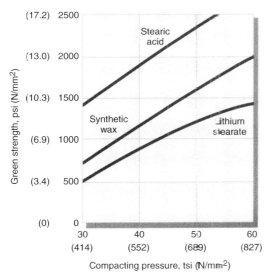

Fig. 3.14 Effect of lubricant on green strength of 316L. Source: Ref 34

because it represents a compromise between the other two extremes and because it has relatively clean burn-off characteristics. The latter is addressed in Chapter 4.

Recent data on binder-augmented lubricants for low-alloy steel powders claim that it is feasible to significantly increase green strength

without any loss in compressibility and ejection load requirements (Ref 30). Application of this technology to stainless steel powders requires careful control of the debinding process in order to achieve low residual carbon contents.

3.4 Contamination, Copper Sulfate and Ferroxyl Tests

Contamination. Even small amounts of iron contamination can have a disastrous effect on the corrosion resistance of sintered stainless steel parts. Such contamination with iron or iron-base powders may originate at the powder producer or the part manufacturer. As little as 10 ppm of iron powder contamination in 316L was found to decrease the corrosion life (based on the appearance of the first rust spot in 5% NaCl aqueous solution at room temperature) by 50%. Therefore, the utmost cleanliness, separate production facilities, and dedicated equipment have become common for the production of stainless steel powders. Active iron or iron-base particles form galvanic couples with the passive stainless steel and corrode anodically in preference to the stainless steel. Figure 3.15 shows this type of corrosion for iron particles embedded in the surface of a pressed-and-sintered 316L part. Rusting occurs within minutes after exposure to the testing solution.

The buildup of the initial corrosion product forms a crevice in which oxygen depletion causes acidification of the solution inside the part and further corrosion.

Experiments with −325 mesh iron powder contamination in 316L parts have shown that sintering at 1260 °C (2300 °F) or higher will

Fig. 3.15 Small circles of rust around iron particles embedded in the surface of sintered type 316L stainless steel after testing in 5% aqueous NaCl. Source: Ref 31

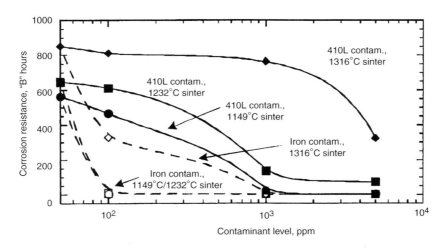

Fig. 3.16 Corrosion resistance of H_2-sintered 316L as a function of contaminant type, contaminant level, and sintering temperature. Reprinted with permission from MPIF, Metal Powder Industries Federation, Princeton, NJ

result in the full homogenization, that is, disappearance, of the contaminant.

Samal and Terrell (Ref 32) (Fig. 3.16) found that contamination of 316L with a ferritic (410L) stainless steel reduced the corrosion resistance of 316L only moderately up to approximately 100 ppm of contamination, and high-temperature sintering at 1316 °C (2400 °F) was highly beneficial in minimizing the loss in corrosion resistance of 316L that was contaminated with 1000 ppm of coarse (100/150 mesh) 410L powder.

Copper Sulfate Test. Iron or iron alloy particles present in stainless steel powder or on the surface of a sintered part can be revealed by placing the powder or part in a concentrated solution of copper sulfate, $CuSO_4$. The dissolved copper plates out on the iron particles within minutes and can easily be seen under a low-magnification microscope. The powder should be tested in the unlubricated condition, because lubricant will prevent the solution from wetting the powder.

Ferroxyl Test. The so-called ferroxyl test is described in ASTM A 380, "Practice for Cleaning and Descaling Stainless Steel Parts, Equipment and Systems." It has recently been adapted for the detection of contamination of stainless steel powder with iron or low-alloy steel powders (Ref 33). This test uses hexacyanoferrate (II/III) solutions with different chloride contents. Immersion of a sintered stainless steel part into such a solution develops,

usually within minutes, a blue precipitate known as Turnbull's blue, in accordance with:

$$K^+(aq) + Fe^{2+}(aq) + [Fe(CN_6)]^{3-}(aq)$$
$$\rightarrow KFe^{III}Fe^{II}(CN)_6(s)$$

The ferroxyl test is superior to the previously described copper sulfate test for detecting contamination with less noble materials in that it is not affected by the presence of a lubricant. Also, unlike the copper sulfate test, in the ferroxyl test, the blue precipitates, formed in the interior of a green part or in the powder mass, grow rapidly in size, thus making it easy to examine. The high sensitivity of the ferroxyl test, is commensurate with the fact that even traces of iron or low-alloy steel contamination can ruin the corrosion resistance of sintered stainless steel parts.

The ferroxyl test also can reveal metallurgical weaknesses or defects due to improper sintering conditions (section 9.1.4 in Chapter 9, "Corrosion Testing and Performance"). However, agreement with salt spray testing and electrochemical measurements is only moderate. Recommended solutions and development times are shown in Table 3.2.

Testing for contamination should be conducted on powders and on green parts prior to sintering, because identification of corrosion problems after sintering is more difficult and more expensive.

Table 3.2 Recommended test solution strengths and development times for Turnbull's blue

Material (300-series SS)	Contaminant powder	Solution strength, %		Development time, min
		$K_3Fe(CN)_6$	NaCl	
Powder, not lubed	Fe or 410L	0.1	0.1	<3
Powder, lubed(a)	Fe or 410L	0.25	0.25	120–180
Green parts(b)	Fe or 410L	0.05	0.05	5–20
Green parts(b)	Fe	0.1	0.1	<3

(a) Rinsed in acetone. (b) Density: 6.6 g/cm³, lubed

REFERENCES

1. A.J. Yule and J.J. Dunkley, *Atomization of Melts*, Clarendon Press, Oxford, 1994
2. E. Klar and J.W. Fesko, Gas and Water Atomization, *Powder Metallurgy*, Vol 7, Metals Handbook, 9th ed., American Society for Metals, 1984, p 26
3. N. Dautzenberg and H. Gesell, Production Technique and Properties of Austenitic Cr-Ni Stainless Steel Powders, *Powder Metall. Int.* Vol 8 (No. 1), 1976, p 14–17
4. T. Tanaka, S. Nakabaya, and T. Takeda, Property of Alloy Powder by Droplet Drawing High Pressure Water Atomizing, *Proc. Spring Meeting of the Japan Society of Powder and Powder Metallurgy*, June 1–3, 1999 (Tokyo, Japan), p 41–46
5. S. Okamoto, T. Sawayama, and Y. Seki, Kobe Steel Advances Water Atomized Powders, *Met. Powder Rep.*, March 1996, p 28–33
6. M. Kikukawa, S. Matsunaga, T. Inaba, O. Iwatsu, and T. Takeda, Development of Spherical Fine Powders by High-Pressure Water Atomization Using Swirl Water Jet, *Proc. of 2000 Powder Metallurgy World Congress*, Part 1, Nov 12–16, 2000 (Kyoto, Japan), p 363–366
7. P.U. Gummeson, Modern Atomizing Techniques, *Powder Metall.* Vol 15 (No. 29), 1972, p 67–94
8. T. Allen, *Particle Size Measurement*, 3rd ed., Chapman and Hall, 1981, p 136
9. Apparatus for Atomizing Molten Metal, U.S. Patent 2,956,304, 1960
10. Apparatus for Producing Metal Powder, U.S. Patent 3,309,733, 1967
11. R.C. Leyton and O. Andersson, High Density Sintered Stainless Steel with Close Tolerances, *Advances in Powder Metallurgy and Particulate Materials* Vol 7, MPIF 2002, p 118–126
12. Y. Okura and T. Kono, Developing Stainless Steel Powders to Meet Market Demands, *Met. Powder Rep.*, Vol 48 (No. 3) 1993, p 46–48
13. C. Schade, R. Causton, and T. Cimino-Corey, Bulk Melts Bring Flexibly Better Metal Powders, *Met. Powder Rep.*, (No. 07), July/Aug 2003, p 34–44
14. E.E. Stansbury and R.A. Buchanan, *Fundamentals of Electrochemical Corrosion*, ASM International, 2000
15. T. Takeda and K. Tamura, *Trans. Natl. Res. Inst. Met.* (Jpn.), Vol 8, 1966, p 74–75
16. J.J. Dunkley, Atomization, *Powder Metal Technologies and Applications*, Vol 7, *ASM Handbook*, ASM International, 1998
17. T. Kato and K. Kusaka, On the Recent Development in Production Technology of Alloy Powders, *Mater. Trans.*, *JIM*, Vol 31, (No. 5) 1990, p 363–374
18. K. Kusaka, T. Kato, and T. Hisada, Influence of S, Cu, and Sn Additions on the Properties of AISI 304L Type Sintered Stainless Steel, *Modern Developments in Powder Metallurgy*, MPIF, Vol 16, 1984, p 247–259
19. L. Nyborg, P. Bracconi, and C. Terrisse, "Physical Chemistry of Sintering of Stainless Steel Powder", Special Interest Seminar, 1998 World Congress (Granada, Spain)
20. J. Reinshagen and A. Neupaver, Principles of Atomization, *Physical Chemistry of Powder Metals Production and Processing*, W.M. Smith, Ed., TMS/AIME, 1989, p 16
21. D.H. Ro and E. Klar, Corrosion Behavior of P/M Austenitic Stainless Steels, *Modern Developments in Powder Metallurgy*. Vol 13, 1980, p 247–287
22. R.M. Larsen and K.A. Thorsen, "Removal of Oxygen and Carbon During Sintering of Austenitic Stainless Steels", presented at PM World Congress (Kyoto, Japan), 1993
23. D.P. Ferris, Surface Analysis of Steel Powders by ESCA, *Int. J. Powder Metall. Powder Technol.*, Vol 19 (No. 1), 1983, p 11–19

24. E. Klar and E.K. Weaver, Process for Production of Metal Powders Having High Green Strength, U.S. Patent 3,888,657, 1975

25. C. Aslund and C. Quichaud, Metallic Powder for Producing Pieces by Compaction and Sintering, and Process for Obtaining this Powder, U.S. Patent 5,460,641, 1995 (assigned to Valtubes)

26. N. Dautzenberg, Eigenschaften von Sinterstählen aus Wasserverdüsten Unlegierten und Fertiglegierten Pulvern (Properties of Sintered Steels from Water Atomized Elemental and Alloyed Powders), *Second European Symposium on Powder Metallurgy*, May 8–10, 1968 (Stuttgart, Germany), Section 6–18, p 1–27

27. R.M. German and A. Bose, *Injection Molding of Metals and Ceramics,* Metal Powder Industries Federation, Princeton, NJ, 1997, p 71

28. G. Schulz, Melt Film Gas Atomization—The Key to Cheaper MIM Powders, *Second European Symposium on Powder Injection Molding*, Oct 18–20, 2000 (Munich, Germany), EPMA, p 265–272

29. Y. Liu, K. Minagawa, H. Kakisawa, and K. Halada, Hybrid Atomization: Processing Parameters and Disintegration Modes, *Int. J. Powder Metall.*, Vol 39 (No. 2), 2003, p 29–37

30. L. Tremblay, F. Chagnon, and Y. Thomas, Enhancing Green Strength of P/M Materials, *Advances in Powder Metallurgy and Particulate Materials*, MPIF, 2000, Part 3, p 129

31. E. Klar, Corrosion of Powder Metallurgy Materials, *Corrosion*, Vol 13, *Metals Handbook*, 9th ed., American Society for Metals, 1987

32. P.K. Samal and J.B. Terrell, "Effect of Contaminant Level and Sintering Temperature on the Corrosion Resistance of P/M 316L Stainless Steel", P/M Conf. (New Orleans, LA), MPIF, 2001

33. E. Klar and P.K. Samal, On Some Practical Aspects Related to the Corrosion Resistance of Sintered Stainless Steels, *Proceedings of PM '94, Powder Metallurgy World Congress*, June 6–9, 1994 (Paris, France), p 2109–2112

34. Metal Powder Industries Federation (MPIF), Princeton, NJ. Unpublished data

CHAPTER 4

Compacting and Shaping

POWDER METALLURGY (PM) employs many methods for compacting and shaping of powders. The vast majority of stainless steel powders, however, are consolidated by one of four methods: rigid die compaction, metal injection molding, extrusion, and hot isostatic pressing. Conventional PM, based on rigid die compaction, is the most popular process, due to its compatibility with low-cost water-atomized powders and its versatility with regard to component shape and size—all of which generally lead to a reasonable cost. However, this process is not suitable for producing pore-free components. Significant tonnages of gas-atomized stainless steel powders are used in the manufacture of seamless tubing by extrusion. Metal injection molding and hot isostatic pressing techniques are commonly used for the production of full or near-full density products. Although relatively small tonnages of stainless steel powders are used in injection molding, the dollar value of injection-molded stainless steel parts is quite high due to the high unit value of these components.

Readers interested in some of the newer and/or lesser used techniques of powder shaping, such as roll compaction, tape casting, rapid prototyping, spray forming, and controlled powder deposition, are referred to *Powder Metal Technologies and Applications,* Volume 7, *ASM Handbook,* 1998 (Ref 1). Nonetheless, when using these techniques, particularly with water-atomized stainless steel powders and in the presence of binders and/or open porosity, the processing precautions described in Chapter 5, "Sintering and Corrosion Resistance," need to be observed so that the corrosion resistance is not impaired.

4.1 Rigid Die Compaction

This section covers some of the significant aspects of rigid die compaction, or the conven-

tional PM technology, as it applies to stainless steel powders. The types of presses and ancillaries employed for compaction of PM stainless steels are basically the same as those used for compaction of iron and low-alloy steel powders. Because information on these types of compacting equipment is covered in detail in a number of books on PM (including Ref 1), those details are not covered here.

4.1.1 Basics of Powder Compaction and Tooling

Phenomenology of Powder Compaction. Compaction of metal powder in a rigid die can be viewed as a process that comprises a number of stages, with some degree of overlap among them. During the initial stage, densification progresses via rearrangement of powder particles, leading to filling up of large voids and breaking up of particle bridges. In this case, the applied pressure needs mainly to be sufficient to overcome the internal friction in the powder mass. Densification in this stage is aided by the presence of lubricants and the smoothness of the powder particles. In the next stage, elastic deformation of the particles becomes a contributor to the process of densification, as particles continue to reposition and reorient themselves. As the applied pressure is further increased, plastic deformation occurs locally at the interparticle contact points, leading to interlocking of protruding asperities on the particle surfaces. In the next stage, plastic deformation becomes widespread, accompanied by shearing, generation of new oxide-free surfaces, and cold welding of contacting surfaces. Shearing is the result of asymmetrically opposed forces, and, as such, irregularly shaped particles lead to a greater degree of shearing and cold welding. In this stage, significant changes occur in the shape of the powder particles, accompanied by large-scale reduction in porosity. Because plastic deformation leads to strain hardening, the pressure required for incremental densification becomes larger and larger as compaction progresses. Green strength of the compact is the result of both interlocking of the rough, irregular surface features of individual particles and the cold welding of contacting surfaces due to shear. Figure 4.1, adapted from German (Ref 2) and Bocchini (Ref 3), schematically shows the various stages of densification of a metal powder in a rigid die.

While irregularly shaped powder particles are desirable from the point of view of green strength, they require a relatively greater amount of deformation in order to close up interparticle voids; this in turn entails a greater degree of strain hardening. Therefore, these powders require higher compacting pressures. Smooth and rounded powder particles, on the other hand, densify readily, beginning with the early stages of compaction. However, these do not develop high green strengths with commonly available compacting pressures. Fine powders require greater compacting pressures because of their larger surface area, greater rate of work hardening, and finer grain size. Green strength and green density are also influenced by the type and amount of lubricant used.

Tool Materials Selection. The critical properties of materials selected for tooling include high compressive strength, wear resistance, toughness, impact resistance, low coefficient of friction, and antigalling characteristics. Two types of materials are generally found suitable for making tools and dies for compaction of stainless steel powders: carbides

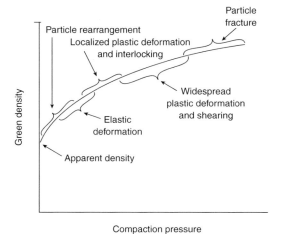

Fig. 4.1 Sketch showing various stages in metal powder compaction. Source: Ref 2, 3

and tool steels (in addition, a zirconia-base ceramic has found limited application as PM tooling material). Carbides typically comprise tungsten carbide with 12 to 14% Co, and, compared to tool steels, they offer higher strength, greater wear resistance, and lower coefficient of friction. Carbides, however, are relatively brittle and do not tolerate any deflection. In tool designs incorporating teeth, blades, or projections (such as for keyways), tool steels (or tool steel inserts) are preferred over carbides. Similarly, for punches and core rods, tool steels are preferred over carbides. While tool steels do not offer wear resistance and antigalling characteristics as good as carbides, they resist chipping and cracking relatively well. In selecting a tool steel (and heat treatment), a compromise must be made between wear resistance and toughness. The higher the hardness of the tool steel, the higher its wear resistance and compressive yield strength but the lower its toughness. Standard grades of tool steel often selected in these applications are A2, D2, D3, S7, and M2. Specially designed PM tool steels that have precisely controlled ratios of carbides to the high-speed steel matrix are the CPM type 3V, M4, 9V, 10V, 15V, T15V, and REX 121 (CPM is the trade name of Crucible Materials Corp.). Surface treatments can be provided to tool steel dies and punches in order to enhance their resistance to galling, seizing, wear, corrosion, and erosion and also to reduce the coefficient of friction. These surface treatments include chemical vapor deposition (TiC, TiN), physical vapor deposition (TiAlN, CrN, TiN, TiCN), nitriding, ion nitriding, and ion implantation (nitrogen, cobalt, titanium + carbon, tin, chromium) (Ref 4).

4.1.2 Compaction of Stainless Steel Powders

General Characteristics. Stainless steel powders intended for PM processing must be sufficiently irregular in shape in order to exhibit good green strength; at the same time, they must deform and densify readily under pressures that are compatible with commercial compacting presses and tooling. The powders must also exhibit good flow properties in order to fill the die cavity in a reasonably short period of time. In comparison to iron and low-alloy steel powders, stainless steel powders generally exhibit lower green strengths, which often lead to relatively slower rates of compaction in order to avoid damage to green parts. Nevertheless, it is possible to formulate stainless steel powders possessing green strengths in the neighborhood of 15.2 MPa (2200 psi), at a 552 MPa (40 tsi) compaction pressure, while still exhibiting very good compressibility and flow rate. In comparison to iron and low-alloy steel powders, stainless steel powders exhibit lower compressibilities, thus requiring much higher compaction pressures to reach the same green density. Stainless steel powders are relatively more abrasive to the tooling. For these reasons, carbide tooling is most often preferred.

Selection of optimal green density in a given application is dependent on a number of factors. These include the desired sintered density and properties, anticipated dimensional change, capabilities of the press and tooling, and the requirements, if any, placed by secondary processes. Also, one must not raise the green density so high that it hinders delubrication during subsequent processing. For alloys prone to crevice corrosion in neutral saline environments, sintered densities in the range of approximately 6.70 to 7.10 g/cm^3 should be avoided (section 5.2.2 in Chapter 5, "Sintering and Corrosion Resistance"). In applications that demand good mechanical properties, including good dynamic properties and/or low interconnected porosity, sintered densities in the range of 7.20 to 7.35 g/cm^3 are found to be optimum.

Austenitic stainless steel powders exhibit somewhat better compressibility, compared to ferritic stainless steel powders, due to their superior ductility. Compacting pressures employed for stainless steel powders largely fall in the range from 552 to 760 MPa (40 to 55 tsi).

Lubricant Effects. The primary reason for using a lubricant is to aid in the ejection of the green compact from the die and to reduce die wear. Lubricants also help with reducing interparticle friction, thereby lowering the pressure needed to achieve the desired green density. A good lubricant extends the particle rearrangement stage of the compaction process and leads to a more uniform density distribution. Lubricant addition generally produces an adverse effect on the green strength, because it reduces surface-to-surface contacts. The relative effects of various lubricants on green strength and green density vary widely, depending on the lubricant composition, particle size, particle morphology, and the amount used. In general, a lubricant that is fine and has the ability to disperse well will

coat the powder surface uniformly and more completely. This type of lubricant will generally lead to a higher green density but will lower green strength. The optimal amount of lubricant necessary for most applications falls in the range of 0.5 to 1.0% by weight. A larger amount may be selected if the part thickness is in excess of 2.5 cm (1 in.) or if the compaction pressure is higher than normal.

Currently, the two most commonly used lubricants for stainless steel powders are ethylene bisstearamide (EBS) and lithium stearate. The former is a wax and is popularly known in the industry as Acrawax C, while the latter is a metallic soap. A number of proprietary modifications of the latter are available. The basic characteristics of EBS are its ability to be removed easily during delubrication and its beneficial effect on the green strength of the powder. Lithium stearate, on the other hand, has a beneficial effect on green density, which usually translates to a high sintered density. Removal of lithium stearate lubricant requires more care in delubrication compared to EBS, and the residues of delubrication may be undesirable from an operational cleanliness point of view. Often, a mixture of the two lubricants is selected in order to arrive at a suitable compromise. A study by Reinshagen and Mason (Ref 5) examined the effect of various ratios of EBS and lithium stearate on the compacting properties of 316L. Figures 4.2 and 4.3 show these effects for three compacting pressures. As the relative amount of EBS is increased, green strength is increased and the green density is decreased. They found the apparent density of the mix to increase linearly from 2.85 to 3.22 g/cm^3 as the

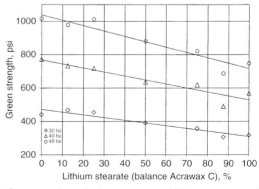

Fig. 4.2 Effect of relative amounts of lithium stearate and Acrawax C on the green strength of 316L, compacted at 414, 552, and 662 MPa (30, 40, and 48 tsi), respectively. The total amount of the two lubricants was 1.0% by weight in all cases. Source: Ref 5. Reprinted with permission from MPIF, Metal Powder Industries Federation, Princeton, NJ

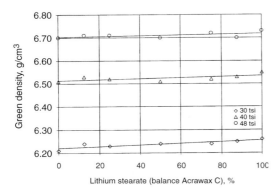

Fig. 4.3 Effect of relative amounts of lithium stearate and Acrawax C on the green density of 316L, compacted at 414, 552, and 662 MPa (30, 40, and 48 tsi), respectively. The total amount of the two lubricants was 1.0% by weight in all cases. Source: Ref 5. Reprinted with permission from MPIF, Metal Powder Industries Federation, Princeton, NJ

relative amount of lithium stearate was increased from 0 to 100%.

Bergkvist (Ref 6) investigated the effects of admixing lithium stearate, EBS (of a brand called Amide Wax), and Kenolube on the compacting and sintered properties of 409L stainless steel. Kenolube is a proprietary lubricant containing 2% Zn as metallic soap. Lubricant mixes investigated are shown in Table 4.1, along with the resulting compacting properties at a compaction pressure of 650 MPa (47 tsi). The median particle sizes were 22 μm for Kenolube, 29 μm for EBS, and 5 μm for lithium stearate.

Mixes 1 and 2 (1.0% Kenolube and 0.9% Kenolube + 0.1% lithium stearate) exhibited good flow rates, high green strengths, and intermediate green densities. Mix 4 (1% lithium stearate) had the lowest green strength and the highest green density; the mix made with 1% EBS showed the lowest green density. The results of other sets of tests carried out at 500 and 800 MPa (36 and 58 tsi) paralleled those

obtained in tests with 650 MPa (47 tsi) compaction. The difference in the green density due to lithium stearate versus EBS observed in this study is somewhat larger than those observed by others. Typically, the difference is found to be approximately 0.04 g/cm³ in the range of compaction pressures from 414 to 828 MPa (30 to 60 tsi). It is likely that the discrepancy is due to the relatively coarser particle size of the grade of EBS used in this study.

Dimensional change was very similar for all six mixes, and thus, the sintered densities paralleled the green densities. Interestingly, mixes containing greater than 0.1% lithium stearate showed higher sintered strengths for the same sintered density. The gains in ultimate tensile strength, elongation, and impact strength due to the presence of lithium stearate were remarkably greater than what would be expected from the increased sintered density (Table 4.2 and Fig. 4.4). The sintered strengths of samples that were processed with 1% lithium stearate were markedly higher than those of all other sample groups. These results suggest that the presence of lithium stearate, even in small amounts, enhances bonding of stainless steel particles during sintering. A possible explanation may be that lithium, which is popularly used as a flux for brazing, promotes surface diffusion during sintering. In this study, sintering was carried out at 1350 °C (2462 °F) for 30 min in a 100% hydrogen atmosphere.

Reinshagen and Mason (Ref 5) compared the effects of twelve lubricants, most of which were proprietary formulations, on the green density and green strength of 316L stainless steel powder. In all cases, the total lubricant content was kept at 0.75%. Three compacting pressures were used: 414, 552, and 662 MPa (30, 40, and 48 tsi). The results of this study are summarized in Fig. 4.5. The highest green

Table 4.1 Effect of lubricant type on compacting properties of 409L powder

Mix no.	Lubricant, wt%	Hall flow, s/50 g	Apparent density, g/cm³	Green strength MPa	Green strength psi	Green density, g/cm³	Ejection force MPa	Ejection force psi
1	1% Kenolube	28.8	2.91	14.3	2070	6.50	9.1	1316
2	0.9% Kenolube + 0.1% lithium stearate	30.2	2.93	13.5	1957	6.51	9.8	1421
3	0.6% Kenolube + 0.4% lithium stearate	35.0	3.01	11.7	1692	6.52	10.4	1508
4	1% lithium stearate	43.5	3.03	9.6	1387	6.54	10.3	1494
5	1% EBS (a)	44.2	2.73	14.8	2138	6.43	9.9	1435
6	0.25% lithium stearate + 0.75% EBS (a)	40.0	3.00	10.0	1455	6.52	10.7	1552

Compacting pressure: 650 MPa (47 tsi). (a) EBS, ethylene bis stearamide

Table 4.2   Effect of lubricant type on the sintered properties of 409L

Mix no.	Lubricant, wt%	Sintered density, g/cm³	Dimensional change, %	Ultimate tensile strength		Elongation, %	Impact energy	
				MPa	ksi		J	ft. lbf
1	1% Kenolube	7.32	−4.5	367	53.2	10.9	200	147
2	0.9% Kenolube +							
	0.1% lithium stearate	7.35	−4.7	363	52.7	11.9	237	175
3	0.6 Kenolube +							
	0.4% lithium stearate	7.37	−4.7	381	55.2	12.5	278	205
4	1% lithium stearate	7.38	−4.7	388	56.3	13.4	>300	>221
5	1% EBS (a)	7.27	−4.6	362	52.5	11.3	185	136
6	0.25% lithium stearate +							
	0.75% EBS (a)	7.40	−4.8	391	56.7	12.9	281	207

All compaction at 650 MPa (47 tsi). (a) EBS, ethylene bis stearamide

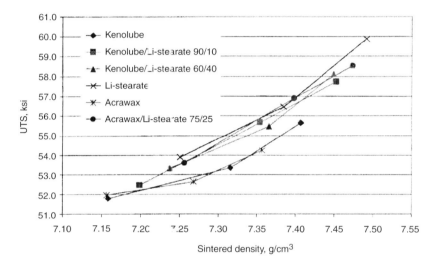

Fig. 4.4 Ultimate tensile strength (UTS) of sintered 409L as a function of lubricant type and sintered density. Source: Ref 6. Reprinted with permission from MPIF, Metal Powder Industries Federation, Princeton, NJ

strength was obtained when no lubricant was employed. The second highest green strength was obtained with a polyethylene lubricant, which typically has a coarse particle size. This was followed by various mixes comprising Shamrock grades S-400, C640-X831, and C640-X830. These mixes showed significantly higher green strengths but had adverse effects on the green density. Most of the other lubricants, including Acrawax C and lithium stearate, showed relatively higher green densities and lower green strengths. Both of these studies indicate that opportunities do exist to enhance one or more of the properties of a stainless steel powder by careful selection of the type and amount of lubricant used.

Identification of Lubricants. In some situations, it may be necessary to positively identify the type and amount of lubricant present in a powder mix. Identification of a lubricant can be made by means of a differential scanning calorimeter. In this test, measurement is made of the heat absorbed by the sample as it passes through its melting point, as well as any other phase changes. Because each lubricant has a characteristic thermogram (heat flow versus temperature profile), comparison of the thermogram of the powder sample against the characteristic thermograms of known lubricants permits identification of a lubricant, or any combination of lubricants, present in the powder mix. Details of the technique, along with thermograms of a number of popular PM lubricants, have been published by Cronin and Berry (Ref 7). Figure 4.6 shows the thermograms of Acrawax C and lithium stearate. The relative quantities of the lubricants present can be determined from the areas of their inverted peaks. An alternate method of determining the quantity of an admixed lubricant is by analysis

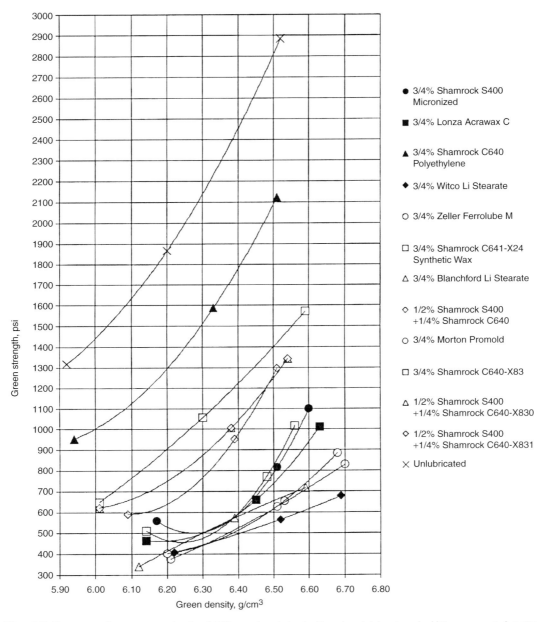

Fig. 4.5 Green strength versus green density of 316L powder admixed with various lubricants and additives compacted at 414, 552, and 662 MPa (30, 40, and 48 tsi), respectively. Source: Ref 5. Reprinted with permission from MPIF, Metal Powder Industries Federation, Princeton, NJ

of total carbon content of the metal powder, provided that only one type of lubricant is present.

Die Wall Lubrication. Much progress has been made in recent years to develop commercially viable die wall lubrication methods for the compaction of iron and low-alloy steel powders. It is anticipated that when this technology is perfected, it can also be applied to stainless steel powders. The main benefits of

die wall lubrication are elimination (or minimization) of the delubrication step as well as achievement of higher green strengths. The benefits of die wall lubrication are mainly realized at high compacting pressures. At low compacting pressures, the green density achieved with die wall lubrication is lower than that achieved with admixed lubrication. As the compaction pressure is increased beyond what

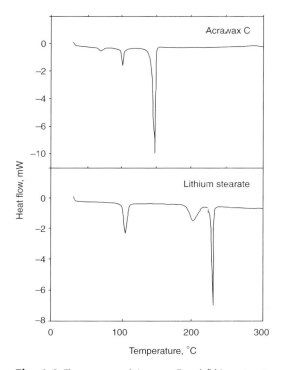

Fig. 4.6 Thermograms of Acrawax C and lithium stearate determined by differential scanning calorimetry. Source: Ref 7. Reprinted with permission from MPIF, Metal Powder Industries Federation, Princeton, NJ

is known as the transition pressure, die wall lubrication results in an increased green density, compared to admixed lubrication. In practice, admixing a small amount of a lubricant, typically 0.125%, is found to be beneficial with the die wall process, because it reduces interparticle friction and optimizes densification.

Double Pressing-Double Sintering is sometimes used to achieve high sintered densities. The first sintering is carried out at a relatively low temperature, which produces sufficient ductility for the second pressing operation. The second sintering is carried out at a conventional sintering temperature. Takeda and Tamura (Ref 8) studied the rate of densification in repressing using three austenitic stainless steels. Results obtained on 316L samples are shown in Fig. 4.7 as iso-density curves for various combinations of pressure used for the first and second pressing operations. Figure 4.8 shows the porosities for various pressing and repressing sequences. After the first compaction, samples were sintered in vacuum at 1050 °C (1922 °F) for 30 min, then repressed. As expected, a threshold pressure has to be met in repressing in order for any densification to occur. This threshold pressure is

determined by the compressive yield strength of the material, which in turn is dependent on the density achieved in the first pressing.

Warm Compaction involves compaction of a powder above room temperature by heating both the die and the powder, typically to approximately 150 °C (300 °F). The process takes advantage of the reduced yield strength of the process material at the higher temperature. It is essential that the lubricant used in the process is capable of withstanding the higher operating temperatures, and as such, it determines the highest permissible operating temperature. Warm compaction has been in commercial practice for iron and low-alloy steel powders for over 10 years. Its applicability to stainless steels has been demonstrated on a laboratory scale, as described subsequently. In order for the technology to be commercially successful, improvements achieved in green and sintered properties must offset the additional process cost. Economic considerations will include projected production volumes and capital investment.

Currently, a number of researchers have demonstrated the applicability of warm compaction to stainless steel powders. Gasbarre (Ref 9) determined an increase in the green density of 316L from 6.81 to 7.09 g/cm^3, under a compaction pressure of 690 MPa (50 tsi), as a result of warm compaction. In similar trials, the green density of 434L was found to increase from 6.51 to 6.71 g/cm^3. In these studies, the

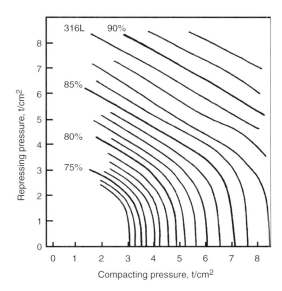

Fig. 4.7 Effects of various combinations of compacting and repressing pressures on the final density of PM 316L. Source: Ref 8

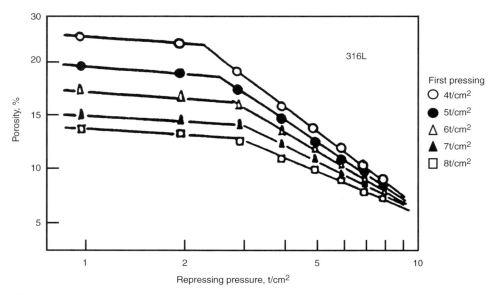

Fig. 4.8 Effects of various combinations of compacting and repressing pressures on the porosity of PM 316L. Source: Ref 8

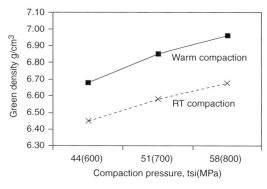

Fig. 4.9 Green densities of 409L obtained under various compaction pressures using warm compaction and room-temperature (RT) compaction. Source: Ref 10. Reprinted with permission from MPIF, Metal Powder Industries Federation, Princeton, NJ

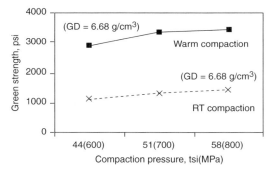

Fig. 4.10 Green strengths of 409L obtained under various compaction pressures using warm compaction and room-temperature (RT) compaction. GD, green density. Source: Ref 10. Reprinted with permission from MPIF, Metal Powder Industries Federation, Princeton, NJ

specimens were in the shape of a thin-walled ring having an outer diameter of 15.5 mm (0.61 in.) and a height of 12.5 mm (0.5 in.). Control samples, pressed at room temperature, contained 0.75% lubricant, while the warm compacted samples contained 0.2% admixed lubricant. In these warm compaction tests, all powders were heated to 79 °C (175 °F), while the temperature of compaction was 204 °C (400 °F) for 316L and 260 °C (500 °F) for 434L.

Leyton and Andersson (Ref 10) evaluated the benefits of warm compaction using 409L powder. Significant improvements in green density and green strength were observed, as shown in

Fig. 4.9 and 4.10. The large increase in green strength should permit the use of higher apparent density, more rounded, and possibly lower-oxygen-content stainless steel powders. The potential benefits of this strategy are discussed in sections 3.1.2 and 3.1.3 in Chapter 3, "Manufacture and Characteristics of Stainless Steel Powders." The increase in sintered density was less pronounced at the intermediate and high sintering temperatures, as shown in Fig. 4.11. The gain in sintered strength (ultimate tensile strength) was minimal (at 4%), but the impact energy increased by 40%. For warm compaction, both powder and tooling were heated to 110 °C (230 °F).

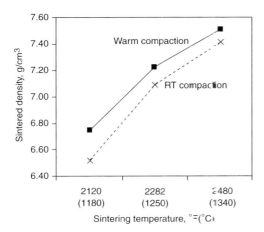

Fig. 4.11 Sintered densities of warm compacted and room-temperature (RT) compacted 409L as a function of sintering temperature. Source: Ref 10. Reprinted with permission from MPIF, Metal Powder Industries Federation, Princeton, NJ

4.1.3 Dimensional Change

Dimensional change is the linear measure of any expansion or contraction taking place in a green compact during sintering. Dimensional change characteristics must be taken into consideration in tool design, because they have significant influence on the final dimensions of the component. For practical reasons, dimensional change is determined by comparing the dimension of the die cavity with the corresponding dimension of the sintered part (typically, dimensional measurement tests use a long dimension of the die cavity that lies perpendicular to the direction of pressing). In doing so, it does not make any adjustments for the springback that normally occurs when the green compact is ejected from the die. Nonetheless, this method of measurement permits direct use of dimensional change data in tool design. A secondary benefit of this method is the avoidance of potential errors arising from the measurement of individual green compacts. Because this method of testing does not involve actual measurement of the green compact, a precise estimation of sintered density is not possible from the dimensional change data. Estimation of sintered density is further complicated by the fact that uniaxially compacted parts do not undergo dimensional change in an isotropic manner. Height changes are not usually measured, because they have no critical bearing on tool design.

Sintering of stainless steel almost always results in the shrinkage of the part, and hence, dimensional change values are typically expressed as negative numbers and often converted to a percentage of the original die dimension. Ferritic stainless steels undergo a greater degree of shrinkage compared to austenitic stainless steels. This is attributed to the greater rate of diffusion of atoms in the more open body-centered cubic (bcc) lattice as compared to the face-centered cubic (fcc) lattice. It should be kept in mind that in some cases, the crystal structure of the alloy at the sintering temperature may be different from that at room temperature. Takeda and Tamura (Ref 11) have determined the densification characteristics of stainless steel powder compacts possessing alpha, gamma, and gamma plus delta phase structures as functions of porosity, particle size, and sintering time. They have also noted that the rate of sintering (densification) of austenitic stainless steels is enhanced by the presence of bcc delta phase in the austenitic matrix.

In the sintering of PM parts, one of the important goals of the parts producer is to consistently reproduce that dimensional change that has been designed or factored into the size of the compacting die, in order to be able to meet print size specifications. This is normally done by advance testing of the powder under closely monitored conditions. A compacting die is then designed and manufactured on the basis of such data, with the anticipation that future lots of powder will exhibit the same dimensional change characteristics as used in the design phase, and the process parameters will essentially remain unchanged.

Dimensional change is a reflection of the degree of sintering, and hence, processes that involve a high degree of sintering will result in a greater amount of dimensional change. With large changes in overall dimensions, greater variations in dimensions may be experienced from part to part as well as within a part. Variation in the dimensional change within a part, which is usually caused by green density variations, can lead to distortion of the part.

Factors Affecting Dimensional Change. Dimensional change is influenced by a large number of material- and process-related parameters, not all of which can be monitored or controlled in commercial processing. Decisions at the plant level are based on "intelligent" experimentation, because formulae based on fundamental parameters are inadequate to predict with sufficient accuracy the complex changes taking place during sintering. The accuracy of computer models for predicting dimensional change is still an order of magnitude below what is required in practice (Ref 12). Nevertheless, it is useful to have a qualitative understanding of the variables that have an effect on dimensional change. Hausner (Ref 13) has summarized the effects of a number of variables that affect dimensional change during sintering.

In practice, adjustments to dimensional change are often accomplished empirically. Small disparities in the powder characteristics or other influences, sometimes of unknown origin, can usually be compensated for by minor adjustments in compacting pressure, sintering temperature, sintering time, and/or composition or amount of lubricant. These adjustments can be based on the sintering characteristics shown in a powder manufacturer's technical brochures. Another practice for modifying dimensional change during sintering is to increase or decrease the amount of fines, that is, the amount of the −325 mesh (<45 μm) powder fraction. This approach relies on the large surface area of this powder fraction that provides much of the driving force for sintering. Addition of a small amount of copper decreases the shrinkage of stainless steels. A more drastic approach to controlling dimensional change involves the adjustment of stainless steel constituents within their specified ranges.

Knowledge of the significance of the major contributors to dimensional change is useful from the points of view of controlling and troubleshooting the manufacturing process.

Powder-Related Factors. The rate of sintering of a PM compact is significantly influenced by the specific surface area of the powder used. The greater the specific surface area, the more rapid the rate of sintering and greater the dimensional change. The specific surface area of a powder is a function of its particle topology and particle size distribution. An irregularly shaped powder (i.e., with lower apparent density) will result in greater shrinkage compared to a powder having a rounded shape. Similarly, a powder having a finer particle size (or a larger fraction of fine particles) will result in greater shrinkage compared to a coarser powder. Because the specific surface area of a powder is inversely proportional to the square of its particle diameter, contribution to sintering and shrinkage becomes more significant from the subsieve size fraction of the powder (−325 mesh, or <45 μm, in diameter) and more so from its superfines fraction (<20 μm in diameter). The effect of the amount of the −325 mesh fraction of a stainless steel powder on its dimensional change is illustrated in Fig. 4.12 for a number of sintering temperatures and times (Ref 14).

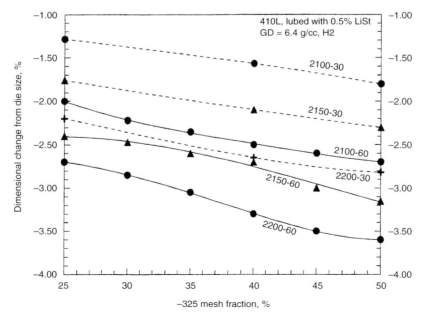

Fig. 4.12 Effect of the percentage of −325 mesh fraction on the dimensional change of 410L powder for sintering in hydrogen at various temperatures (2100, 2150, and 2200 °F) and for two sintering times (30 and 60 min). GD, green density. Source: Ref 14. Reprinted with permission from MPIF, Metal Powder Industries Federation, Princeton, NJ

These data demonstrate the need for close control of the amount of subsieve size particles in a stainless steel powder in order to keep dimensional change in a manageable range. It must be kept in mind that the presence of excessive amounts of superfine particles in a powder can lead to premature die scoring, because these particles can become lodged in the clearance between die and punch during processing.

Oxides present on the surface of a powder can have a noticeable effect on the rate of sintering, especially if the dewpoint of the sintering atmosphere is not low enough. The progress of sintering will be relatively slow until these oxides are reduced, resulting in a decrease in dimensional change. Because of this, the effect of surface oxides on the dimensional change is more pronounced at the lower sintering temperatures, such as 1149 °C (2100 °F), compared to the higher sintering temperatures, such as 1316 °C (2400 °F). Water-atomized stainless steel powders typically contain from 1800 to 3500 ppm oxygen, much of which is concentrated at the surface of powder particles in the form of oxides of silicon, manganese, and chromium. From the point of view of dimensional control, the lot-to-lot oxygen variation should be held to within approximately 800 ppm.

Compaction-Related Factors. Dimensional change is significantly influenced by the green density of the compact. The amount of shrinkage decreases with increasing green density. Figure 4.13, taken from Hirschhorn (Ref 15), illustrates the effect of green density on the rate of sintering. A lower green density leads to a greater change in density during sintering, thus

resulting in a higher dimensional change. Nonuniform green density can lead to nonuniform shrinkage within the part, resulting in distortion. Similarly, any part-to-part differences in overall green density (due to problems such as poor powder flow rate, variations in the apparent density, etc.) can lead to variations in the part-to-part sintered dimensions. In general, use of a hydraulic press for compaction results in more uniform green density, both from part-to-part and within a part, compared to a mechanical press.

Sintering-Related Factors. Dimensional change is significantly influenced by the sintering temperature, time, and atmosphere. Sintering temperature is usually the most important process parameter that determines dimensional change and sintered properties. In high-temperature sintering (typically above 1232 °C, or 2250 °F), a change in the sintering temperature of as little as 15 °C (27 °F) can result in a marked shift in dimensional change.

Sintering in hydrogen produces greater shrinkage compared to sintering in a nitrogen-bearing atmosphere. This is attributed to two factors. First, compared to nitrogen-bearing atmospheres, hydrogen is more effective in reducing surface oxides, and so it removes the surface oxides more rapidly. This results in a longer effective sintering period. Secondly, sintering in a nitrogen-bearing atmosphere (which usually involves cooling in a similar atmosphere) leads to the absorption of significant amounts of nitrogen. Some of this nitrogen forms chromium nitrides during cooling. Because chromium nitride has a significantly lower density compared to the stainless steel matrix, its presence tends to cause a slight expansion of the part. Heavily nitrided PM stainless steels can even exhibit net growth instead of shrinkage.

In comparison to sintering temperature, sintering time has a somewhat smaller but still significant influence on dimensional change. Experimental studies by German (Ref 16) and Takeda and Tamura (Ref 17) have shown that linear shrinkage in the sintering of austenitic stainless steel is proportional to the square root of sintering time. Data presented in Fig. 4.12 are also found to be in agreement with this rule, despite the fact that these are for a ferritic stainless steel. In this case, for most of the sample groups, doubling of the sintering time is found to increase shrinkage by a factor of $\sqrt{2}$ (or by approximately 40%).

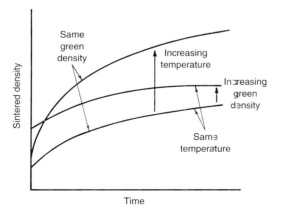

Fig. 4.13 Schematic representation of the effects of green density, sintering temperature, and sintering time on sintered density. Source: Ref 15. Reprinted with permission from MPIF, Metal Powder Industries Federation, Princeton, NJ

A high carbon potential of the sintering furnace and/or high carbon content of the part, resulting from inadequate delubrication, can affect the rate of sintering of ferritic grades of stainless steel. The high carbon content could transform most of the ferrite to austenite at the temperature of sintering, thus resulting in a lower rate of sintering.

Statistical Analysis of Variables. McMahon and Reen (Ref 18) have statistically evaluated 24 lots of water-atomized 316L stainless steel powders from different powder manufacturers. The lubricated powders were pressed to various densities and sintered for 3 h at 1204 °C (2200 °F) in dry hydrogen. The shrinkages were calculated from the differences between the sintered and die dimensions. The percent linear shrinkage was found to be represented by the following equation:

Percent linear shrinkage = $11.76 - 0.234$ (%Cr) $- 0.103$ (%Ni) $- 0.465$ (%Mo) $+ 9.15$ (%N) $- 0.119$ $(CR)^2 + 0.0749$ $(NI)^2 + 1.186$ $(MO)^2 + 0.0505$ $(CR \times NI) + 0.659$ $(CR \times MO) - 0.0662$ $(NI \times MO)$ $- 1.20$ $(AD^*) + 0.0621$ $(AD)^2 - 0.0196$ (CP) $+ 0.00202$ $(CP)^2 - 0.0243$ $(CR \times AD) + 0.0011$ $(CR \times CP) + 0.0184$ $(AD \times CP) - 0.203$ $(N \times CP)$

$R^2 = 89.8\%$

Standard error = 0.27%

where:

%Cr, %Ni, %Mo, %N = percentage of elements

CP^* = compacting pressure, tsi

AD^* = apparent density, g/cm^3

$CR = (\%Cr - 17.23)$

$NI = (\%Ni - 12.27)$

$MO = (\%Mo - 2.44)$

$AD = (AD^* - 3.02)$

$CP = [(CP^* - 35) \div 5]$

$N = (\%N - 0.035)$

The effects for several constituents, including chromium, nickel, and molylodenum, within the AISI composition range are substantial and can be used to modify the shrinkage characteristics of a powder.

Myers et al. (Ref 19) found the surface chromium content to significantly increase shrinkage during sintering, while bulk manganese and bulk carbon had the opposite effect. The effect of surface chromium can probably be explained in terms of activated sintering, because much of the surface chromium, present as Cr_2O_3, becomes reduced during sintering in dry hydrogen. The effect of bulk manganese could be due to the tendency of manganese (Section 3.1.3 in Chapter 3, "Manufacture and Characteristics of Stainless steel Powders") to favor SiO_2 formation on the particle surfaces during water atomization, which in turn impairs the sintering process.

Dimensional Change—Interpretation of Test Data. As discussed previously, dimensional change is influenced by a large number of materials- and process-related variables. In most situations, a good number of these variables are difficult to monitor and control. As a result, under seemingly similar test conditions, tests run in two different sintering furnaces may produce appreciably different results. Similarly, stainless steel powders having the same nominal composition, similar particle size distribution, and similar apparent density, but produced at different manufacturing facilities, may exhibit different dimensional change values in a side-by-side test. When designing a new process, it is therefore advisable to establish at least an approximate value of dimensional change by using a powder sample from the selected powder source and then sintering in the furnace selected for the job, using test parameters that are close to the targeted process parameters.

For the purposes of powder evaluation, dimensional change testing typically involves the use of a reference powder sample to be tested side by side with the test lot powder, so that some adjustment can be made to the raw data to correct for possible variations in the sintering run. This is especially critical for the testing of ferritic grades of stainless steel powder, because small differences in the carbon or nitrogen contents of a sample can influence its crystal structure (relative amounts of bcc versus fcc) at the sintering temperature. Similarly, tests involving a nitrogen-bearing sintering atmosphere are prone to variations in dimensional change arising from the possible differences in the amount of nitrogen absorbed in the sample during cooling.

4.2 Powder Injection Molding of Stainless Steel

Metal powder injection molding (MIM) is a technology suitable for the high-volume manufacture of small (largest dimension typically less than 100 mm, or 4 in.) and complex shapes. It evolved from the well-known technology of injection-molded plastic parts. It is used where combinations of shape capability, dimensional tolerances, and cost make it superior to alternate fabrication techniques. Many MIM uses involve complex parts of high-performance materials, which, with alternate fabrication techniques, would require extensive and precise grinding, machining, and/or drilling. Much of the following treatment is adapted from the chapter "Powder Injection Molding" by German in *Powder Metal Technologies and Applications,* Volume 7, *ASM Handbook,* 1998 (Ref 20).

4.2.1 Powders for MIM

Approximately 20 years ago, the metal powders used for MIM consisted mostly of fine iron and carbonyl nickel powders and fine fractions of atomized stainless steel powders scalped from coarse powders intended for other uses. Much progress has been made in the technology of powder manufacture to support the needs of the growing MIM industry. High-pressure atomization, hybrid gas-water atomization and special gas-atomization techniques (Chapter 3, "Manufacture and Characteristics of Stainless Steel Powders") with good yields of fine (<20 μm) powder now offer a much wider choice of powders at significantly reduced cost.

Spherical powders give desirably low viscosity mixtures as well as isotropic shrinkage and high fill density, while slightly irregular powders are less prone to slumping during debinding and early stages of sintering. Slightly irregular powders also provide the so-called brown strength or the compact strength subsequent to debinding, which is desirable in many situations. These and other opposing requirements suggest that a spheroidal powder, that is, a nearly but not fully spherical powder, the particles of which have an aspect ratio of approximately 1:1.2 and a packing density of approximately 60% of theoretical, is considered most desirable (Ref 21). Alternately, some MIM fabricators use blends of gas- and water-atomized powders to obtain an optimal combination of properties. Pascoli et al. (Ref 22) have studied several injection molding characteristics as well as the mechanical properties of MIM parts prepared from mixtures of gas- and water-atomized 316L stainless steel powders. Nevertheless, the water-atomized component of such powder blends still has an oxygen content that is over an order of magnitude larger than that of the inert gas-atomized component, and hence, its negative effect on the mechanical and corrosion properties (Chapters 7 and 9) of the final MIM part should be taken into consideration.

Master alloys that are blended with carbonyl iron powder are also in use and are said to provide improved surface finish and shape definition at reduced cost.

In continuing efforts to lower the cost of MIM powders, development efforts in the past few years have been devoted to the use of less expensive coarser powders and the feasibility of liquid-phase and supersolidus sintering.

4.2.2 Feedstock

In the most common version of MIM, a fine (<20 μm), near-spherical prealloyed metal powder or a mixture of elemental powders are combined with an organic binder, typically a multicomponent thermoplastic polymer, and palletized to form the feedstock. The latter is available from major chemical companies. The metal powder must be of small particle size to accomplish sintering to nearly full density. High packing densities are also desirable, to minimize the amount of binder necessary to fill all the voids of the powder.

The rheological characteristics of a powder-binder mixture are of critical importance. Successful feedstock requires a carefully balanced ratio of powder and binder for shape retention during debinding. High shear-rate mixing of binder and powder assures feedstock homogeneity, which is important for minimizing defects and distortion during sintering (Ref 23). Figure 4.14 shows examples of pelletized feedstock. Table 4.3 lists compositions of several commercially available feedstocks, including two examples of stainless steel feedstock.

Fig. 4.14 Feedstock pellets and worms for molding. Source: Ref 20

Table 4.3   Examples of powder injection molding feedstock

Powder	Binder, wt%	Solids loading, vol%	Density, g/cm^3	Molding temperature °C	Molding temperature °F	Viscosity Pa.s	Strength MPa	Strength ksi
4 μm Fe	55PW-45PP-5SA	61	5.12	150	300	19	22	3.2
4 μm Fe-2Ni	90PA-10PE	58	4.52	180	360	190	20	2.9
10 μm stainless	55PW-45PP-5SA	67	5.60	130	265	100	15	2.2
15 μm stainless	90PA-10PE	62	5.33	190	375	80	20	2.9
12 μm tool steel	90PA-10PE	62	5.33	190	375	180	20	2.9
1 μm W-10 Cu	60PW-35PP-5SA	64	11.41	135	275	55	6	0.87

PA, polyacetal; PE, polyethylene; PP, polypropylene; PW, paraffin wax; SA, stearic acid. Source: Ref 20

The most commonly used feedstock system in Europe is the BASF polyacetal binder that uses catalytic debinding (Ref 24). The feedstock is injection molded into the desired shape, typically at a temperature below 200 °C (392 °F), in a plastic injection molding machine. The binder provides the necessary viscosity for the feedstock to flow into the cavity under hydrostatic pressure (>60 MPa, or 8700 psi). Cooling channels in the die accelerate cooling of the feedstock, so that the shape of the molded part is retained after it is ejected from the die cavity.

4.2.3   Tooling and Molding

Tooling used is similar to that used for injection molding of plastics. Sprues and runners are recycled for maximum feedstock utilization. The tool materials used range from easy-to-machine soft alloys, such as aluminum, to hard and wear-resistant tool steels and cemented carbides. The choice of material depends on the number of molding cycles as well as some other factors. Tooling made of tool steels typically has a surface roughness of 0.2 μm (8 μin.) and a hardness of at least 30 HRC. A hard tool set can mold up to one million parts, whereas soft tool materials have a life of only 1000 to 10,000 parts. Frequently, tool sets are designed with multiple cavities, leading to high production rates. The dimensions of a tool cavity are oversized to accommodate the large, ideally isotropic shrinkage (typically 15%) taking place during sintering. Undercuts and holes perpendicular to the direction of molding are possible through the use of side-actuated cores or inserts. Internal and external threads are also possible.

Molding Machines. The most common types of molding machines are reciprocating screw, hydraulic plunger, and pneumatic. Adequate feedstock pressurization is essential for minimization of defects arising from shrinkage during cooling. Thus, low-pressure pneumatic molding machines are used only for small components that can tolerate small internal flaws. Horizontal reciprocating screw machines are suitable for high-volume production. During a molding cycle, the screw initially rotates and compresses the feedstock. Then, for injection, the screw moves forward like a plunger. A high injection rate assures die fill before the feedstock cools.

Molding rate depends on cavity size and filling and cooling times, which range from a few seconds to over a minute. For the feedstock to flow and ensure complete mold filling, the molding temperature needs to be above the softening point of the binder. Excessive molding temperature, on the other hand, causes binder degradation, flashing, powder-binder separation, and prolongs the cooling period. Table 4.4 lists typical ranges of powder injection molding parameters and, as an example, the parameters used for molding a trigger guard for a rifle.

4.2.4   Debinding

Removal of the binder from the injection-molded part is accomplished in the debinding step of the process. Many debinding variants exist. Thermal, solvent, or catalytic-phase erosion debinding (or a combination of these) now require only hours, instead of days, to remove a major portion of the binder. The use of a multicomponent binder is

Table 4.4   Typical powder injection molding parameters

Parameter	Typical range	Trigger
Barrel temperature, °C	100–200	160
Nozzle temperature, °C	80–200	180
Mold temperature, °C	20–100	40
Screw rotation speed, rpm	35–70	35
Peak injection pressure, Mpa	0.1–130	20
Packing pressure, Mpa	0–10	8
Fill time, s	0.2–3	0.6
Packing time, s	2–60	3
Cooling time, s	18–45	20
Cycle time, s	8–360	37

Source: Ref 20

key to rapid progressive and stage-controlled debinding. The residual binder is removed during sintering. Nonetheless, section thickness of an injection-molded part is usually limited to approximately 50 mm (2.0 in.) in the interest of economic debinding times. The smallest section thickness can be 0.5 mm (0.02 in.). Residual stresses from incomplete or improper debinding can lead to cracking during the sintering cycle (Ref 22). Fast debinding techniques permit their integration into the sintering cycle. Solvent and catalytic debinding techniques provide superior dimensional control.

Mathew and Mastromatto (Ref 25) describe a water-based agar binding system for stainless steels that they claim provides clean and rapid debinding with the ability to use soft tooling and low molding pressures. The metal powder is mixed with water, agar (a polysaccharide derived from seaweed), and a gel strength-enhancing agent, for example, calcium borate and zinc borate, to form feedstock pellets. This approach uses only 2 to 3% binder, which permits fabrication of large and thick parts. Debinding is combined with sintering.

Table 4.5 shows a comparison of various debinding techniques for parts with section thickness of 10 mm (0.4 in.) and made from a 5 μm steel powder.

4.2.5 Sintering

Sintering conditions for injection-molded stainless steels are similar to those used for conventionally pressed-and-sintered stainless steels (Chapter 5, "Sintering and Corrosion Resistance"), except that residual binder removal should be accomplished prior to the closure of open porosity in the interest of low carbon content. For the low-oxygen-content gas-atomized powders, this must be done with great care in order to achieve the required low carbon contents required for good corrosion resistance.

For the hybrid water/gas-atomized powders that have higher oxygen contents, sintering in hydrogen or vacuum can be controlled to cause the oxygen from the powder to react with residual carbon from the binder (Chapter 5).

Kyogoku et al. (Ref 26) have compared the effects of several microstructural factors on the mechanical properties of high-temperature (1400 °C, or 2552 °F), vacuum-sintered, injection-molded parts made from water- and gas-atomized 304L stainless steel. The parts had been debound to give identical sintered densities of 98% for both materials. Parts made from the water-atomized powder (0.39% oxygen) had a 16% higher yield strength, which was attributed to dispersion strengthening by silicon oxides; their fatigue strength, however, was marginally inferior. For both materials, the relationship between grain size and yield strength followed the Hall-Petch relation. Also, both pores and precipitates showed Oswald ripening and satisfied the Lifshitz-Wagner equation. The authors had not determined corrosion resistance of the parts.

The hydrostatic nature of the injection molding process provides a relatively gradient-free and isotropic density distribution of the feedstock throughout the molded part. Thus, with a uniformly compounded feedstock, this feature permits a distortion-free, uniform shrinkage during sintering, even at very high sintering temperatures. Nonetheless, only small changes in section thickness are desirable to further improve dimensional accuracy. Dimensional tolerances are typically within 0.3% of a target. Experience has shown that the majority of problems related to dimensional control can be traced back to the molding conditions.

The MIM parts have typical sintered densities of 95 to 98% of theoretical. They can be densified to theoretical density by hot isostatic pressing.

The overall complexity of the MIM process suggests that the prospective user of the component and the part manufacturer discuss the

Table 4.5 Comparison of debinding techniques and times

Binder system	Debinding techniques	Conditions	Time
Wax-polypropylene	Oxidation	Slow heat 150 °C, hold heat to 600 °C in air	60 h
Wax-polyethlene	Wicking	Slow heat to 250 °C, hold, heat to 750 °C in hydrogen	4 h
Wax-polymer	Supercritical	Heat in freon vapor at 10 °C/min to 600 °C under 10 MPa pressure	6 h
Wax-polyethlene	Vacuum extraction	Slow heat while passing low-pressure gas over compacts, heat to sintering temperature	36 h
Water-gel	Vacuum sublime or freez dry	Hold in vacuum to extract water vapor from ice	8 h
Oil-polymer	Solvent immersion	Hold in ethylene dichloride at 50 °C	6 h
Water-gel	Air drying	Hold at 60 °C	10 h
Polyacetal polyethlene	Catalytic debind	Heat in nitric acid vapor at 135 °C	4 h

Note: Section thickness, 10 mm; particle size, 5 μm; solids loading, 6 vol%. Source: Ref 20

manufacturing approaches early on in the design phase of the component.

4.2.6 Process Criteria and Design Guidelines

Table 4.6 summarizes typical fabrication ranges of metal injection molded parts; Table 4.7 shows a list of component design guidelines. Some of these criteria will broaden with an increase in manufacturing experience. The single most critical requirement is that of a reduced and uniform section thickness, as mandated by the debinding process. Table 4.8 summarizes

Table 4.6 Fabrication ranges for metal injection molding

Attribute	Minimum	Maximum	Typical
Thickness, mm	0.2	25	10
Thickness variation	None	100×	2×
Longest dimension, mm	2	1,000	100
Tolerance, % (standard deviation)	0.03	2	0.3
Number of dimentional specifications	20	1,000	100
Mass, g	0.02	20,000	40
Material	Simple element	Composites	Alloys
Properties	Unimportant	Highest attainable	Handbook
Cost per part, $	0.20	400	2
Production quantity per year	200	20,000,000	150,000

Source: Ref 20

Table 4.7 Nominal metal injection molding component design guidelines

Restrictions
No inside closed cavities
Corner radius greater than 0.075 mm
Smallest hole diameter 0.1 mm
Weight range 0.02 g to 20 kg
No undercuts on internal bores
2° draft on long parts
Minimum thickness 0.2 mm

Desirable features
Gradual section thickness changes
Weight less than 100 g
Assemblies in one piece
Small aspect ratio geometries
Largest dimension below 100 mm
Wall thickness less than 10 mm
Flat surfaces for support

Allowed design features
Holes at angles to one another
Stiffening ribs
Protrusions and studs
D-shaped and keyed holes
Hexagonal, square, blind, and flat bottom holes
Knurled and waffle surfaces
External or internal threads
Part number or identification in die

Source: Ref 20

Table 4.8 Typical tolerances for powder injection molded components

Characteristic	Typical	Best possible
Angle, degrees	2	0.1
Density, %	1	0.2
Weight, %	0.4	0.1
Dimension, %	0.3	0.05
Absolute dimension, mm	0.1	0.04
Hole diameter, %	0.1	0.04
Hole location, %	0.3	0.1
Flatness, %	0.2	0.1
Parallelism, %	0.3	0.2
Roundness, %	0.3	0.3
Perpendicularity	0.2% or 0.3°	0.1% or 0.1 °
Average roughness, μin.	10	0.4

Source: Ref 27

minimum and typical tolerances possible with powder injection molding without secondary operations.

Figure 4.15 shows recommended designs for MIM components and suggestions for improved processing (Ref 28). Figure 4.16 compares different production technologies in terms of production quantity and geometric complexity (Ref 29). Thus, the graph indicates that MIM competes mainly with die casting and precision casting and, to a lesser degree, with conventional (press-and-sinter) PM.

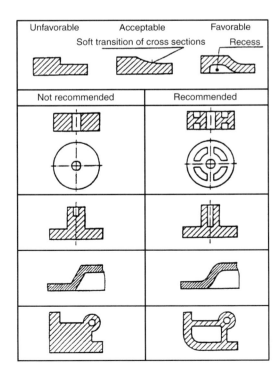

Fig. 4.15 Recommended designs for metal injection molded components. Source: Ref 28

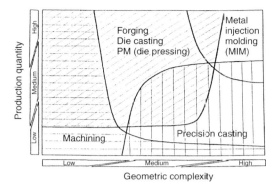

Fig. 4.16 Economic comparison of various production technologies in terms of production volume and geometric complexity. Source: Ref 29

Table 4.9 Comparative mechanical properties of conventionally produced wrought (cold worked and amealed) and PM extruded stainless steel tubes

Grade	Type(a)	Number of samples	Yield strength, 0.2% offset		Tensile strength		Elongation, %
			MPa	ksi	MPa	ksi	
Type 304L	C	84	302	44	582	84	57
	PM	18	325	47	609	88	58
Type 304	C	133	321	46	600	87	57
	PM	72	350	51	660	96	55
Type 316L	C	90	319	46	604	88	53
	PM	128	336	49	632	92	52
Type 316	C	134	306	44	584	85	64
	PM	125	346	50	649	94	51
Type 904L	C	49	334	48	651	94	45
	PM	112	382	55	681	99	43

(a) C, conventional production; PM, powder metallurgy (extruded). Source: Ref 30

4.3 Extrusion of PM Stainless Steels

Seamless stainless steel tubes and associated products are made commercially by Dynamet Anval, Sweden, by extrusion of gas-atomized stainless steel powders (Ref 29, 30). The coarse (approximately 0.5 mm, or 0.02 in.) inert-gas-atomized powder (Section 3.2 in Chapter 3, "Manufacture and Characteristics of Stainless Steel Powders") is canned in carbon steel capsules (fill density >70%), then cold isostatically compacted into a billet of 85 to 90% of theoretical density. The billet is then hot extruded in two stages at 1200 °C (2192 °F) in a conventional glass-lubricated extrusion press. The capsule material is removed by decladding. With extrusion ratios of 20 to 30, seamless tubes 10 to 15 m (33 to 50 ft) long are produced. Advantages of the PM process over the conventional wrought billet route include:

- Improved homogeneity of microstructure and lower slag content
- Improved control of composition
- Improved corrosion and mechanical properties due to reduced segregation of alloying elements
- Reduced grain size in the heat-affected zone after welding
- The ability to make difficult compositions, including titanium-stabilized ferritic and austenitic stainless steels
- Greater flexibility in production (greatly reduced throughput time eliminates the need to stock a wide range of finished tubing in the more expensive grades)

Standard grades include most of the common austenitic stainless steels as well as some special austenitic, ferritic-austenitic, and ferritic stainless steels. Nitrogen contents are somewhat higher (900 versus 500 ppm for wrought type 316). Slightly higher yield and tensile strengths, without a loss in elongation, are attributed to the aforementioned differences in comparison to wrought products (Table 4.9).

Most mechanical properties at elevated temperature are practically identical to those of conventionally produced materials. The creep properties are generally better because of the finely dispersed oxide particles. A titanium-stabilized ferritic stainless steel had a markedly improved impact ductility, which was attributed to small TiCN particles lying in the grain boundaries and preventing grain growth.

No difference between PM-extruded and conventional material has been found with respect to the resistance to intergranular corrosion in tests based on ASTM A 262 (practices C and E). Also, the resistance to pitting attack, as measured by the pitting corrosion breakthrough potential, is superior for several PM grades compared to the corresponding conventional grades (Ref 29).

Uses and applications include highly alloyed stainless steels for very corrosive environments, offshore oil rigs, and chemical plants.

4.4 Hot Isostatic Pressing of Stainless Steels

Gas-atomized powders, with their spherical particle shape and low oxygen contents, are used to produce a variety of relatively simple shapes by hot isostatic compaction. Although such parts have been commercially available since the 1960s (Ref 29), their documentation is sketchy. The applications are predominantly nonautomotive and include (Ref 31):

- Oil/gas drilling: valve components
- Aerospace/aircraft: turbine components
- Defense: ordnance
- Marine: diesel engine components
- Architectural: sputtering targets
- Chemical: pump bodies

Dynamet Anval manufactures flanges, tube hollows, billets for wire rolling, valve components, valve seats, gates, separator rotors, and so on by this process. These are produced mainly in the higher-alloyed stainless steels, which cannot be easily obtained through ingot metallurgy. Hot forging difficulties that can occur with some stainless steels are absent in the PM process. The PM materials also have fine grain size and more uniform material properties throughout the component.

Hjorth and Ericksson (Ref 32) reported about the use of hot isostatically pressed duplex and austenitic stainless steels in the oil, gas, and petrochemical industries. Valve bodies, fittings, and large manifolds for piping systems are produced in a cost-effective manner. Figure 4.17 shows a hot isostatically pressed valve body made of an austenitic stainless steel.

Bodycote Powdermet AB, Sweden, manufactures pressure vessel components, valve bodies, swivels, and thick-wall fittings by hot isostatic pressing of ferritic, austenitic, and duplex ferritic-austenitic stainless steels.

ASTM standard A 988 (1998) summarizes the specifications for several hot isostatically pressed martensitic, austenitic, and austenitic-ferritic stainless steels.

Biancaniello et al. (Ref 33) reported a pitting potential increase of 600 mV over that of wrought 316L for high chromium/high nitrogen (28–30Cr, 0.8–1.0N, ≥12Ni, ≤15Mn, ~2Mo, 0.02C) austenitic stainless steels prepared by alloy melting under nitrogen followed by nitrogen atomization and hot isostatic compaction. The improvement was attributed to the high level of nitrogen. In addition to the vastly improved pitting resistance, the alloys also had dramatically improved yield strength and other excellent mechanical properties. The high levels of chromium and manganese served to keep the large amount of nitrogen in solution.

Fig. 4.17 Hot isostatically pressed valve body in austenitic stainless steel. Weight: 2 t (2.2 st)

REFERENCES

1. *Powder Metal Technologies and Applications,* Vol 7, *ASM Handbook,* ASM International, 1998, p 313–436
2. R.M. German, *Powder Metallurgy Science,* 2nd ed., MPIF, Princeton, NJ, 1994, p 205
3. G.F. Bocchini, "The Warm Compaction Process—Basics, Advantages and Limitations," Paper 980334, SAE International Congress and Expo. (Detroit, MI), 1998
4. R. Phillips, "Tooling for Molding Stainless Steel Parts," presented at PM Stainless Steel Seminar, March 1–2, 2000 (Durham, NC), MPIF, Princeton, NJ
5. J.H. Reinshagen and R.P. Mason, An Evaluation of Methods for Improving the Green Properties of PM Stainless Steels, *Advances in Powder Metallurgy and Particulate Materials,* ed. W. Eisen, S. Kassam, Vol 7, MPIF, Princeton, NJ, 2001, p 7-121 to 7-134
6. A. Bergkvist, Stainless Steel Powders for High Density Applications, *Advances in Powder Metallurgy and Particulate Technology,* ed. W. Eisen, S. Kassam, Vol 3, MPIF, Princeton, NJ, 2001, p 3-251 to 3-262

7. T. Cronin and D.F. Berry, Identification of Lubricants in Metal Powder Mixes by Differential Scanning Calorimeter, *Advances in Powder Metallurgy and Particulate Materials,* ed. E. Andreotti, P. McGeehan, Vol 1, MPIF, Princeton, NJ, 1990, p 295–306

8. T. Takeda and K. Tamura, Pressing and Sintering of Cr-Ni Austenitic Stainless Steels, *J. Jpn. Soc. Powder Met.,* Vol 17 (No. 2), 1970, p 70–76

9. G.P. Gasbarre, Jr., "Warm Compaction and Die Wall Lubrication," PM Advances, Innovations and Emerging Technologies Seminar, March 2002 (Cincinnati, OH), Sponsored by MPIF, Princeton, NJ

10. R.C. Leyton and O. Andersson, High Density Sintered Stainless Steel with Close Tolerances, *Advances in Powder Metallurgy and Particulate Materials,* ed. V. Arnhold, C.-L. Chu, W. Jandesha, Jr., H. Sanderow, Vol 7, MPIF, 2002, p 7-127 to 7-126

11. T. Takeda and K. Tamura, Densification of Stainless Steel Powder Compacts Powder During Sintering, *J. Jpn. Soc. Powder Met.,* Vol 17 (No. 5), Jan 1971, p 28–36

12. R.M. German, Computer Modeling of Sintering Processes, *Int. J. Powder Metall.,* Vol 38 (No. 2), 2002, p 48–66

13. H.H. Hausner, *Handbook of Powder Metallurgy,* Chemical Publishing Company, Inc., New York, NY, 1982, p 168–206

14. P.K. Samal, "Sintering of Stainless Steels," presented at PM Stainless Steel Seminar, March 1–2, 2000 (Durham NC), Sponsored by MPIF, Princeton, NJ

15. J.S. Hirschhorn, *Introduction to Powder Metallurgy,* American Powder Metallurgy Institute, Princeton, NJ, 1969, p 204

16. R.M. German, The Sintering of 304L Stainless Steel Powder, *Metall. Trans. A,* Vol 7, Dec 1976, p 1879–1885

17. T. Takeda and K. Tamura, Pressing and Sintering of Cr-Ni Austenitic Stainless Steel Powder, *J. Jpn. Soc. Powder Met.* Vol 7, 1970, p 22–28

18. D.J. McMahon and O.W. Reen. The Prediction of Processing Properties of Metal Powders, *Modern Developments in Powder Metallurgy,* Vol 8, MPIF, 1974, p 41–60

19. N. Myers, R.K. Enneti, L. Campbell, and R. German, Effects of Chemistry Variations on Dimensional Control of 316L Stainless Steel, *Advances in Powder Metallurgy and Particulate Materials,* ed.

V. Arnhold, C.-L. Chu, W. Jandesha, Jr., H. Sanderow, Vol 7, MPIF, Princeton, NJ, 2002, p 7-127 to 7-133

20. R.M. German, Powder Injection Molding, *Powder Metal Technologies and Applications,* Vol 7, *ASM Handbook,* ASM International, 1998, p 355–364

21. R.M. German and A. Bose, *Injection Molding of Metals and Ceramics,* MPIF, Princeton, NJ, 1997, p 57

22. S. Pascoli, P.A.P. Wendhausen, and M.C. Fredel, Keeping to Form in the Powder Business, *Met. Powder Rep.,* Vol 57 (No. 3), March 2002, p 32–37

23. J.A. Sago, J.W. Newkirk, and G.M. Brasel, The Effects of MIM Processing Control Parameters on Mechanical Properties, *Advances in Powder Metallurgy and Particulate Materials,* ed. R. Lawcoch, M. Wright, Vol 8, MPIF, 2003, p 8-217 to 8-233

24. F. Petzoldt and T. Hartwig, Overview on Binder and Feedstock Systems for PIM, *Second European Symposium on Powder Injection Molding,* Oct 18–20, 2000 (Munich, Germany), EPMA, p 43–50

25. B.A. Mathew and R. Mastromatto, Metal Injection Moulding for Automotive Applications, *Met. Powder Rep.* Vol 57 (No. 3), 2002, p 20–23

26. H. Kyogoku, S. Komatsu, M. Shinzawa, D. Mizuno, T. Matsuoka, and K. Sakaguchi, Influence of Microstructural Factors on Mechanical Properties of Stainless Steel by Powder Injection Molding, *Proc. 2000 PM World Congress* (Kyoto, Japan), K. Kosugo and H. Nagai, Ed., Part I, Japan Society of Powder and Powder Metallurgy, p 304–307

27. D.S. Hotter, PIM Breathes Life into Medical Products, *Mach. Des. (Eng. Mater.)* Oct 9, 1997, p 78–81

28. H. Cohrt, "Metal Injection Molded Components for Automotive Applications," Workshop Notes: Designing for High Performance PM Automotive Components, PM 97 (Munich, Germany), EPMA

29. The Production of Stainless Steel Tube and Associated Products by Powder Metallurgy, *Stainless Steel Ind.,* Vol 9 (No. 52), 1981

30. C. Tornberg, "The Manufacture of Seamless Stainless Steel Tubes from Powder," Paper 8410-013, presented at the 1984 ASME International Conference on

New Developments in Stainless Steel Technology (Detroit, MI), American Society of Mechanical Engineers, 1984, p 1–6

31. A.J. Clayton, Non-Automotive Markets for PM Applications: Opportunities and Challenges, *Advances in Powder Metallurgy and Particulate Materials,* ed. H. Ferguson, D. Whychell, Jr., Vol 10, MPIF, Princeton, NJ, 2000, p 10-73 to 10-78

32. C.G. Hjorth and H. Eriksson, New Areas for HIPing Components for the Offshore and Demanding Industries, *Hot Isostatic Pressing, Proc. Int. Conf. Hot Isostatic Pressing,* ASM International, May 20–22, 1996, p 33–38

33. F.S. Biancaniello, D.R. Jiggets, M.R. Stoudt, R.E. Ricker, and S.D. Ridder, Suitability of Powder Processed High Nitrogen Stainless Steel Alloys for High Performance Applications, *Advances in Powder Metallurgy and Particulate Materials,* ed. V. Arnhold, C.-L. Chu, W. Jandesha, Jr., H. Sanderow, Vol 3, MPIF, Princeton, NJ, 2002, p 3-198 to 3-207

CHAPTER 5

Sintering and Corrosion Resistance

THE PRIMARY GOAL in stainless steel sintering is to obtain good corrosion resistance along with good mechanical properties and adequate dimensional tolerances. Most aspects of sintering have a bearing on corrosion resistance; therefore, in the following, sintering is discussed with an emphasis on its effect on corrosion resistance.

In wrought stainless steels, superior corrosion resistance is of paramount importance, because mechanical properties similar to and even superior to stainless steels can be obtained much less expensively with conventional carbon steels. However, over several decades, despite modest corrosion resistances, commercial sintered stainless steels found niche applications (for example, office machine parts, lock parts, mirror mounts, some appliance parts, etc.) where sintered stainless steels were able to compete with wrought or cast stainless steels because their corrosion properties met the moderate requirements. Also, powder metallurgy (PM) parts offered their typical advantages: good material utilization and low-cost net shape fabrication (no machining costs).

From the 1950s until the mid-1980s belt furnaces were the dominant method of industrial sintering of stainless steels in North America. Maximum sintering temperature was approximately 1150 °C (2100 °F), and furnace atmosphere was dissociated ammonia (DA). The lower-cost atmosphere and the higher strength levels possible with sintering in DA were attractive, but it was also more difficult to achieve good corrosion properties in DA than in hydrogen or vacuum. Hence, the gradual shift to hydrogen and vacuum sintering, or the use of a $90H_2$-$10N_2$ atmosphere, during the past 10 years. There was also a shift toward high-temperature (>1205 °C, or >2200 °F) sintering.

The majority of studies on the corrosion resistance of sintered stainless steels still lack a full description of the experimental conditions employed. The most frequently omitted process parameters are the dewpoint of the sintering atmosphere and the cooling rate after sintering. Because these and other parameters are of critical importance to corrosion properties of sintered stainless steels, only publications providing critical processing data and/or permitting unambiguous conclusions are reviewed in the context of corrosion-resistance properties.

If sintering conditions are conducive to the development of good corrosion resistance, good mechanical properties usually follow. The reverse is not necessarily true. Each sintering atmosphere has its own peculiarities with regard to stainless steels, mainly because each responds differently to a number of chemical reactions involving the interstitials carbon, nitrogen, and oxygen. The details of these reactions largely determine the corrosion and dynamic mechanical properties of sintered parts. The many misconceptions about sintering stainless steels (Table 1.1 in Chapter 1, "Introduction") arise in large part from a lack of appreciation of the importance of these chemical reactions and from ignoring their differences for the various sintering atmospheres. Even though the extent of these reactions typically varies from only several hundred to a few thousand parts per million, they are of critical importance. Viewing the sintering atmosphere as mainly an inert cover to protect parts from oxidation, typical of the early years, grossly misjudges its importance.

In wrought stainless steel technology, oxygen, carbon, and nitrogen are controlled at the refining stage of the production process; in PM, they are controlled during powder manufacture and sintering. Excessive amounts of carbon and nitrogen can give rise to the formation of chromium carbides and chromium nitride, with negative effects on corrosion resistance. These precipitates can be identified metallographically and through special corrosion tests. Furthermore, they resemble the corresponding phenomena in wrought stainless steels. However, precipitates of silicon dioxide that form during cooling after sintering usually do not show up in a metallographic cross section and are normally absent in properly finished wrought stainless steels.

5.1 Sintering Furnaces and Atmospheres

Notwithstanding the importance of powder selection, the sintering process is of even greater importance for the successful processing of stainless steels. It encompasses many more elements, from furnace type and atmosphere choice to process parameter choices. All of these influence the quality of a sintered part. To the extent that these elements are common to general PM processing, their treatment in the following is only cursory. For a detailed general treatment of both practice and fundamentals of sintering, the reader is advised to consult the literature sources suggested at the beginning of Chapter 1, "Introduction." However, those elements and parameters that have a special bearing on sintered stainless steels, both regarding their mechanical properties and, more so, their corrosion-resistance properties, are treated in detail.

Sintering Furnaces. Most commercial sintering of stainless steel parts is performed in continuous mesh belt conveyor furnaces at temperatures up to approximately 1150 °C (2100 °F). Pusher, walking beam, and vacuum furnaces are used for higher temperatures up to approximately 1345 °C (2450 °F). In recent years, ceramic belt furnaces have also been introduced for high-temperature sintering. The higher temperatures are favored for improved mechanical and corrosion

properties. Vacuum and other high-temperature furnaces began to be more widely used in the 1980s, as a result of increasing demands on magnetic and corrosion-resistance properties. Although some sintering furnaces for carbon steel parts now have gas quench capability in their cooling zones, permitting so-called sinter hardening, most industrial furnaces for stainless steels presently lack this feature, despite its benefits in minimizing reoxidation in the cooling zone and reducing the risk of sensitization. In this regard, vacuum furnaces, with their readily available gas quench features, are advantageous. Among the belt furnace types, so-called humpback furnaces (Fig. 5.1) (Ref 1), give lower dewpoints. Their inclined entrance and exit zones retain the lighter hydrogen better than the more common horizontal furnaces.

It is the inferior control of dewpoint and slower cooling after sintering in many commercial furnaces, compared to laboratory furnaces, that has led to one of the half-truths (Table 1.1 in Chapter 1) about the corrosion resistance of stainless steel parts, namely, that PM parts possessing good corrosion resistance can be produced in laboratory furnaces but not in industrial furnaces.

Sintering Atmospheres. Typical sintering atmospheres for stainless steels include hydrogen, hydrogen-nitrogen mixtures, dissociated ammonia, and vacuum. Because a low-dewpoint capability is important for both hydrogen and hydrogen-nitrogen atmospheres, there is a widespread belief throughout the industry that the use of cryogenic nitrogen in hydrogen-nitrogen mixtures makes it easier to attain the required low dewpoints because of the dryness of cryogenic nitrogen. However, the reducibility criterion for nitrogen-containing hydrogen demands lower dewpoints than those for pure hydrogen (Fig. 5.15).

Dissociated ammonia with dewpoints of approximately –45 °C (–50 °F) was the most

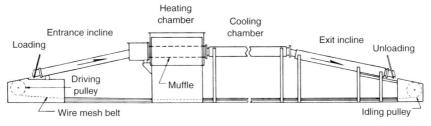

Fig. 5.1 Schematic of a humpback mesh belt furnace. Source: Ref 1. Reprinted with permission from MPIF, Metal Powder Industries Federation, Princeton, NJ

widely used sintering atmosphere until the mid-1980s. In the interest of high-strength parts and low-cost sintering atmospheres, some stainless steel parts were even sintered in N_2-H_2 atmospheres containing as little as 3% H_2. As is shown later, such parts had very low corrosion-resistance requirements. Vacuum sintering of stainless steels is conducted with low pressures (1000 to 3000 μm Hg) of argon or nitrogen to minimize chromium losses due to the high vapor pressure of that element at elevated temperature (section 5.2.5 in this chapter).

5.2 Sintering of Stainless Steels

Prior to actual sintering, PM parts are delubricated either as part of the sintering process or in a separate step. The importance of delubrication for stainless steels had been underrated for many years, with the result that parts possessed excessive carbon contents due to lubricant decomposition. If the furnace temperature rises above approximately 540 to 650 °C (1000 to 1200 °F) before most of the lubricant has had time to volatilize, lubricant decomposition and carbon absorption by the part will take place rather than lubricant volatilization. While vacuum furnaces can readily cope with carbon absorption (section 5.2.5 in this chapter), lowering carbon levels in hydrogen and hydrogen-nitrogen atmospheres is more difficult. This subject is discussed in section 5.2.3 in this chapter.

5.2.1 Fundamental Relationships

The development of properties of stainless steels as a function of density and sintering temperature is similar to those of carbon steels. Figures 5.2 (Ref 2) and 5.3 (Ref 3) illustrate some of the basic relationships between sintering temperature, sintered density, and properties of parts. Corresponding explicit data for sintered stainless steels are given in Chapter 7, "Mechanical Properties."

In addition to the effect of density, sintered steels differ from similar wrought steels in that they usually possess smaller grain size. High-temperature-sintered ferritic stainless steels, however, possess large grain sizes. Also, inclusions and second phases are distributed throughout the matrix rather uniformly, and the oxygen contents are often an order of magnitude higher than those in wrought stainless steels. The

latter arises from the oxidation during water atomization of a stainless steel powder (section 3.1.5 in Chapter 3, "Manufacture and Characteristics of Stainless Steel Powders"). Typical commercial sintering conditions remove or reduce only a small portion of this oxygen.

In order for mechanical properties to develop properly and in a reasonable amount of time, it is critical that atomic diffusion during sintering is not impeded by the oxide layers of the water-atomized powder particles. In the case of plain iron powders, such oxides are readily reduced by any of the common sintering atmospheres. In the case of water-atomized stainless steels, much drier (lower-dewpoint) atmospheres are necessary for reduction. Residual oxides, sometimes termed acid insolubles, can reduce the mechanical properties of a sintered part. Dautzenberg and Gesell (Ref 4) showed that the ultimate tensile strength of a sintered austenitic stainless steel was increased by 30% when its oxygen content was decreased from 1.0 to 0.2%. Although

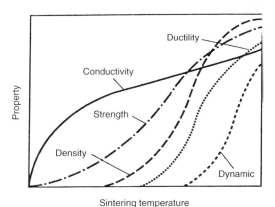

Fig. 5.2 Variation in compact properties with degree of sintering, as represented by sintering temperature. Source: Ref 2. Reprinted with permission from MPIF, Metal Powder Industries Federation, Princeton, NJ

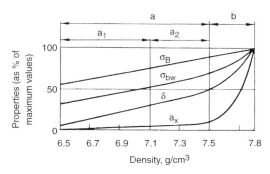

Fig. 5.3 Correlation of process-dependent density and important properties. Source: Ref 3

acceptable mechanical properties (acceptable in this case means that sintered parts will meet the properties specified in Metal Powder Industries Federation, or MPIF, and ASTM standards) are obtainable with the common industrial sintering atmospheres and with sintering times of only 30 min, sintering under conditions that lower the amount of oxygen (oxides) clearly and significantly improves the dynamic mechanical properties of sintered stainless steels, that is, fatigue and impact strength.

For the same fundamental reason, properties of fully dense parts, made from gas-atomized powders with low oxygen contents and/or low contents of undesirable interstitials, can be superior to their wrought counterparts because of their lower levels of interstitials and the more uniform (isotropic) distribution of these interstitials within the matrix. Superior mechanical and corrosion-resistance properties have been documented for fully dense PM stainless steels, PM superalloys, and PM aluminum alloys (Ref 5).

For vacuum sintering at high temperature (>1200 °C, or >2192 °F), the addition of graphite to the stainless steel powder prior to compacting can produce oxygen contents of less than 300 ppm, with carbon contents of less than 0.03% (section 5.2.5 in this chapter) and a 40% improvement in impact strength (Ref 6).

In hydrogen-nitrogen atmospheres, lower dewpoint and higher hydrogen content give better reduction of oxides. Lower compact density will also produce lower oxygen contents because of faster diffusion of the reducing gas and reaction products (H_2O, CO, CO_2). Longer sintering time, of course, will also result in more reduction.

Differences in the mechanical properties of sintered stainless steels as well as in their dimensional change (whether from lot to lot or from producer to producer) are mainly caused by differences in the amount and distributions of the interstitials, oxygen, carbon, and nitrogen. These in turn arise from differences in processing. With good sintering practice, homogenization of the microstructure takes place quite rapidly. This is described in more detail in section 5.2.3 in this chapter.

While it is possible to obtain good corrosion resistance in any of the common sintering atmospheres, each atmosphere demands its own controls. It is therefore convenient to discuss this subject individually for each sintering atmosphere. However, the control of sintered density and how it affects corrosion resistance is common to all types of sintering and is therefore addressed now.

5.2.2 Effect of Sintered Density on Corrosion Resistance

The corrosion resistance of stainless steels can differ widely, depending on the testing environment. Different mechanisms of corrosion have been correlated with certain environments.

Acidic Environment. Testing of sintered stainless steels in acids, that is, H_2SO_4, HCl, and HNO_3, shows that corrosion resistance, measured as weight loss, improves significantly with increasing density (Fig. 5.4) (Ref 7). This relationship is confirmed elsewhere (Ref 8–10).

The detrimental effect of pores is attributed to two factors: first, to the large internal surface areas of sintered parts, which, at the typical densities of many structural parts (i.e., 80 to 85% of theoretical), are still 2 orders of magnitude larger than their exterior geometric surface areas and therefore can be subject to increased general corrosion; second, to a lack of passivation within the pores of a sintered part. Open-circuit measurements (section 9.1.3 in Chapter 9, "Corrosion Testing and Performance") of wrought stainless steels in an acidic environment show that the potential typically increases with time (Ref 11). This can be interpreted as passivation and/or healing of active areas. In contrast, sintered stainless steels often exhibit decreasing potential, indicating activation of the surface. Itzhak and Aghion (Ref 12) and Raghu et al. (Ref 11) interpret the declining open-circuit potential of sintered stainless steels as gradually increasing

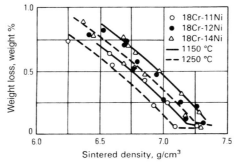

Fig. 5.4 Relationship between sintered density and weight decrease of three austenitic stainless steels in 40% HNO_3. Source: Ref 7. Reprinted with permission from MPIF, Metal Powder Industries Federation, Princeton, NJ

activation as the acid penetrates the pores. This is accompanied by hydrogen evolution on the surface of a part. The main reaction taking place is $2H^+ + 2e^- \rightarrow H_2$. Thus, corrosion in an acidic environment can be viewed as the operation of a hydrogen concentration cell between the external surface of a part and its internal pore surface. The surface of the pores acts as the anode and the engineering surface as the cathode. Metal dissolution occurs primarily in the interior of the material. After 40 h, the activation process comes to an end and the potential increases.

Neutral Chloride-Containing Environment. Corrosion resistance in neutral saline solutions has been found to decline with increasing density (Fig. 5.5) (Ref 13).

The parts of Fig. 5.5 had been prepared from typical –100 mesh compacting-grade powders. The decline of corrosion resistance is moderate at low density but becomes very steep at a relative density of approximately 80 to 84% of theoretical, depending on pore size, pore morphology, and possibly on residual oxygen content. The fact that some specimens in Fig. 5.5 are capable of bridging the low corrosion-resistance gap suggests that the effect is a borderline one and that it may disappear by increasing the intrinsic crevice and pitting resistance of an alloy. In fact, some of the specimens of Fig. 5.5, as a result of carbon-assisted vacuum sintering (section 5.2.5 in this chapter), had very low oxygen contents, comparable to wrought stainless steels. Also, corrosion-resistance/density curves for 317L, a somewhat more corrosion-resistant material because of higher chromium and molybdenum contents, appear to possess a greater number of specimens bridging the corrosion-resistance gap.

Potential-time curves of sintered stainless steels in a neutral saline solution exhibit similar behavior to those in an acidic environment; that is, they also may be characterized by the potential decreasing with time, indicating activation rather than passivation.

Raghu et al. (Ref 14) have performed cyclic potentiodynamic polarization studies on sintered 316L prepared from narrow sieve fractions. Densities varied from 37 to 71%, and testing was performed in 3% NaCl. The difference potential, ΔE, which is a measure for a material susceptibility to crevice corrosion, increased with decreasing pore size (Fig. 5.6)

The detrimental effect of pores is very strong for small pores up to approximately 20 μm (as determined by the bubble point test method for filters) and thereafter becomes much less pronounced. The important variable in this case is pore size rather than porosity.

These results are best explained by assuming the operation of an oxygen concentration cell, which establishes itself in accordance with the mechanism shown in Fig. 5.7 (Ref 15) and Fig. 5.8 (Ref 16).

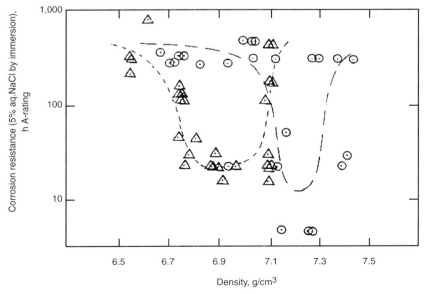

Fig. 5.5 Effect of density on corrosion resistance of 316L parts. Δ, pressed and sintered only; o, pressed, sintered, re-pressed, and annealed. Source: Ref 13. Reprinted with permission from MPIF, Metal Powder Industries Federation, Princeton, NJ

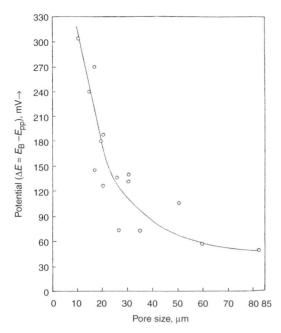

Fig. 5.6 Effect of pore size on size of hysteresis(*E*) for sintered 316L in 3% NaCl (27 °C, or 81 °F). Source: Ref 14. © NACE International 1989

The overall reaction involves the dissolution of metal, M (immersed in aerated saline solution), and the reduction of oxygen to hydroxide in accordance with:

Oxidation

$$M \Rightarrow M^+ + e$$

Reduction

$$O_2 + 2H_2O + 4e \Rightarrow 4OH^-$$

As a result of limited diffusion within the pore space of a part, oxygen within that space becomes depleted and oxygen reduction ceases. However, as shown in Fig. 5.7, metal dissolution continues within the pore space. The latter creates a positive charge (M^+) within the pore space, which is neutralized by the migration of chloride ions into the pore space. The increased metal chloride concentration within the pore space undergoes hydrolyzation into insoluble hydroxide and free acid according to:

$$M^+Cl^- + H_2O = MOH \Downarrow + H^+Cl^-$$

The free acid increases metal dissolution, which in turn increases migration, representing an accelerating, autocatalytic process. As the corrosion within the pore space increases, the

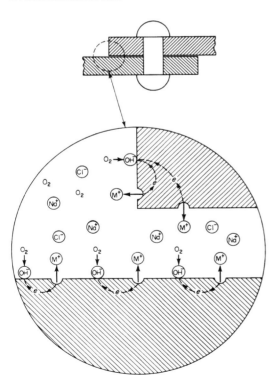

Fig. 5.7 Crevice corrosion mechanism—initial stage. Source: Ref 15. Reprinted with permission, Fontana, *Corrosion Engineering*, 2d ed. © The McGraw-Hill Companies, Inc., 1978

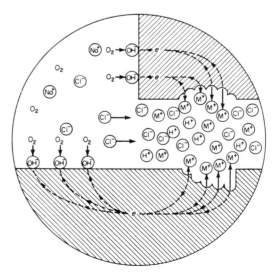

Fig. 5.8 Crevice corrosion mechanism—later stage. Source: Ref 16. Reprinted with permission from MPIF, Metal Powder Industries Federation, Princeton, NJ

rate of oxygen reduction on the internal pore surfaces also increases. This cathodically protects the external surfaces, which explains why during crevice corrosion the attack is localized within the porous or shielded areas, while the remainder suffers little or no damage.

Pore morphology of a sintered part is affected by powder particle shape, particle size distribution, compacting pressure, amount of shrinkage during sintering, re-pressing, and so on. For a stainless steel part made from a typical –100 mesh compacting-grade stainless steel powder, minimum corrosion resistance, as measured by immersion in 5% NaCl, appears at a relative density of approximately 87 to 90%, again depending on pore morphology. Past the minimum, corrosion resistance increases again. The increase past the minimum is attributed to the disappearance of pores as sintered structural parts approach the region of closed-off porosity at approximately 92% of theoretical density. A more uniform density distribution in a sintered part, such as is obtainable with isostatic pressing, or, more practical, with warm compaction, may reduce the width of the crevice-corrosion density regime.

It should be emphasized that corrosion resistance as shown in Fig. 5.5 was measured by the time it took for the development of rust, based on visual assessment. Weight change measurements are not reliable for assessing corrosion resistance of sintered parts that were tested in a neutral environment.

It is not clear why the corrosion resistance at the higher densities past the corrosion-resistance minimum does not approach that of its wrought counterpart. Both pore morphology and residual oxygen content may play a role. In fact, as is shown in section 5.3 on liquid-phase sintering of stainless steels, a high-density, boron-containing, liquid-phase-sintered 316L

had a chloride (immersion in 5% aqueous NaCl) corrosion resistance similar to wrought 316L, while other high-density-sintered stainless steels of the same composition but without boron had much lower corrosion resistances. The boron may have scavenged and redistributed the residual oxygen of the sintered material, with the formation of less detrimental borosilicates.

Conflicting with the aforementioned results, short-term potentiodynamic polarization tests by Lei et al. (Ref 17) pointed to a beneficial effect of density in a saline environment. The controversy was resolved when Maahn and Mathiesen (Ref 18) observed that in short-term polarization tests, there was not enough time for the time-consuming buildup of localized attack within pores (Table 5.1) (Ref 19).

While the corrosion resistance related to the outer surfaces, given by i_{peak} and i_{pass}, in general improves with increasing density, with E_{pit} remaining unchanged, more relevant long-term exposure techniques, such as E_{stp} and salt spray testing (NSS1 and NSS2) (Chapter 9, "Corrosion Testing and Performance"), show increasing susceptibility to crevice corrosion with increasing density and increasing oxygen content. Thus, by using slow, stepwise polarization (section 9.1.3 in Chapter 9), the expected relationship—a decrease of the stepwise initiation potential, equivalent to deteriorating corrosion resistance—was observed.

In the past, many instances of corrosion of sintered stainless steels were interpreted as crevice corrosion because of porosity, when in fact they were clearly the result of incorrect or suboptimal sintering that produced metallurgical defects that gave rise to intergranular or galvanic corrosion or that, because of an excessive vacuum (section 5.2.5 in this chapter), led to severe chromium depletion of the surfaces of

Table 5.1 Effect of density and oxygen content on corrosion resistance of hydrogen-sintered 316L

Specimens sintered at 1250 °C (2282 °F), 120 min in pure hydrogen

Compaction pressure		Sintered density, g/cm^3	Open pores, %	O, ppm	I_{peak}(a), µA/cm^2	I_{pass} (a), µA/cm^2	E_{pit} (a), mV SCE	E_{stp} (b), mV SCE	NSS1, h	NSS2
MPa	ksi									
295	43	6.34	19.4	340	31	20	475	0	>1500	9
390	57	6.62	15.5	1260	18	19	425	−100	985	7
490	71	6.86	12.3	970	25	15	475	−75	36	4
540	78	6.94	10.8	1900	18	15	500	−200	60	3
590	86	7.02	9.7	1410	21	14	450	−125	28	2
685	99	7.13	7.6	2150	9	7	500	−225	48	2
785	114	7.23	5.7	2040	7	7	475	−200	24	2

(a) 0.1%Cl⁻, pH 5, 30 °C (86 °F), 5 mV/min. (b) 5% NaCl, 30 °C (85 °F), 25 mV/8 h

sintered parts. Mathiesen and Maahn (Ref 20) have used image analysis on 316L parts, sintered in hydrogen under various conditions of time and temperature, to obtain a wide range of sintered densities. Considering the pores as cylindrical holes, they expressed the severity of corrosion in pores as:

$$S = i_a \cdot 1/d$$

where d is the pore diameter, and i_a is the corrosion rate in the passive state. Figure 5.9 shows a plot of the visual rating of corrosion (10 = no rust; 1 = 50% rust) after a 1500 h salt spray test versus the aforementioned severity value.

In the same investigation, Mathiesen and Maahn show that stepwise pitting potential (0.5% Cl) decreases with increasing sintered density of 316L (Table 5.2).

The numbers within the body of Fig. 5.9 refer to the tables of the paper in which the various experiments are described. It is apparent that corrosion resistance begins to deteriorate rapidly at a density of approximately 6.6 g/cm³ (82% of theoretical). The authors attribute the deterioration of corrosion resistance with increasing density to both a critical pore geometry and to impeded reduction of oxides.

Figure. 5.10 (Ref 15) shows the results of a crevice-corrosion test in accordance with ASTM G 48 wrought 316L and sintered 316L. The density of the sintered 316L was 6.8 g/cm³ (85% of theoretical), that is, well within the steep decay region for a compacting-grade material.

Interestingly, the sintered part showed only a mild attack in comparison to the severely corroded wrought stainless steel of the same

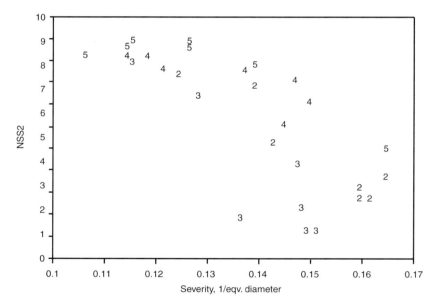

Fig. 5.9 Visual rating after 1500 h salt spray test versus severity value calculated as the reciprocal of average pore diameter. Reprinted with permission from MPIF, Metal Powder Industries Federation, Princeton, NJ

Table 5.2   Effect of density for 316L cylindrical specimens sintered at 1250 °C (2282 °F), 120 min in pure hydrogen

Green density, g/cm³	Density(a), g/cm³	Open pores(a), %	Average pore diameter(b), μm	Roundness(b)	NSS1, h	NSS2	Ferroxyl test, Cl⁻: 0.5%	E_{stp}(c) mV SCE, Cl⁻: 0.1%	0.5%
5.80	6.40	19.3	9.5	0.73	1336	9	0	350	150
5.91	6.51	17.8	8.8	0.74	>1500	10	0	250	150
6.05	6.65	15.9	8.0	0.72	>1500	10	0	275	100
6.19	6.75	14.4	8.7	0.79	>1500	10	0	250	125
6.25	6.83	13.4	8.0	0.74	>1500	10	0	300	100
6.38	6.93	11.8	7.3	0.75	1168	9	0	325	100
6.44	7.01	10.7	6.1	0.71	192	5	0	300	50

(a) Measured by oil impregnation technique. (b) Measured by image analysis. (c) Stepwise polarization designation. Evidently, because the entire PM part, on account of its porosity at a relative density of 85%, already represents a system of interconnected crevices, any additional crevice

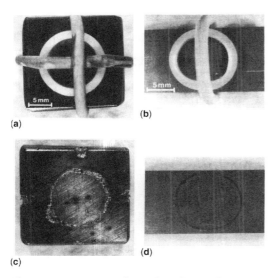

(a)

(b)

(c)

(d)

Fig. 5.10 Comparison of wrought and sintered type 316L stainless steels before and after testing in 10% aqueous FeCl₃. (a) Assembled crevice-corrosion test specimen of wrought type 316L (100% dense). (b) Assembled crevice-corrosion test specimen of sintered type 316L (85% dense). (c) Wrought specimen after test showing severe attack at four crevices under rubber bands and synthetic fluorine-containing resin ring. (d) Sintered specimen after test showing slight attack under synthetic fluorine-containing resin ring. Source: Ref 15. Reprinted with permission from MPIF, Metal Powder Industries Federation, Princeton, NJ

in accordance with ASTM G 48 test seems to have only a minor effect. This relationship is analogous to the lower notch sensitivity of sintered parts in comparison to wrought parts. The authors also found sintered type 304L and 316L stainless steels to be less susceptible to crevice corrosion than wrought 316L, on the basis of the areas of the hysteresis loops of their cyclic polarization curves (Ref 15).

The reduced crevice sensitivity of sintered stainless steels may be attributed to their interconnected pores, which facilitate oxygen diffusion through and from neighboring pores. As such, it appears that the pore space surrounding a crevice should be taken into account in assessing its susceptibility to crevice corrosion. Oxygen diffusivity within the pore space of a sintered part, as a measure for its capability to transport oxygen to its internal surfaces in order to maintain passivity, appears to be a better characterization for its resistance to crevice corrosion than an average pore diameter. A permeability or diffusivity number takes into account the entire pore space, including its tortuosity. Characterization of the pore space, through mercury porosimetry would also appear to provide more relevant characterization than an average pore diameter. In mercury porosimetry (Fig. 5.11) (Ref 21), the measured pore sizes represent the bottlenecks between neighboring pores rather than pore diameters themselves. It is the totality of these bottlenecks, rather than

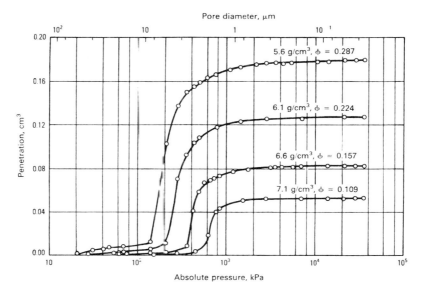

Fig. 5.11 Mercury porosimetry curves of sintered steel parts of varying densities. Green skeletons were sintered at 1093 °C (2000 °F) for 20 min. Total porosity ϕ is determined from sample weight and dimensions. Source: Ref 21

the actual pore diameters or pore volumes, that exert the greater influence on a part's capability to facilitate gas diffusion through the pore space. According to Fig. 5.12, the majority of the pore bottlenecks of a sintered steel part (made from a compacting-grade, −100 mesh, water-atomized powder), of a relative density of 80 to 84%, are 4 to 5 μm in size, and at a relative density of 87 to 90%, approximately 2 μm. Assigning greater importance to the bottlenecks would also explain the shift of the crevice-corrosion minimum to a higher density as a result of repressing (Fig. 5.6). In pressing or repressing, densification comes about first through the collapse of the larger pores (Ref 22), whereas in sintering, it is the small pores and the connections between large pores, that is, the aforementioned bottlenecks, that, because of their small curvatures and greater surface energies, become active. Thus, repressing simply increases the density of a part without greatly affecting its bottleneck pores or its diffusivity characteristics, and hence, the shift of maximum corrosion to a higher density.

According to Maahn et al. (Ref 19), the corrosion attack in a ferric chloride test (ASTM G 48) may develop within the pores beneath the surface of a part. In this case, the superior surface appearance of a sintered part may be misleading,

and interior examination and testing for mechanical property degradation is appropriate.

For a better assessment of the effect of pore morphology on crevice-corrosion resistance in the low-density range below approximately 80% of theoretical, optimally sintered parts with oxygen contents below approximately 200 ppm should be evaluated. Such parts could be prepared with carbon-assisted optimal vacuum sintering, or, easier, by optimal gravity sintering of a low-oxygen-content, inert-gas-atomized stainless steel powder, or by warm compaction and sintering of such a powder.

Molins et al. (Ref 23) have investigated the influence of several finishing operations on the corrosion resistance of sintered 316L as measured potentiodynamically according to ASTM B 627 in a solution of 0.1 N NaCl and 0.4 N NaClO$_4$ (Fig. 5.12).

Worsening passivation due to tumbling was interpreted as due to smearing of pores and potential contamination from additives. The best and most significant improvement resulted from operations that sealed surface porosity: grinding, turning, and shot blasting.

It should be stressed again that the effectiveness and/or ranking of such treatments will depend on the quality of sintering. Thus, while any such results may be relevant and practical

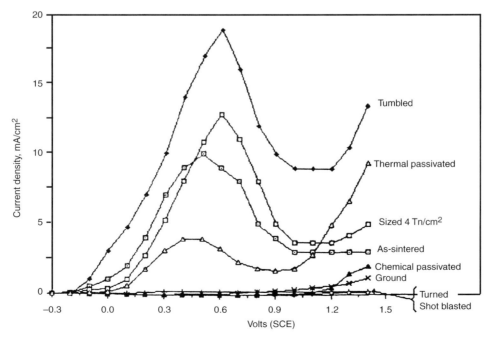

Fig. 5.12 Potentiodynamic curves of 316L stainless steels as a function of surface finishing treatment. Reprinted with permission from MPIF, Metal Powder Industries Federation, Princeton, NJ

for an individual parts producer, they differ for each parts producer. Only when sintering conditions approach optimal should results and ranking be generalized.

There have been attempts to decrease or alleviate crevice corrosion in sintered stainless steels by impregnating the pores with a resin, by metallurgical modification of the pore surfaces, or by the use of higher-alloyed stainless steels, particularly those containing higher concentrations of molybdenum. The authors found resin impregnation beneficial only in cases where the stainless steel parts had been improperly sintered and therefore had an initial low corrosion resistance. In optimally sintered parts, that is, parts that endured a long exposure to the testing solution, the tested resins separated from the pore surfaces, and the testing liquid was able to seep into the spaces. In several instances, resin impregnation also introduced ferrous contamination and unacceptable galvanic corrosion. The approaches based on surface modification and higher-alloying additions (Chapter 6, "Alloying Elements, Optimal Sintering, and Surface Modification in PM Stainless Steels") show promising results. Another promising approach to avoiding the problem of long-term corrosion in a neutral salt solution by the presence of crevice-sensitive pores is to make use of liquid-phase sintering and to achieve sintered densities greater than 7.4 g/cm^3 (section 5.3 in this chapter).

5.2.3 Sintering of Stainless Steels in Hydrogen

Hydrogen has now become the most widely used atmosphere for sintering stainless steels. In the interest of good corrosion resistance, the primary goal in processing is to lower the oxygen content of the green part as much as possible, to prevent reoxidation in the cooling zone of the furnace, and to maintain a low carbon content of approximately 0.03% in austenitic stainless steels (0.02% for high-nickel contents), and preferably, still lower for ferritic stainless steels. Apart from the sintering temperature, the two most critical parameters are the dewpoint of the hydrogen atmosphere (a measure of the water content of the atmosphere) and the cooling rate after sintering.

Oxygen Control during Sintering. For elevated-temperature metallurgical reactions, equilibrium data are very informative because of the greater ease with which reactions take place at high temperatures. For sintering of stainless

steels in hydrogen, the equilibria of interest are usually shown as so-called redox curves. Such curves show, as a function of temperature, at what water content of the sintering atmosphere a metal becomes oxidized. In scientific literature, the water content is usually shown in terms of water pressure, P_{H_2O}; in technical engineering-type literature, it is often shown in terms of dewpoint of the atmosphere, because of the easy way to determine dewpoints. The two scales can be converted into each other via temperature-pressure data for steam (Fig. 5.13) (Ref 24).

The dewpoint, τ, may also be calculated by the following equation (Ref 25):

$$\tau = -273 - A/(\ln P_{H_2O} - B) \ [°C]$$

where A = 6128 and B = 17.335 for $-100\ °C \leq \tau \leq 0\ °C$ and for $10^{-8} \leq pH_2O \leq 6 \times 10^{-3}$

Figure 5.14 (Ref 19) shows such redox curves, calculated from thermodynamic data, for several of the pure, high-oxygen-affinity elements present in stainless steels as well as for some of these elements present as solid solutions in stainless steel.

Figure 5.15 (Ref 25) shows redox curves for pure metals and their oxides against both the partial pressure ratio P_{H_2}/P_{H_2O} and the dewpoint, as well as for H_2-N_2 mixtures.

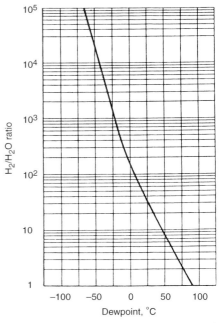

Fig. 5.13 Relationship between ratio of H_2/H_2O and dewpoint. Source: Ref 24

Partial pressures of water and temperatures to the left of a redox curve indicate that the oxide of that particular metal is stable under such conditions, whereas under conditions that lie to the right of that curve, the pure metal is stable.

Because reduction equilibria depend on the ratio of the partial pressures of hydrogen and water, that is, p_{H_2}/p_{H_2O}, and not on the absolute water pressure, as dewpoint does, the dewpoints for oxide reduction of hydrogen-nitrogen mixtures differ from those of pure hydrogen. The minimum temperature at which a metal oxide can be reduced in a hydrogen-nitrogen mixture of a given dewpoint can be derived from Fig. 5.15 by drawing a horizontal line at the height of that dewpoint in the left part of the figure.

At the intersection of this horizontal line with the gas mixture of the atmosphere, a perpendicular line is drawn up to the curve for pure hydrogen. From this intersection, a horizontal line is extended into the right part of Fig. 5.15. All oxides above this line can be reduced, whereas all oxides below this line are stable. The example in Fig. 5.15 shows that with a dewpoint of –35 °C (–31 °F), it is possible to reduce Cr_2O_3 at a temperature of approximately 1000 °C (1830 °F) or higher in pure hydrogen (dashed line), whereas an H_2-N_2 atmosphere with 95% N_2 requires a minimum temperature of almost 1600 °C (2912 °F) (solid line) or a dewpoint of almost –60 °C (–76 °F). For dissociated ammonia and atmospheres containing lesser amounts of nitrogen, this dew-

point correction is relatively small. Figure 5.14 shows that very low dewpoints are required, or only very small amounts of water vapor can be tolerated, if stainless steels are to be kept from becoming oxidized during sintering. Also, this requirement is easier to fulfill as the sintering temperature increases. Furthermore, when an element is present in the form of an alloy, its activity is decreased, and it is easier to keep it from becoming oxidized than if the same element is present as a pure metal. Of the most oxidation-prone constituents in conventional stainless steels— manganese, chromium, and silicon—silicon exhibits the greatest affinity to oxygen. Also, for silicon, the difference between pure and alloyed states is particularly large and explains, according to Larsen, why silicon dioxide in sintered stainless steels can be reduced under commercial sintering conditions, a fact that often had been doubted in earlier years. Another scenario for the successful reduction of SiO_2 at reasonable dewpoints is based on the reduction of SiO_2 to volatile SiO followed by the iron-catalyzed reduction of SiO to silicon (Ref 16).

Figure 5.16 shows the Auger composition-depth profile of a 316L part sintered in hydrogen at 1260 °C (2300 °F).

A comparison with Fig. 3.10, which shows the same profile for water-atomized 316L in the green condition, demonstrates the enormous degree of reduction of SiO_2. Furthermore, the width of the oxygen profile of Fig. 5.16 is narrow, approximately 30 atomic layers (~50 Å), and more akin to the thickness of a passive film. This material's corrosion resistance was excellent. Figure 5.17 (Ref 26) shows a similarly

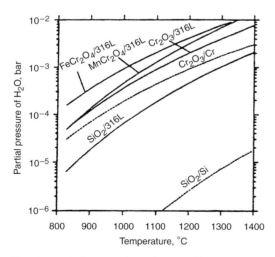

Fig. 5.14 Redox curves for oxides in equilibrium with 316L in H_2 at atmospheric pressure. Source: Ref 19. Reprinted with permission from MPIF, Metal Powder Industries Federation, Princeton, NJ

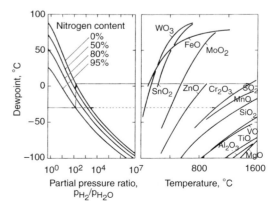

Fig. 5.15 Dewpoint for various hydrogen-nitrogen mixtures in equilibrium with metal/metal oxide. Source: Ref 25

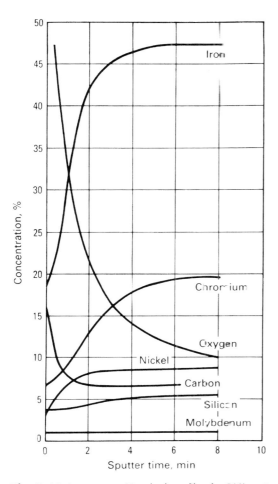

Fig. 5.16 Auger composition-depth profile of a 316L part sintered in hydrogen at 1260 °C (2300 °F)

excellent profile for a high-temperature (1295 °C, or 2363 °F) vacuum-sintered 316L part.

The absence of silicon in Fig. 5.17 appears to be due to the authors' use of the nonscanning mode of Auger analysis, because they were able to confirm its presence when they were using the scanning mode.

Figure 5.18 shows the profiles of a 316L part sintered at 1120 °C (2048 °F) in hydrogen with a dewpoint of –35 °C (–31 °F).

Based on Fig. 5.18, these conditions are marginal for SiO_2 reduction, and it is therefore not surprising that the corrosion resistance of this material was quite inferior. Also, the width of the Auger oxygen profile of this material is wider than that of 316L reduced at higher temperatures.

Table 5.3 (Ref 19) further illustrates how a marginal dewpoint (–35 °C, or –31 °F) for 316L parts sintered in hydrogen at 1120 and 1250 °C

(2048 and 2282 °F) affects the electrochemical passivation characteristics as well as the long-term-exposure corrosion resistance in 5% aqueous NaCl in comparison to a much lower dewpoint of –70 °C (–94 °F).

Microstructures. Thus, dewpoint value, together with temperature, largely determines if adequate particle bonding will take place during sintering. Interparticle bonding can readily be ascertained through metallography. Figure 5.19 shows the microstructure of a 316L stainless steel part sintered in hydrogen for 30 min at 1093 °C (2000 °F). Prior-particle boundaries and angular pores are evident as a result of insufficient sintering.

Figure 5.20 shows the polished cross section of undersintered 304L, with many oxides in the grain boundaries.

In contrast, well-sintered 316L (Fig. 5.21), shows good interparticle bonding, well-rounded pores, and narrow and precipitate-free grain boundaries in the austenitic structure.

Figure 5.22 shows the same attributes for a well-sintered ferritic stainless steel, 434L, except for the absence of twin boundaries, which are characteristic of face-centered austenitic stainless steels.

Figure 5.23(a) (Ref 27) shows the surface and Fig. 5.23(b) the cross section of 316L vacuum sintered for 1 h at 1150 °C (2102 °F). The oxide particles seen in Fig. 5.23(a) are typically less than 1 μm in diameter.

After exposure to $FeCl_3$, the oxide particle sites develop corrosion pits (Fig. 5.24b) (Ref 27).

It was only recently recognized that, in order to develop excellent corrosion properties in sintered stainless steels, not only a thorough reduction of oxides but also prevention of reoxidation after sintering is required. The parts in Fig. 5.25 (Ref 28), sintered under various conditions, had good static mechanical properties, but they had a broad range of corrosion resistances as measured by submersion in a saline solution. Corrosion increased with increasing oxygen content of the sintered part.

The detrimental effect of oxygen on corrosion resistance was confirmed by Maahn and Mathiesen (Ref 18). The pitting potential, a measure of a steel's resistance to pitting, declined with increasing oxygen content in 316L.

Kinetic Considerations. Using gas and mass spectrometry analysis during sintering of stainless steels in hydrogen and under vacuum, at various temperatures, and with additions of graphite to the powder, Tunberg et al. (Ref 6)

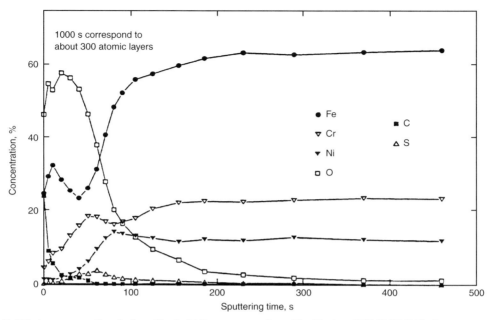

Fig. 5.17 Auger composition-depth profile of a 316L part vacuum sintered for 30 min at 1295 °C (2363 °F). Oxygen content was 0.20%. Source: Ref 26

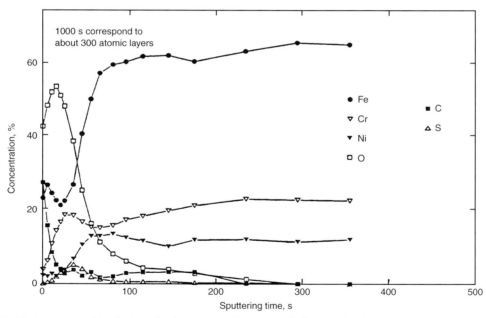

Fig. 5.18 Auger composition-depth profile of 316L sintered for 30 min in hydrogen with a dewpoint of –35 °C (–31 °F). Oxygen content was 0.24%

and Larsen and Thorsen (Ref 29, 30) showed that the various reduction reactions occurred in fairly close agreement with the equilibrium gas concentrations calculated from thermodynamic data. More specifically, the dominant reactions for sintering in hydrogen are:

$$2H_2 + SiO_2 = 2H_2O + Si_{316L}$$

$$2C + SiO_2 = 2CO + Si_{316L}$$

$$C + 2H_2 = CH_4$$

Table 5.3 Corrosion properties of 316L steel sintered in hydrogen with a dewpoint of –35 or –70 °C (–31 or –94 °F) at different combinations of time and temperature

Dewpoint, °C	i_{peak}(a), μA/cm²		i_{pass}(a), μA/cm²		E_{pit}(a), mV SCE		NSS1, h		NSS2	
	–35	–70	–35	–70	–35	–70	–35	–70	–35	–70
1120 °C/30 min	150	10	29	11	250	375	36	>1500	5	9
1250 °C/30 min	105	7	20	12	325	325	288	1260	4	8
1120 °C/120 min	120	10	25	10	325	375	48	1272	5	7
1250 °C/120 min	83	4	19	9	325	500	24	96	1	7

(a) 0.1%Cl⁻, pH 5, 30 °C (86 °F), 5 mV/min. Source: Ref 19

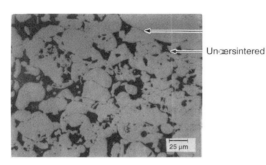

Fig. 5.19 Undersintered 316L (unetched) revealing prior-particle boundaries and angular pores

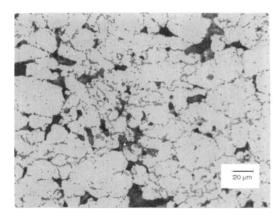

Fig. 5.20 Polished cross section of undersintered 304L revealing oxides in grain boundaries

and for sintering in vacuum:

$$2C + SiO_2 = 2CO + Si_{316L}$$

However, due to low equilibrium pressures of the reaction products, it can take a long time to remove oxygen from the parts. Increased part density also slows down the reactions.

Larsen and Thorsen (Ref 30) have shown that carbon is a much more effective reducing agent than H_2 for the reduction of oxides, and carbon removal is much faster in vacuum. Tunberg et al.

Fig. 5.21 Well-sintered 316L (etched) revealing interparticle bonding, twin boundaries, rounded pores, and precipitate-free grain boundaries

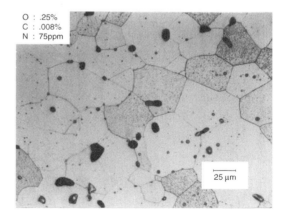

O : .25%
C : .008%
N : 75ppm

Fig. 5.22 Well-sintered 434L (etched) revealing interparticle bonding, rounded pores, and precipitate-free grain boundaries

(Ref 6) were able to obtain a tenfold reduction in oxygen content (from 0.31 to 0.03%) by adding 0.19% C to a 304L stainless steel powder and by vacuum sintering at 1200 °C (2192 °F) for 1 h. Removal of surface oxides led to improved interparticle bonding, as reflected in markedly improved dynamic mechanical properties (elongation, impact strength). The

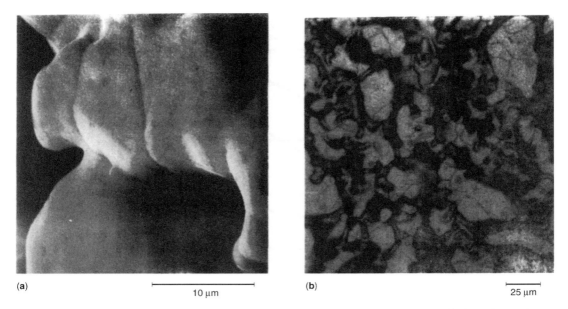

(a)   10 μm   **(b)**   25 μm

Fig. 5.23 Microstructures of vacuum-sintered 316L. (a) SEM. (b) Light microscopy. After exposure to FeCl₃, the oxide particle sites develop corrosion pits (Fig. 5.24b). Source: Ref 27. Reprinted with permission from MPIF, Metal Powder Industries Federation, Princeton, NJ

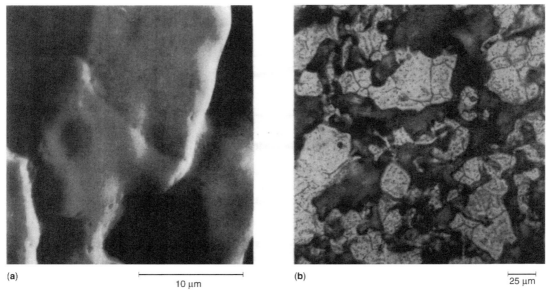

(a)   10 μm   **(b)**   25 μm

Fig. 5.24 (a) SEM and (b) light microscopy microstructures of vacuum-sintered 316L after exposure to 6% FeCl₃. Reprinted with permission from MPIF, Metal Powder Industries Federation, Princeton, NJ

investigators attributed the superior oxygen reduction during vacuum sintering to the faster removal of carbon monoxide from the pores. The faster chemical reaction rates for vacuum sintering and the low oxygen levels achievable with the addition of an appropriate amount of carbon should be of considerable commercial interest (section 5.2.5 in this chapter).

Figure 5.26 (Ref 28) shows the very strong effect of part size on weight loss during sintering of 316L transverse-rupture specimens in dry hydrogen. The H_2O and CO account for the major portion of the total weight loss. Rate-controlled by the slow transportation of reaction products through the tortuous pores, oxides near the surface of a part are reduced first, followed

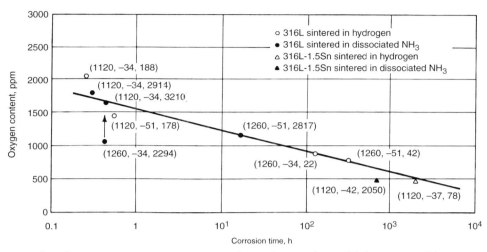

Fig. 5.25 Effect of oxygen content on corrosion resistance of sintered 316L and tin-modified 316L (sintered density: 6.65 g/cm³; cooling rate: 75 °C/min, or 135 °F/min. Values in parentheses are sintering temperature (°C), dewpoint, (°C), and nitrogen content (ppm), respectively. Time indicates when 50% of specimens showed first sign of corrosion in 5% aqueous NaCl. Source: Ref 28. Reprinted with permission from MPIF, Metal Powder Industries Federation, Princeton, NJ

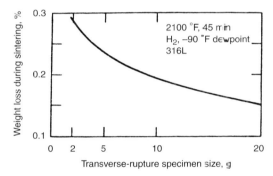

Fig. 5.26 The effect of transverse-rupture specimen size on weight loss during sintering in hydrogen (density of specimens: approximately 6 g/cm³). Source: Ref 28. Reprinted with permission from MPIF, Metal Powder Industries Federation, Princeton, NJ

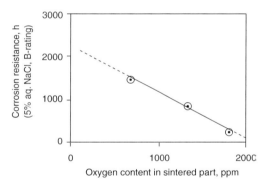

Fig. 5.27 Corrosion resistance of 316L stainless steel parts sintered under various conditions under exclusion of defects, except for residual oxides

by the reduction of oxides farther inside a part. Again, this suggests that there exists an oxygen content gradient in sintered stainless steels, with the oxygen content increasing from the surface to the interior of a part. Part supports (ceramic bodies, metallic belts) and part spacing will also affect the diffusion of gases into and out of the sintered bodies and therefore their reaction rates with interior oxides.

In their tests with 316L parts, Samal and Klar et al. (Ref 31, 32) carefully eliminated all known corrosion defects, such as contamination, grain-boundary carbides and nitrides, reoxidation on cooling, and the crevice-sensitive density range around 6.9 g/cm³, except for residual oxygen (oxides). A reduction of the

oxygen content of the sintered parts from approximately 1900 to approximately 1300 ppm through increasing the sintering temperature from 1138 to 1316 °C (2080 to 2400 °F) improved the saline (5% NaCl) corrosion resistance by 400%, (Fig. 5.27). Vacuum-sintered 316L parts with oxygen contents of approximately 700 ppm showed a 700% improvement over the low-temperature hydrogen-sintered parts. Sintered 316L parts with still lower oxygen contents (200 to 300 ppm) are expected to have a yet higher corrosion resistance, with reduced or no evidence of crevice corrosion, but with general corrosion, as measured by I_{pass}, reflecting the larger effective surface areas of such parts. Such low-oxygen parts could be made from a thermally agglomerated (Ref 33), low-oxygen-content, gas-atomized 316L powder or, more practically, by warm compaction (Chapter 4, "Compacting and Shaping") of a low-oxygen-content, gas-atomized powder in combination with, if necessary, an appropriate binder.

Oxygen Control during Cooling. The importance of a low dewpoint during the sintering of stainless steels has been described previously. However, it is also important to control the cooling conditions after sintering if maximum corrosion resistance is to be achieved. For illustration, Lei and German (Ref 34) subjected a wrought 304L stainless steel specimen, together with PM parts, to sintering in dry hydrogen (dewpoint ≤ –35 °C or ≤ –31 °F) for 60 min at 1250 °C (2282 °F). The corrosion rate of the wrought stainless steel (as measured by potentiodynamic scanning in 3.5% saltwater) after exposure to sintering increased by a factor of 100. Electrochemical testing of the exterior surfaces of the sintered PM stainless steels showed similar degradations. No cooling rates were disclosed in these experiments, but the authors had observed second-phase inclusions on the surfaces of the wrought stainless steel after its simulated sintering cycle. They attributed the large decrease in corrosion resistance of both the sintered and wrought specimens to chromium losses from the surfaces due to chromium evaporation. It is more likely, however, that the culprit was reoxidation during cooling, with the formation of spheroidal oxides on the exposed surfaces. Figure 5.28 (Ref 13) shows a 316L stainless steel part that had first been vacuum sintered to reduce the oxygen content to approximately 700 ppm (Ref 27) and then allowed to cool in hydrogen (dewpoint –40 °C, or –40 °F) from 1127 °C (2061 °F) at a cooling rate of 187 °C/min (337 °F/min). The oxide particles formed during cooling were spherical and measured between 0.5 and 2.0 μm in diameter. To the naked eye, the surface brightness of such parts is not affected by this type of reoxidation. Auger line analysis (Fig. 5.29)

10 μm

Fig. 5.28 Spheroidal SiO_2 particles formed on 316L part on cooling. Source: Ref 13

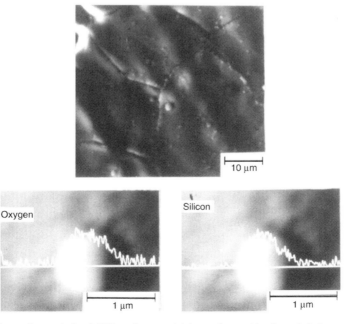

10 μm

Oxygen

Silicon

1 μm   1 μm

Fig. 5.29 SEM and Auger line analysis of 316L surfaces containing surface oxides formed during cooling. Reprinted with permission from MPIF, Metal Powder Industries Federation, Princeton, NJ

identified the particles as consisting predominantly of silicon oxide. Chromium was absent.

After exposure to aqueous $FeCl_3$ for 7 h, the oxide particles had formed corrosion pits. The chloride corrosion resistance had decreased to a very low value. As shown in Fig. 5.32, 316L loses approximately 99% of its corrosion resistance in 5% saltwater when cooled under the aforementioned conditions. SS-100 parts, a higher-chromium, higher-nickel austenitic stainless steel, treated identically, exhibited very little pitting and had lost only approximately 50% of their chloride corrosion resistance. The oxide particles of the SS-100 material were of spherical and triangular shape.

Redox curves and cooling-rate relationships provide insight on why it is important to control the cooling process. The redox curves in Fig. 5.30 (Ref 19) show two sintering scenarios.

In scenario 1, with a dewpoint of –40 °C (–40 °F), the sintered parts, as they enter the cooling zone and their temperature decreases, will begin to become oxidized at approximately 1070 °C (1958 °F), the temperature at which the dewpoint of –40 °C (–40 °F) crosses the redox curve. In scenario 2, with a lower dewpoint of –60 °C (–76 °F), oxidation is delayed until the parts have cooled to a lower temperature of 960 °C (1760 °F). It is clear that the parts become more oxidized under scenario 1 than under scenario 2.

The importance of a fast cooling rate for minimizing reoxidation is self-evident. It is also obvious that a high dewpoint requires a faster cooling rate than a lower dewpoint, because of the higher concentration of water vapor in a higher-dewpoint atmosphere and because reoxidation starts at a higher temperature. Sands et al. (Ref 35) suggested a maximum water content of 50 ppm (corresponding to a dewpoint of –48 °C, or –54 °F) for slow cooling in hydrogen. Figure 5.31 illustrates these relationships in a semiquantitative scheme.

Upper Critical Cooling Temperatures. Figure 5.32 shows the upper critical cooling temperatures for 316L, that is, the lowest temperatures where rapid cooling must begin in order to avoid sensitization. The required cooling rates, as a function of dewpoint, are also shown in that figure.

The curves marked with percentage figures indicate, semiquantitatively, how rapidly corrosion resistance (in 5% NaCl) deteriorates with decreasing cooling rate. Thus, a part cooled at 400 °C/min (720 °F/min) in a dewpoint atmosphere of

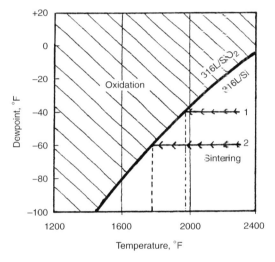

Fig. 5.30 Redox curves and sintering scenarios for 316L in H_2 at atmospheric pressure (schematic). Source: Ref 19. Reprinted with permission from MPIF, Metal Powder Industries Federation, Princeton, NJ

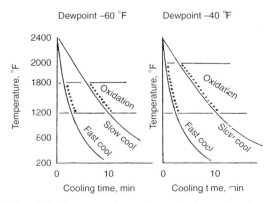

Fig. 5.31 Temperature-time cooling profiles for two dewpoints, showing schematically approximate reoxidation regimes for 316L

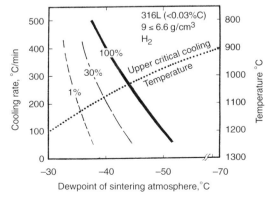

Fig. 5.32 Upper critical cooling temperature and ISO corrosion-resistance curves (%) for H_2-sintered 316L (5% NaCl by immersion; B rating) (schematic)

–40 °C (–40 °F) attains its full or optimal corrosion resistance for the particular oxygen content of the part, but only approximately 30% when the cooling rate is reduced to 200 °C/min (360 °F/min). The data in Fig. 5.32 were obtained from 316L parts sintered in hydrogen at various temperatures, dewpoints, and cooling rates. Their densities were under 6.6 g/cm³, and their oxygen contents ranged from approximately 1300 to 2200 ppm. It is not known how severely oxygen content affects the position or shape of these curves. The upper critical cooling temperature in Fig. 5.32 is derived from the redox curve for 316L. It denotes the temperature at which cooling must start, at the latest, for avoiding reoxidation.

Thus, with rapid cooling, the extent of silicon dioxide formation is minimized and corrosion resistance maximized. As the part cools, the oxidation rate of silicon decreases, and at sufficiently low temperatures, the oxide layer formed contains less silicon and more chromium; in other words, it contains the elements that have a beneficial effect regarding the formation of the passive layer.

Misconceptions regarding the correct dewpoint for sintering stainless steels still linger in the industry (Ref 36). It is clear from the preceding

that the basis for sintering stainless steels is the more demanding redox equilibria for the silicon contained in a stainless steel and not those for chromium (Fig. 5.14).

Figure 5.33 (Ref 13) shows cooling rate/dewpoint curves for three austenitic stainless steels. The 316LSC is a tin-copper-modified 316L; SS-100 is a high-chromium, high-nickel stainless steel. As mentioned earlier, higher-alloyed steels appear to be less sensitive, or more forgiving, because lower cooling rates are sufficient to keep the surfaces free from reoxidation.

Lei et al. (Ref 38) confirm the important effect of cooling rate; 304LSC sintered at 1250 °C (2282 °F), for 45 min in H_2 or 83%H_2-17%N_2 (dewpoint $\leq$–35 °C, or –31 °F) had passive current densities that increased by over 2 orders of magnitude when the cooling rate was changed from fast to slow. Such a big change cannot be attributed to a change in internal surface area as a result of different cooling rates. The latter is only small. A more probable interpretation is reoxidation of the outer surfaces as a result of slow cooling, in accordance with the examples shown in Fig. 5.32. The large increase of the passive current density due to the formation of surface oxides illustrates the effect of metallurgical defects on electrochemical characteristics. As

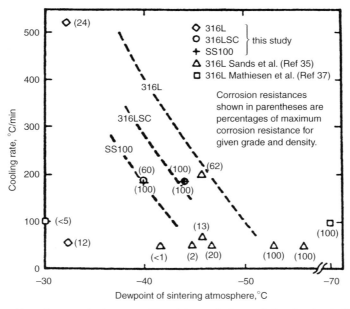

Fig. 5.33 Cooling rate/dewpoint curves for three austenitic stainless steels. Source: Ref 13. Reprinted with permission from MPIF, Metal Powder Industries Federation, Princeton, NJ

mentioned earlier, many investigators in the past have failed to account for such defects and then concluded, mistakenly, that their low corrosion-resistance properties were due to the presence of pores, crevice corrosion, or both.

It is appropriate to mention that the two automotive large-volume uses of sintered stainless steels—antilock brake sensor rings and exhaust flanges—are based on high-temperature sintering (>1200 °C, or >2192 °F) in a low-dewpoint atmosphere of hydrogen but without accelerated cooling. Their oxygen contents are typically between 1500 and 2000 ppm. This illustrates that maximum corrosion resistance is not always necessary for successful use. Furthermore, in applications where some surface wear occurs readily, or in a strong enough acidic environment, shallow surface defects such as oxides can disappear with time.

For wrought stainless steels, it is known that rough surfaces from rolling and drawing operations, or from pickling and passivation treatments, increase the tendency for pitting, as do surface oxides formed during annealing and welding operations. Only in the so-called bright annealing of wrought thin-gage stainless steel sheet are processing conditions similar to those employed in PM processing, namely, rapid cooling in low-dewpoint hydrogen with limited amounts of nitrogen (Ref 39).

Little and conflicting data are available for the chemical cleaning of sintered stainless steels (section 9.1 in Chapter 9, "Corrosion Testing and Performance"). Much of the published information relates to stainless steel parts that had relatively low corrosion properties. Any benefits from cleaning may not necessarily apply to properly sintered parts. Also, the presence of pores makes cleaning in solutions difficult, because of the capillary forces of the pores that tend to retain the cleaning liquid.

Carbon Control: Delubrication and Sintering Conditions. A vast amount of literature exists on the subject of chromium carbide precipitation in wrought stainless steels. Most of these data are applicable to PM stainless steels and are used here where relevant. For sintered stainless steels, proper delubrication is important for keeping carbon levels below where they can cause sensitization.

Thermodynamics and Kinetics Background. Intergranular corrosion, one of the various forms of corrosion in stainless steels, arises from excessive amounts of carbon, which can

form chromium-rich carbide precipitates at grain boundaries. Because these chromium-rich carbides ($M_{23}C_6$) have a higher chromium content than the alloy, chromium in the surrounding matrix, that is, next to the grain boundaries (Fig. 5.34) (Ref 40), is depleted to below the level necessary to maintain passivation. Chromium carbide itself is not susceptible to rapid corrosion.

Figure 5.35 (Ref 41) illustrates chromium carbide precipitates in sintered 316L for various carbon levels and typical (i.e., slow) commercial cooling conditions.

Austenitic Stainless Steels. At low carbon levels (Fig. 5.35a), the austenitic structure reveals desirable clean and thin grain boundaries and ample twinning; at intermediate carbon levels (Fig. 5.35b), so-called necklace-type chromium-rich carbide precipitates (if present, nitrogen can participate in the precipitation) are visible in the grain boundaries; and at high carbon levels (Fig. 5.35c), the grain boundaries are heavily decorated with continuous precipitates. The latter two cases give rise to various degrees of intergranular corrosion.

In austenitic stainless steels, chromium carbide precipitation occurs in the temperature range of 816 to 538 °C (1500 to 1000 °F). Carbon, due to its small atomic size, diffuses rapidly in the steel matrix. Hence, during cooling from an elevated temperature, for example, after sintering in a typical belt, pusher, or walking beam furnace, any carbon present in excess of the limit of solubility

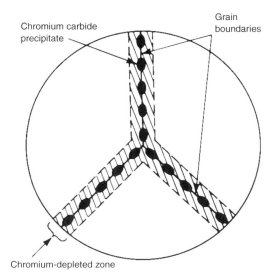

Chromium carbide precipitate

Grain boundaries

Chromium-depleted zone

Fig. 5.34 Schematic of sensitization. Source: Ref 40. Reprinted with permission from McGraw-Hill

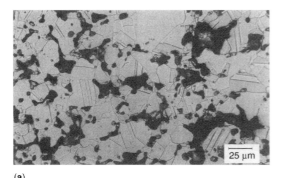

(a)

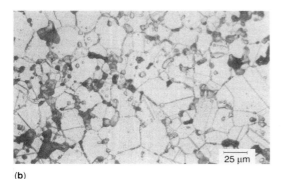

(b)

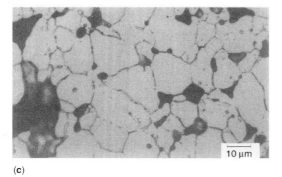

(c)

Fig. 5.35 Microstructures of type 316L stainless steel sintered in hydrogen at 1150 °C (2100 °F) (glyceregia). (a) Carbon is 0.015%; thin and clean grain boundaries. (b) Carbon is 0.07%; necklace-type chromium-rich carbide precipitates in grain boundaries. (c) Carbon is 0.11%; continuous chromium-rich carbide precipitates in grain boundaries. Source: Ref 41

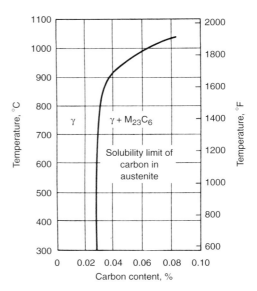

Fig. 5.36 Solid solubility of carbon in austenitic stainless steel. Source: Ref 42

can easily migrate out of the matrix to the grain boundaries, where it would combine with chromium to form chromium carbide. Figure 5.36 (Ref 42) shows the limit of solubility of carbon in an austenitic stainless steel as a function of temperature. Based on these data, the maximum amount of carbon in an austenitic stainless steel should be approximately 0.03%.

The diffusion rate of chromium in an austenitic matrix is not rapid enough to make

up for the chromium lost due to chromium carbide precipitation. If, however, the sintered part is cooled very rapidly, carbon atoms will not be able to diffuse out of the matrix and form chromium carbide precipitates but instead will be held in solution. Hence, by cooling rapidly from an elevated temperature, sensitization can be either prevented or minimized in an alloy containing carbon in excess of its limit of solubility.

Critical cooling rates necessary for the prevention of sensitization in wrought stainless steels are commonly depicted as time-temperature-sensitization (TTS) diagrams. Figure 5.37 (Ref 42) is an example of a set of TTS diagrams for five 18Cr-9Ni austenitic stainless steels with different carbon contents.

According to Fig. 5.37, a steel containing 0.08% C must be cooled through the sensitization range in less than approximately 30 s in order to avoid chromium carbide precipitation. A steel containing only 0.03% C, however, may be cooled through the same temperature range in approximately 50 min without risking sensitization. For constituents that either increase or decrease the tendency for carbide precipitation, see Ref 43 and 44.

Figure 5.38 (Ref 45) shows the effects of carbon content and cooling rate on intergranular corrosion for hydrogen-sintered 316 parts that had been prepared with various amounts of lubricants and with various delubrication conditions to produce parts that possessed a wide range of carbon contents, from 0.01 to 0.11%.

The curve separating the sensitized from the sensitization-free parts represents the critical cooling rates necessary for various carbon contents to avoid intergranular corrosion. The curve was derived from the TTS curves of Fig. 5.37 by drawing the tangents from 1260 °C (2300 °F), the sintering temperature used in this experiment, to the time minima of the various carbon-level curves. This allows one to calculate an average cooling rate for each carbon level. Good agreement between wrought and sintered stainless steel data confirms the applicability of wrought stainless steel data to sintered stainless steels.

The most reliable means to prevent sensitization in austenitic stainless steels is to restrict the carbon content to 0.03% maximum. Alloys thus modified are designated as "L" grades. In wrought stainless steels, "L" grades are recommended for applications requiring welding and/or thermal cycling. For PM stainless steels, "L" grades are recommended if sintering is performed in typical commercial sintering furnaces with their slow cooling rates. All stainless steel powders destined for conventional compaction and sintering are also of the L-grade designations, because of their superior compacting properties. However, as mentioned in section 3.1.3 in Chapter 3, "Manufacture and Characteristics of Stainless Steel Powders," for 304L and 316L with high nickel contents close to their upper limits, safe maximum carbon contents are only approximately 0.02%.

Ferritic Stainless Steels. The phenomenon of intergranular corrosion in ferritic stainless steels differs somewhat from that in austenitic stainless steels. The limit of solubility of carbon is much lower in ferritic stainless steels, and the diffusion rates of interstitials are much higher due to their body-centered cubic (bcc) structure. As illustrated in Fig. 5.39 (Ref 46), these characteristics require very fast cooling rates in order to prevent sensitization.

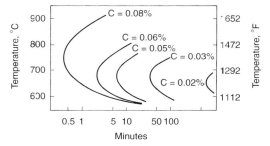

Fig. 5.37 Time-temperature-sensitization diagrams for five 18Cr-9Ni austenitic stainless steels with different carbon contents. Source: Ref 42. Reprinted with permission of John Wiley & Sons, Inc.

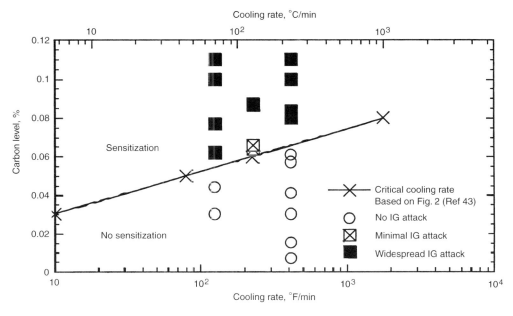

Fig. 5.38 Effect of carbon content and cooling rate on intergranular corrosion of hydrogen-sintered 316. IG, intergranular. Source: Ref 45. Reprinted with permission from MPIF, Metal Powder Industries Federation, Princeton, NJ

Ferritics with intermediate levels of interstitials also exhibit serious loss of ductility in addition to loss of corrosion resistance. As a result, it becomes necessary to have either a very low carbon plus nitrogen level of approximately 0.02% or to use a strong carbide former (stabilizer), such as niobium, titanium, or tantalum, which forms a more stable carbide in preference to chromium carbide, thereby preventing sensitization. With the introduction of argon-oxygen decarburization (AOD) and with vacuum and electron beam refining, it became possible to produce wrought stainless steels possessing much lower levels of interstitials. Before the advent of AOD, the ease with which wrought ferritic stainless steels (i.e., 430 and 434) could be sensitized, together with ductile-to-brittle transitions occurring above ambient temperature, had limited their use. The addition of stabilizers has been a common practice for wrought ferritic alloys intended for applications requiring welding or exposure to elevated temperatures. The control of interstitials to very low levels, with and without the use of stabilizers, has led to the development of the high-chromium superferritics with good toughness, stress-corrosion resistance, and general corrosion resistance.

Among sintered PM stainless steels, only one stabilized grade is featured thus far in the MPIF and ASTM standards, namely, ferritic 409L, containing niobium as a stabilizer. However, Samal et al. (Ref 47) have shown that it is possible to obtain the equivalent of sintered superferritics by using niobium as a stabilizer, by sintering (at 1148 °C, or 2100 °F) in a low-dewpoint atmosphere of hydrogen, and by employing rapid

cooling. Although some of the niobium reacts with nitrogen during delubricating in nitrogen and with carbon from the lubricant, leading to increased carbon and nitrogen levels, the amount of niobium was sufficient to precipitate these interstitials as carbides and nitrides and to achieve superior corrosion resistances.

Of the various stabilizers (titanium, tantalum, niobium) used in wrought stainless steels to cope with higher carbon contents and combat sensitization, only niobium has been used thus far with some success (409Nb, 434Nb) in sintered stainless steels. In wrought stainless steels, niobium carbide-stabilized steels have been found to be more resistant to intergranular corrosion than titanium carbide-stabilized steels (Ref 48). Titanium and tantalum, probably because of their higher oxygen affinities, form objectionable surface oxides during water atomization.

In contrast to the face-centered cubic austenitic stainless steels, the diffusion rate of chromium atoms in the bcc ferritic matrix is approximately 100 times faster. Because of this, a ferritic stainless steel can be cured of its sensitized condition by a suitable annealing step between approximately 704 and 954 °C (1300 and 1750 °F). Replenishment of chromium-depleted regions can be satisfactorily achieved and corrosion resistance restored, despite the presence of chromium carbides along the grain boundaries (Ref 49).

Delubrication. In wrought and cast stainless steels, carbon control is accomplished during melting. In sintered PM stainless steels, the carbon content is determined not only by the carbon content of the powder but also by its lubricant and the delubrication and sintering conditions. Of these three, the lubricant contribution is the more complex. It is described in some detail, because it has been the cause for many underperforming sintered stainless steels in the past.

Ideally, for H_2 and H_2-N_2 sintering atmospheres, a lubricant should be completely removed from the part by complete combustion into volatile constituents during delubrication. In practice, however, combustion and volatilization are incomplete, and at least a small portion of the lubricant typically decomposes into carbon and other organic constituents. Lack of control can easily increase the carbon content of a stainless steel to above the 0.03% limit and sometimes to as much as 0.1%. Moyer (Ref 50) has discussed the delubrication and sintering conditions on the

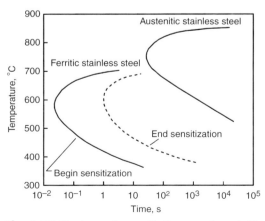

Fig. 5.39 Time-temperature-sensitization curves for austenitic and ferritic stainless steels of equivalent chromium content. Source: Ref 46

efficiency of lubricant removal in 316L powder compacts. Saha and Apelian (Ref 51) described an empirical model and closed-loop control system for delubrication. The degrees of volatilization and decomposition depend on many factors, including part density, heat-up rate, dewpoint, furnace atmosphere, and gas flow rates. In the early years of stainless steel parts production, when much of the sintering was in dissociated ammonia at 1120 to 1150 °C (2050 to 2100 °F), lubricant removal was usually accomplished in the preheat zone. Decomposition of the lubricant caused carbon contents to exceed 0.03%. Because the low-temperature sintering conditions were not conducive to lowering the carbon content during sintering, part fabricators resorted to delubricating the parts separately in air, for approximately 30 min for small parts and longer for larger parts. While this procedure reduced the carbon content to more acceptable levels, it also caused oxidation (Fig. 5.40).

This oxidation is lessened or avoided in dissociated ammonia. Figure 5.41 (Ref 41) shows similar data for Acrawax as a lubricant. In comparison to lithium stearate, Acrawax has cleaner burn-off characteristics, but, as mentioned in Chapter 3, "Manufacture and Characteristics of Stainless Steel Powders," it does not impart the compressibility advantage of lithium stearate.

Oxidation begins before complete lubricant removal, even when delubrication is performed in dissociated ammonia. It appears impossible under these conditions to obtain maximum carbon removal without additional oxidation. Though oxides formed at low temperatures are more easily reduced during sintering in a reducing atmosphere than those formed at very high temperatures during water atomization, the goal is to keep oxidation as low as possible.

Delubrication should always be viewed in the context of sintering. With higher sintering temperatures (>1205 °C , or >2200 °F), the reaction between residual oxygen and carbon is more complete, and delubricating is therefore preferably completed in a reducing atmosphere.

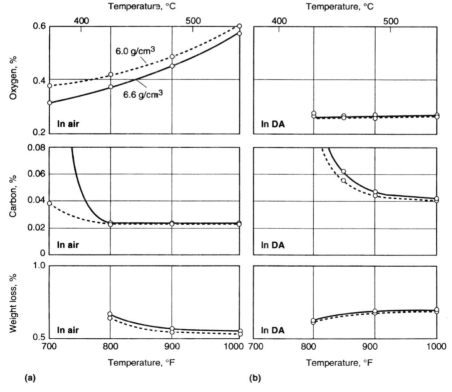

Fig. 5.40 Effect of delubrication temperature on oxygen, carbon, and weight loss of 316LSC parts of two densities (6.0 g/cm³, dashed lines; 6.6 g/cm³, solid lines), lubricated with 1% lithium stearate and delubricated for 30 min in (a) air and (b) dissociated ammonia (DA) (unpublished data)

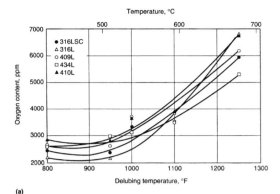

(a)

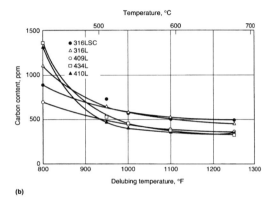

(b)

Fig. 5.41 Effect of delubricating temperature on (a) oxygen content and (b) carbon content of stainless steel parts (6.5 to 6.7 g/cm³), lubricated with 1% Acrawax and delubricated for 30 min in dissociated ammonia. Source: Ref 41

As Fig. 5.41 shows for several stainless steels lubricated with 1% Acrawax and pressed to green densities of 6.5 to 6.7 g/cm³, delubricating in dissociated ammonia prevents any significant oxidation up to 510 to 538 °C (950 to 1000 °F). Although carbon removal under these conditions is not yet at its maximum, and carbon content is still >0.03%, sintering at a higher temperature will lower the carbon content to below 0.03%. It is in part because of these relationships that parts sintered at high temperatures exhibit better corrosion resistances.

Martensitic Stainless Steels. Of the various families of stainless steels, the martensitic stainless steels have the highest carbon contents, sometimes exceeding 1.2%. In this case, the carbon function is to form martensite and primary carbides that endow these steels with their hardness, strength, and abrasion-resistance properties for which they are known and used. Structural requirements limit the chromium content of these steels; thus, their corrosion resistances are limited.

5.2.4 Sintering of Stainless Steels in Hydrogen-Nitrogen Gas Mixtures

The primary goal in sintering stainless steels in H_2-N_2 mixtures is to achieve corrosion resistance equal or superior to sintering in hydrogen, in combination with markedly improved strength. In recent years, there has been a shift from dissociated ammonia to hydrogen and vacuum sintering. This was clearly the result of the increasing emphasis on corrosion resistance and the difficulty in achieving good corrosion resistance with dissociated ammonia. However, as is clear in the following, the shift to hydrogen and vacuum may be unfortunate in view of the lower cost of nitrogen-containing atmospheres and, more importantly, in view of the potency of nitrogen to markedly increase the pitting resistance of a stainless steel at very low cost. In most studies, this beneficial effect of dissolved nitrogen has not been observed, because it was overshadowed by the negative effect of Cr_2N precipitation that causes sensitization and intergranular corrosion. However, as is seen, the use of lower-nitrogen-content atmospheres, such as $90H_2$-$10N_2$ instead of $75H_2$-$25N_2$, coupled with appropriate cooling rates, readily establishes these benefits.

Like carbon, nitrogen has a strong affinity to chromium, and its absorption from the sintering atmosphere can be exploited for increasing strength and hardness of stainless steels. Nitrogen-containing sintering atmospheres are mainly used for austenitic stainless steels, where nitrogen up to approximately 0.3% does not promote sensitization with correct processing. It is therefore superior to carbon as a means of increasing strength, particularly yield strength. Strengthening is caused by the lattice expansion of the γ phase (austenite) from the dissolved nitrogen, as well as the precipitation of finely divided Cr_2N. The latter is also beneficial to the fatigue properties but detrimental to corrosion resistance.

Sintering temperature and dewpoint requirements for effective oxide reduction and interparticle bonding are similar to those of hydrogen sintering. Figures 5.42 (Ref 16) and 5.43 illustrate good and bad Auger composition-depth profiles of 316L parts, sintered in dissociated ammonia under good and unacceptable conditions.

While the part in Fig. 5.42 had excellent corrosion resistance, that in Fig. 5.43 was very inferior. The temperature-dewpoint conditions

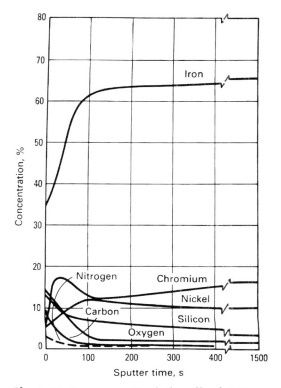

Fig. 5.42 Auger composition-depth profile of 316L part sintered in dissociated ammonia at 1177 °C (2151 °F). Dewpoint –40 °C (–40 °F). Source: Ref 16. Reprinted with permission from MPIF, Metal Powder Industries Federation, Princeton, NJ

for Fig. 5.43 were to the left of the redox curve for SiO₂/316L in Fig. 5.14. As a result, the part picked up oxygen during sintering. The oxygen profile in Fig. 5.43 is extended to 2000 s of sputtering, that is, approximately 30 times the value of a properly sintered material. Under proper reducing conditions, however, the challenge of achieving good corrosion resistance in nitrogen-containing atmospheres is related to the control of nitrogen.

Nitrogen Control during Sintering. For optimal sintering of stainless steels in H_2-N_2 mixtures, that is, for exploiting the beneficial effects of nitrogen for both strengthening and improving corrosion resistance, it is important to understand their equilibrium solubilities with nitrogen.

Thermodynamic Relationships: Nitrogen Solubility of Stainless Steels. According to Zitter and Habel (Ref 52), the solubility of nitrogen in austenitic chromium and chromium-nickel steels is determined, on one hand, by the solution of gaseous nitrogen in the matrix and, on the other hand, by the precipitation of dissolved nitrogen as chromium nitride, Cr_2N. The neglect of the fact that there are two equilibria to consider has led to erroneous data in the literature. The interaction of the two relationships leads to a maximum solubility for nitrogen, which depends only slightly on the chromium content but

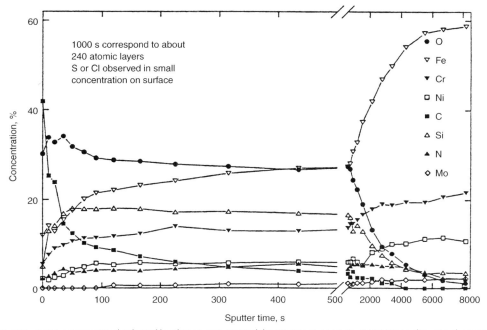

Fig. 5.43 Auger composition-depth profile of 316L part sintered for 20 min at 1110 °C (2030 °F) in dissociated ammonia of dewpoint –30 °C (–22 °F). Oxygen content of sample was 0.39%

which shifts markedly to higher temperatures with increasing chromium content and increasing partial pressure of nitrogen (Fig. 5.44) (Ref 52).

Nickel reduces the nitrogen solubility for the equilibrium with Cr_2N (Ref 53).

For sintering in dissociated ammonia, with chromium contents of 14, 18, and 22%, the maximum solubility temperatures are, according to Fig. 5.44, 947, 1057, and 1163 °C (1737, 1935, and 2125 °F), and their nitrogen solubilities are 0.39, 0.43, and 0.46%, respectively. For sintering a 22% Cr austenitic stainless steel in $90H_2$-$10N_2$, the minimum sintering temperature is approximately 1105 °C (2021 °F). Below this temperature, chromium becomes fully nitrided to form insoluble Cr_2N.

The negative temperature coefficient for solubility is unusual for metals and accounts, in part, as is seen subsequently, for the preference of sintering at higher temperatures in nitrogen-containing atmospheres. On cooling, Cr_2N begins to precipitate at these temperatures of maximum solubility. Above these temperatures, only dissolved nitrogen exists in the solid phase.

Dautzenberg (Ref 54) has shown the effect of nitrogen content on ultimate tensile strength and elongation of 304L stainless steel. The amount of nitrogen absorbed follows known phase equilibria in accordance with Sievert's law; that is, nitrogen absorption is proportional to the square root of the partial pressure of nitrogen in the sintering atmosphere (Ref 55).

The negative effect of nitrogen on ductility and impact strength is also apparent from the mechanical properties tables in Chapter 7, "Mechanical Properties."

Miura and Ogawa (Ref 56) used mechanical alloying of elemental powder mixtures with Fe10N to produce high-nitrogen chromium-nickel and chromium-manganese stainless steel powders with nanostructures having a wide composition range for austenite stability. The addition of AlN or NbN as dispersion agents allowed them to fully consolidate the mechanically alloyed materials by hot rolling near 1173 K (1652 °F), while still retaining nanostructures.

Sinter-Nitrided Martensitic 410. 410L is sometimes sintered in dissociated ammonia to obtain what may be considered an equivalent of the martensitic 410 stainless steel with a carbon content of 0.15% (Ref 57). In comparison to conventional martensitic 410, however, the properties of this so-called sinter-nitrided steel have been found to be somewhat erratic, probably due to the dependence of the amount of absorbed nitrogen on various processing parameters.

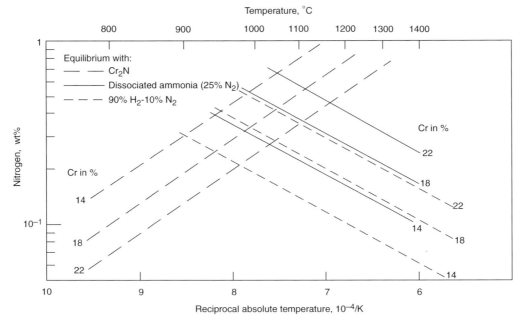

Fig. 5.44 Solubility of nitrogen in chromium-nickel steels in equilibrium with gaseous nitrogen or nitrides, depending on temperature and partial pressure of nitrogen. Source: Adapted from Ref 52

Ferritic stainless steels are not sintered in nitrogen-containing atmospheres, because nitrogen content must be very low to guarantee phase stability and good magnetic and corrosion-resistance properties.

As already mentioned, in recent years, commercial sintering has been experiencing a shift from dissociated ammonia ($75\%H_2$-$25\%N_2$) to $90H_2$-$10N_2$, because of the difficulty in obtaining good corrosion-resistance properties with the former. Strengthening, although decreased, is still possible with $90H_2$-$10N_2$, together with good corrosion resistance. The amount of total nitrogen absorbed determines the degree of strengthening. The amount of nitrogen present in solid solution contributes to the pitting corrosion resistance of a steel; the amount of nitrogen precipitated as Cr_2N during cooling determines its loss in corrosion resistance due to intergranular corrosion.

In polarization studies on PM 316L of various nitrogen contents (tested in 0.5% H_2SO_4), Johannson (Ref 58) has shown the material containing 0.61% N to exhibit unfavorable passivation characteristics (higher i_p), due to the presence of Cr_2N, because of its high nitrogen content. The two materials containing 0.049 and 0.40% N have all or most of their nitrogen dissolved and therefore exhibit lower passivating currents and more favorable (higher) pitting potentials.

Pitting Corrosion Index. For wrought austenitic and duplex stainless steels, the beneficial effect of nitrogen has been found to be interactive with the effects of chromium and molybdenum. Even though a number of other alloying elements can move the pitting potential in the noble direction, an empirical pitting or crevice-corrosion index, also known as pitting resistance equivalent number PREN, has been developed based on these three elements, according to which the pitting potentials of these steels increase with the following compositional parameter (Ref 59):

$$\%Cr + 3.3\%Mo + 16\%N$$

Other techniques for comparing alloy composition resistance of austenitic stainless steels to localized corrosion are based on the critical pitting temperature (CPT) and the critical crevice temperature (CCT), respectively. The former (CPT) involves the determination of the lowest temperature on the test surface at which stable propagating pitting occurs in accordance with ASTM G 150; the latter (CCT) involves the

determination of the maximum temperature at which no crevice attack occurs during a 24 h testing period (Ref 60).

In PM, the beneficial effect of nitrogen is often overshadowed by the negative effects from Cr_2N precipitation and other metallurgical weaknesses, particularly when sintering is performed in dissociated ammonia. It can, however, be exploited by employing rapid cooling after sintering, preferably in a $90H_2$-$10N_2$ gas mixture.

Larsen and Thorsen (Ref 61) report the positive effect of nitrogen for 316L specimens sintered in H_2-N_2 mixtures with N_2 contents from 0 to 25%. The pitting potentials increase with increasing nitrogen content in the sintered specimens.

The positive effect of nitrogen on both corrosion resistance and strength, together with the importance of cooling rate, is also shown in Fig. 5.45 (Ref 32).

Nitrogen Control during Cooling. Under commercial conditions of sintering in nitrogen-containing atmospheres, all nitrogen contained in a green stainless steel powder part, as well as additional nitrogen absorbed during sintering, is completely dissolved in the stainless steel matrix at sintering temperature. Corrosion problems with sintering in nitrogen-containing atmospheres arise on cooling, when some of the dissolved nitrogen, as a result of decreasing solubility, precipitates as Cr_2N unless cooling is very rapid. Also during cooling, additional nitrogen can be absorbed from the sintering atmosphere and precipitated as Cr_2N. The attendant chromium depletion in the surrounding matrix is similar to that known as sensitization in carbon-containing stainless steels. The low-chromium-content areas next to the grain boundaries caused what is termed grain-boundary or intergranular corrosion. Figure 5.46 (Ref 13) shows chromium nitride precipitates obtained under various sintering conditions.

At low concentration (Fig. 5.46a), Cr_2N typically forms precipitates in the grain boundaries, whereas at higher concentrations (Fig. 5.46b), the precipitates tend to form lamellae. The two critical processing parameters for minimizing Cr_2N precipitates are dewpoint of the sintering atmosphere and cooling rate. Figure 5.47 shows three different sintering scenarios, as represented by the dashed lines, for the sintering of an austenitic stainless steel in dissociated ammonia and $90H_2$-$10N_2$.

If cooling is very rapid, the nitrogen contents in the sintered parts will be only slightly higher

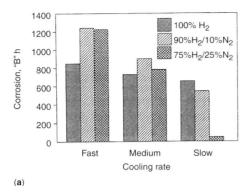

(a)

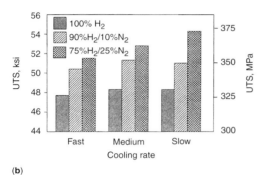

(b)

Fig. 5.45 (a) Corrosion resistances and (b) tensile strengths of 316L specimens, achieved with various sintering atmospheres and cooling rates. UTS, ultimate tensile strength. Source: Ref 32. Reprinted with permission from MPIF, Metal Powder Industries Federation, Princeton, NJ

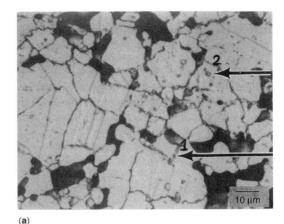

(a)

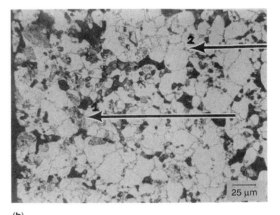

(b)

Fig. 5.46 Chromium nitride precipitates in 316L. (a) Sintered at 1150 °C (2100 °F) in dissociated ammonia; 4500 ppm N_2; Cr_2N precipitates along grain boundaries (1) and within grains (2). (b) Sintered at 1120 °C (2050 °F) in dissociated ammonia and slowly cooled; 6500 ppm N_2; Cr_2N in lamellar form near surface (1) and as grain-boundary precipitate in the interior (2). Source: Ref 13

than the gas equilibrium concentrations given by the upper curves. Because chemical reactions at these high temperatures are quite rapid, some additional nitrogen will be absorbed in accordance with the increasing solubility for nitrogen as the temperature decreases, as well as with decreasing density of a part. Nevertheless, the part sintered in dissociated ammonia not only picks up more nitrogen than the one sintered in $90H_2$-$10N_2$ but also, on cooling, begins to precipitate Cr_2N at a higher temperature, because nitrogen solubility decreases at lower temperatures.

TTS Diagrams. The time-temperature-sensitization diagrams for wrought stainless steels can be used for sintered stainless steels to estimate cooling rate requirements as a function of their nitrogen contents. Figure 5.48 (Ref 62) is an example of such a diagram.

In an early study, Sands et al. (Ref 35) pointed out that 316L sintered in dissociated ammonia required a cooling rate of 200 °C/min (360 °F/min) to prevent nitrogen absorption and precipitation of Cr_2N. More recently, Frisk et al. (Ref 63)

determined in a laboratory study that sintering of 316L in dissociated ammonia at 1250 °C (2280 °F) required cooling rates of >450 °C/min (>810 °F) (Fig. 5.49).

Narrow Dewpoint Window. The higher critical cooling rate of Frisk et al. can probably be ascribed to their much lower dewpoint (–100 °C, or –148 °F) versus –40 to –60 °C (–40 to –76 °F) for Sands, which allows for more rapid nitrogen absorption during cooling, as illustrated in Fig. 5.50 (Ref 64) for the bright annealing of stainless steel strip.

Because nitrogen absorption occurs through diffusion, the presence of nitrides or oxides in the material reduces the rate of nitrogen absorption. A low dewpoint removes surface oxides and facilitates nitrogen absorption from the atmosphere. This explains why nitrogen absorption during cooling increases with decreasing dewpoint of the sintering atmosphere.

The deleterious reactions of increasing nitrogen absorption (with decreasing dewpoint) and increasing oxidation (with increasing dewpoint) during cooling leave a relatively narrow dewpoint window for optimal sintering in dissociated ammonia. Apart from this dewpoint effect, sintering in an atmosphere of $90H_2$-$10N_2$ is a much better compromise that

greatly reduces the high and impracticable cooling rate requirements of dissociated ammonia to more manageable levels, while still benefiting substantially from the strengthening obtainable with the lower nitrogen concentration. Good corrosion resistance for such conditions was reported by Larsen (Ref 26) and Mathiesen (Ref 65) and more recently by Samal et al. (Ref 32).

Figure 5.51 (Ref 28) shows corrosion weight losses in 10% aqueous HNO_3 for austenitic stainless steels sintered under various conditions.

The steep increase in weight loss at a nitrogen content of approximately 3000 ppm has led some investigators to conclude that 3000 ppm represents the upper limit for nitrogen for good corrosion resistance. This conclusion is misleading. It would be better to state that nitrogen levels of up to 3000 ppm are less likely to cause problems, because required cooling rates are easier to manage. Bulk nitrogen analysis cannot reveal to what extent the surface of a part is enriched with nitrogen, whether the nitrogen has reacted with chromium to form chromium nitride, and if sensitization is present or not. Metallographic examination and electrochemical testing according to the double-loop eletrochemical potentiokinetic reactivation (EPR) technique (Chapter 9, "Corrosion Testing and Performance") are the methods of choice for sensitization problems. Mathiesen and Maahn (Ref 67) cite cases for sintered 316L where EPR data indicate the presence of significant sensitization at nitrogen contents of only 2100 ppm.

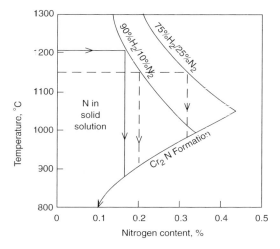

Fig. 5.47 Schematic diagram showing equilibrium nitrogen/ nitride contents for fast cooling from two sintering temperatures, in two sintering atmospheres, for an austenitic stainless steel. A higher sintering temperature, lower nitrogen content of atmosphere, and rapid cooling will lead to a lower total nitrogen content and a smaller percentage of chromium nitrides in the sintered material

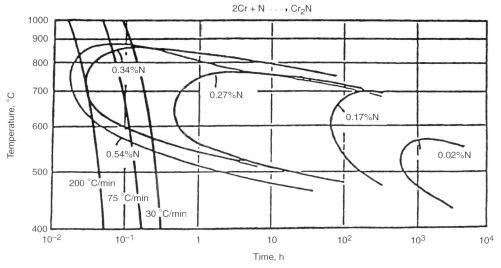

Fig. 5.48 Time-temperature-corrosion diagram for 18%Cr-10%Ni austenitic stainless steel. Source: Ref 62

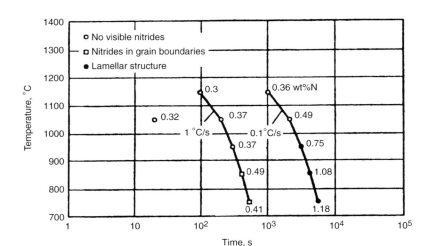

Fig. 5.49 Effect of cooling rate on the presence of chromium nitrides in the microstructure of 316L parts sintered at 1250 °C (2282 °F) in dissociated ammonia. Source: Ref 63. Reprinted with permission from MPIF, Metal Powder Industries Federation, Princeton, NJ

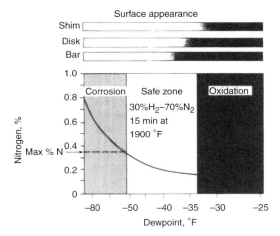

Fig. 5.50 Effect of dewpoint on nitrogen absorption and oxidation of 316L shim, disk, and bar stock annealed for 15 min at 1038 °C (1900 °F) in 30%H₂-70%N₂. It took 2.3, 2.8, and 4.7 min, respectively, to cool the three materials from 1038 to 538 °C (1900 to 1000 °F). Source: Ref 64. Reprinted with permission from MPIF, Metal Powder Industries Federation, Princeton, NJ

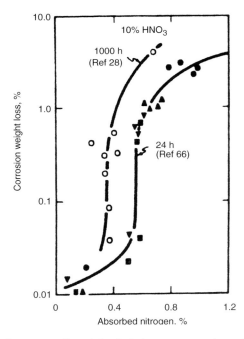

Fig. 5.51 Effect of absorbed nitrogen on corrosion weight losses of austenitic stainless steels in 10% aqueous HNO_3. Source: Ref 28. Reprinted with permission from MPIF, Metal Powder Industries Federation, Princeton, NJ

The MPIF standard 35 (2003) and ASTM standard B 783 (1999) list upper nitrogen contents of 0.6% for the two austenitic stainless steels, 304 and 316, sintered in dissociated ammonia at 1149 °C (2100 °F) (N1 designation) and 1288 °C (2350 °F) (N2 designation). According to Zitter's data, which are corroborated elsewhere (Ref 68), the maximum nitrogen solubilities for these two stainless steels in dissociated ammonia are only approximately 0.3% for sintering at 1149 °C (2100 °F) and lower for sintering at 1288 °C (2350 °F). Nitrogen levels above these

solubility limits will result in Cr_2N precipitation and attendant sensitization for intergranular corrosion. In order to avoid higher nitrogen levels, rapid cooling after sintering must begin above approximately 1057 °C (1935 °F). Better nitrogen specification levels for these two steels would be 0.15 to 0.30%. The 300-series wrought stainless steels in general do not encounter any sensitization problems from nitrogen, because nitrogen levels in these steels are kept below 0.16%, which assures full dissolution of nitrogen.

Chromium enrichment on the surface of a stainless steel part that was sintered in a nitrogen-containing atmosphere can also be gleaned from surface chemical analysis. A pronounced chromium peak, present within the first 100 s of sputtering and corresponding to a depth of approximately 120 Å, is characteristic of sintering in a nitrogen-containing atmosphere with Cr_2N precipitation during cooling (Fig. 5.52).

As dissolved chromium is removed on the surface of nitrogen-accessible pores by precipitation of Cr_2N, a steep dissolved chromium gradient is established that brings about continuing chromium diffusion from the interior to the surface. It is possible to increase this chromium diffusion to the surface to very high levels. Figure 5.52 (Ref 28) shows the profile of a 316L part that was sintered in dissociated ammonia and then slowly furnace cooled.

As mentioned earlier, the positive effect of dissolved nitrogen on corrosion resistance, as documented for wrought stainless steels, applies equally to sintered stainless steels. In polarization studies on 18%Cr-8%Ni stainless steels containing various amounts of nitrogen and tested in a

hydrogen-purged 1 N H_2SO_4 + 0.5 M NaCl solution at ambient, Eckenrod and Kovach (Ref 69) showed that the passive current density decreases and the passive range expands up to 0.25% N.

It is shown in Chapter 6, "Alloying Elements, Optimal Sintering, and Surface Modification in PM Stainless Steels," that tin-modified stainless steels are less sensitive to nitrogen absorption on cooling because of tin's barrier to nitrogen diffusion. Thus, such surface-modified stainless steels can exhibit superior corrosion resistances in nitrogen-containing sintering atmospheres despite only moderate cooling rates.

5.2.5 Sintering of Stainless Steels in Vacuum

Vacuum sintering of stainless steels offers the advantage of high-temperature sintering combined with superior oxide reduction. State-of-the-art furnaces permit continuous prodution, including delubrication (Ref 70) and rapid cooling through gas pressure quenching. Various atmospheres, including hydrogen, may be introduced. Although gas consumption is much less than with atmosphere sintering, capital cost and maintenance of vacuum furnaces are higher. With the aforementioned attributes, however, a vacuum furnace seems to have superior characteristics for the optimal sintering of stainless steels: vacuum sintering produces the lowest levels of interstitials, the prime criterion for maximum magnetic and corrosion-resistance properties. Through judicial addition of carbon to a water-atomized stainless steel powder, the oxygen content of a sintered part can be reduced to levels approaching those of wrought stainless steels. Nevertheless, as is seen later, certain process precautions are necessary.

High Vapor Pressure of Chromium. Stainless steel parts producers recognized early on that vacuum sintering of stainless steels caused the surfaces of sintered parts to be depleted of chromium because of the high vapor pressure of chromium at elevated temperature. This led to a deterioration of general corrosion resistance as well as lower pitting resistance. Therefore, partial pressures of an inert gas were applied during sintering. Over the years, these increased from less than 100 µm of mercury to several hundred. More recently, Klar and Samal (Ref 27) showed that chromium losses continued to decline at 1260 °C (2300 °F) with partial pressures of argon increasing to over 1000 µm of mercury. Significant improvement of corrosion

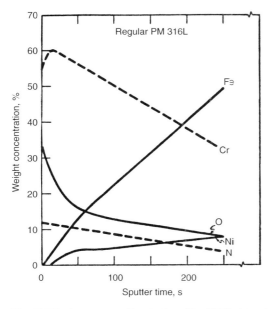

Fig. 5.52 Auger composition-depth profile of 316L sintered in dissociated ammonia and slowly furnace cooled. Note chromium and nitrogen enrichment on surface. Source: Ref 28. Reprinted with permission from MPIF, Metal Powder Industries Federation, Princeton, NJ

resistance also occurred when, toward the end of sintering, the partial pressure of argon was increased to 1 atm for a short time. Presumably, this allowed the parts to replenish surface chromium that had been lost during low-partial-pressure sintering. A short holding period at a lower temperature of approximately 1150 °C (2102 °F) had a similar beneficial effect. In the absence of specific data, the authors recommend a partial pressure of nitrogen or argon of several thousand micrometers of mercury.

Superior Oxygen and Carbon Removal. Despite the absence of an external reducing gas atmosphere, vacuum-sintered stainless steel parts typically have lower oxygen and carbon contents than atmosphere-sintered stainless steels. This is mainly due to the improved reaction between residual carbon and oxides, particularly at high sintering temperatures (>1200 °C, or >2192 °F), (Fig. 5.53) (Ref 71).

However, the typical amount of carbon present in stainless steel powder is usually insufficient to reduce most of its original oxides. These unreduced oxide particles give rise to pitting corrosion as well as to lower mechanical properties. Admixing small amounts of graphite to a stainless steel powder greatly enhanced overall oxide reduction. Using this technique, it should be kept in mind, however, that the carbon content of the sintered part will increase. Thus, the optimal graphite addition is the maximum addition that generates no chromium carbide precipitates in the cooling zone of the sintering furnace. It depends, among other factors, on the

composition of the stainless steel, the oxygen content of the powder, the sintering temperature, and the cooling rate after sintering. With this technique, delubrication can be conducted under conditions that increase or even maximize carbon content, so that less graphite needs to be added.

Beiss (Ref 72) states that good oxide reduction in vacuum-sintered stainless steels occurs without carbon addition, because of the low stability of Cr_2O_3 at temperatures above 900 °C (1652 °F), where it starts to decompose. This may be true for low-silicon-content, gas-atomized stainless steels that may have picked up oxygen during debinding. Water-atomized stainless steels, however, generally have silicon contents approaching 1%, and their major surface oxide is SiO_2, which becomes much more effectively reduced with additions of carbon.

Figure 5.54 illustrates the combined effects of sintering temperature, graphite addition, and partial gas pressure on chromium losses, chromium carbide formation, and oxide reduction, and their effects on chloride corrosion resistance (Ref 27).

In this study, vacuum sintering was performed at 1150, 1260, and 1260 °C (2102, 2300, and 2300 °F), followed by holding at 1150 °C (2102 °F), all with partial pressures of nitrogen and argon of 1300 mm of mercury. The corrosion resistance results are best interpreted as follows:

- *0.08% graphite addition*: Optimal corrosion resistances are due to reduced oxide levels at

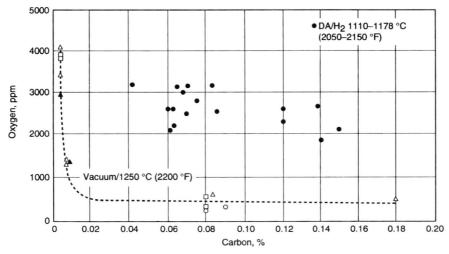

Fig. 5.53 Oxygen versus carbon contents of vacuum- and atmosphere-sintered powder metallurgy austenitic stainless steels of varying compositions. DA, dissociated ammonia. Source: Ref 71. Reprinted with permission from MPIF, Metal Powder Industries Federation, Princeton, NJ

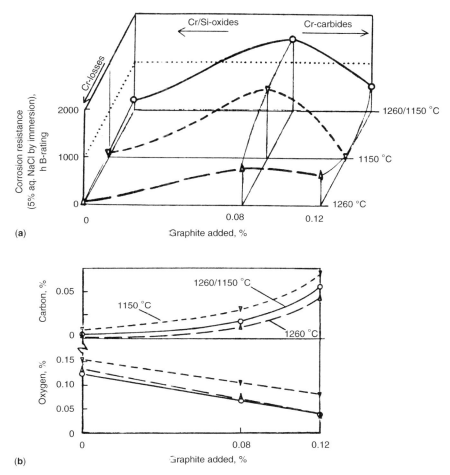

Fig. 5.54 Effect of sintering temperature and graphite addition on (a) corrosion resistance and (b) carbon and oxygen content of vacuum-sintered 316L (green densities 6.6 g/cm³; cooling rate: 30 °C/min, or 54 °F/min). Reprinted with permission from MPIF, Metal Powder Industries Federation, Princeton, NJ

acceptable carbon contents (<0.03%). Chromium losses determine the different corrosion values for the three different sintering conditions.

- *0% graphite addition*: Low-corrosion properties are due to substantial amounts of "original" oxides (from the water-atomization process) plus the effect of chromium losses.
- *0.12% graphite addition:* Low corrosion resistances are the result of chromium carbide formation due to excessive graphite additions and insufficient cooling rates.

For superior corrosion resistance, vacuum-sintered stainless steels should be rapidly cooled in a nonoxidizing gas to prevent the formation of deleterious surface oxides. Cooling in nitrogen will generate chromium nitrides on the surfaces of the parts, and the attendant chromium depletion

will cause the parts to have low corrosion resistance. It should be stressed, however, that metallurgical defects that are limited to shallow surface regions of a part, if removed by dissolution or by wear and tear, may be tolerable for certain uses. As mentioned previously, this would explain the good performance of antilock brake system sensor rings and auto exhaust components that are produced without any accelerated cooling.

5.3 Liquid-Phase Sintering of Stainless Steels

Liquid-phase sintering in PM is often used to achieve improved mechanical properties through higher sintered densities. A more recent interest was the elimination of interconnected porosity in sintered stainless steels, in order to reduce the large internal surface areas that give rise to large corrosion currents as well as to eliminate the density region that, in neutral saline environments, can drastically lower corrosion resistance because of crevice corrosion (section 5.2.2 in this chapter). Of the two common methods for liquid-phase sintering, supersolidus and activated sintering, the former is of lesser interest with stainless steels because of their high solidus temperatures of over 1350 °C (2462 °F) and the narrow temperature windows available for successful sintering.

Activated Sintering and Requirements for Liquid-Phase-Forming Additives. For activated sintering, several liquid-phase-forming additives have been investigated. In wrought stainless steels, liquid-phase-forming elements (for example, antimony, arsenic, boron, phosphorus, sulphur) are kept at very low levels, because even very small amounts of liquid phase lead to hot shortness during hot working and welding. With sintered stainless steels, however, hot working is not practiced, and such precaution is therefore unnecessary. The maximum amount of liquid phase in a sintered part is often determined by the resistance of a part to distortion during sintering (dimensional stability) and can be quite high, depending on the solid-liquid dihedral angle (Fig. 5.55) (Ref 73).

Most literature data on liquid-phase-sintered stainless steels are based on the use of sintering additives that form a persistent liquid phase at commonly used sintering temperatures of 1100 to 1200 °C (2012 to 2192 °F). This led to the desired high densities. In many cases, however, secondary phases were formed during solidification, which impaired corrosion resistance or mechanical properties, particularly dynamic mechanical properties, or both. The secondary phases should have a corrosion resistance similar to or greater than that of the matrix.

When sintered 316L is annealed for extended times at intermediate temperatures (955 to 900 °C or 1751 to 1652 °F), sigma, eta, or Laves phases can develop (Ref 74). The development of such phases was accompanied by increases in the passive currents and characteristic secondary passivating peaks.

Reen (Ref 75) reports the formation of a B-Cr-Ni-Mo-containing secondary phase in boron-containing 316L that depletes the chromium and molybdenum content of the matrix and thereby impairs its corrosion resistance. Maahn et al. (Ref 19) used additions of

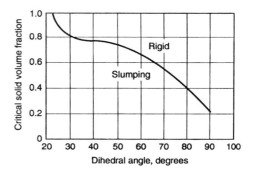

Fig. 5.55 Effect of dihedral angle on the volume fraction for freestanding structural rigidity. Source: Ref 73

boron, BN, CrB, and NiB to 316L and showed improved corrosion properties (Table 5.4), except for high l_r/l_a values in the EPR test (section 9.1.3 in Chapter 9, "Corrosion Testing and Performance"), which reflect chromium and molybdenum depletions around chromium- and molybdenum-rich borides.

Maahn et al. calculated the composition of the boron-containing austenitic (316L) phase during solidification, assuming limited diffusion in the solid state, and found that the last fraction of solidified austenite was significantly decreased in both chromium and molybdenum. Becker et al. (Ref 76) also made use of computer-calculated phase diagrams for identifying prealloy compositions or sintering additives that would give rise to liquid-phase-sintering temperatures between 1100 and 1200 °C (2012 and 2192 °F) and would avoid the problems of chromium depletion due to high-chromium-content intermetallic phases. The prediction of a homogeneous austenitic alloy based on Fe17Cr 12Ni with up to 5% Si was confirmed by microstructural analysis.

In order to compensate for any chromium and/or molybdenum losses of the matrix due to the formation of intermetallic phases, Reen increased both chromium and molybdenum in 316L and came up with SS-100 (section 2.2 in Chapter 2, "Metallurgy and Alloy Compositions"), an austenitic stainless steel powder possessing good compacting properties and which, with optimal processing, in the presence of an open pore structure, had corrosion-resistance characteristics similar to wrought 316L.

By liquid-phase sintering with boron (0.15 to 0.40%), Samal and Terrell (Ref 77) were able to increase the corrosion resistance of PM 316L stainless steel, as determined by immersion testing in 5% aqueous NaCl, to that equaling wrought 316L, with an "A"-rating (no rust or

Table 5.4 Corrosion properties of liquid-phase-sintered 316L stainless steels with addition of boron-base sintering additives

All steels were sintered at 1250 °C (2282 °F) for 60 to 120 min in pure hydrogen.

Additive	Density, g/cm³	Open pores, %	E_{stp}(a), mV SCE	NSS1, h	NSS2	$l_r/l_a \times 1000$
None	6.86	8.2	150	96	7	0.0
0.2% B(–38 μm)	7.83	0.1	500	>1500	10	4.4
1% BN (–63 μm)	7.61	0.2	400	762	9	2.5
1% NiB (–38 μm)	7.67	0.1	525	>1500	10	3.8
1% CrB (–38 μm)	7.64	0.1	550	>1500	10	2.5

(a) 0.1% Cl⁻, pH 5, 30 °C (86 °F), 25 mV/8 h. Source: Ref 19

stain) of over 7000 h. They tentatively attributed this improvement over lower-density 316L to optimal sintering combined with the formation of an almost fully dense and smooth surface layer (Fig. 5.56). This also had been observed by other investigators (Ref 78).

The liquid-phase-sintered specimens had oxygen contents from 0.10 to 0.19%. The fluxing properties of boron appear to redistribute the residual oxides, which are dispersed throughout the stainless steel matrix, in a different, more coagulated form. It is not clear to what extent, if any, this phenomenon contributed to the observed improvement in corrosion resistance.

In their investigation, Samal et al. did not observe any negative effect of boron on the dynamic (room temperature) properties of 316L, although the static strength properties were lower, due to considerable grain growth during liquid-phase sintering. Another side effect with boron was higher carbon contents. Boron scavenges some of the oxygen present in the powder

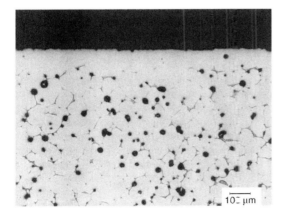

Fig. 5.56 Micrograph of 316L + 0.20% B sample sintered in hydrogen and showing a nearly fully dense outer layer that is 0.076 mm (0.003 in.) deep. This type of layer was observed in all boron-containing samples. Reprinted with permission from MPIF, Metal Powder Industries Federation, Princeton, NJ

to form boron oxide. This reduces the amount of oxygen available for decarburization. The boron oxide forms low-melting, glassy, mixed oxides with SiO_2 (Ref 79).

In view of the discrepancies regarding the effects of boron in liquid-phase sintering of austenitic stainless steels, Samal et al. stressed the usefulness of electrochemical corrosion testing for identifying weaknesses or defects in stainless steel. However, they cautioned against relying too heavily on such testing (without assuring the absence of other metallurgical defects) for determining actual corrosion resistances. This was because of the many complexities with sintered stainless steels, such as indeterminate active surface areas of sintered materials and varying rates of passivation of pore surfaces relative to external surfaces. Corrosion testing methods valid for wrought stainless steels are not necessarily applicable to porous stainless steels without modification or complementary metallographic analysis.

By adding 5% fine (–25 μm) BNi-1 brazing powder (4.5% Si, 3% B) to water-atomized 316L and vacuum sintering at 1200 °C (2192 °F), Sharon and Itzhak (Ref 80) obtained densification to 7.52 g/cm³ and passivation characteristics in 0.5 M H_2SO_4 similar to wrought stainless steel. Transient liquid-phase sintering resulted in a single austenitic phase, eliminating potential mechanical property problems from secondary phases. The boron of the brazing powder additive assists the diffusion rate in the liquid state by reducing the metal oxides in the stainless steel powder. Although the hard vacuum employed during sintering (1.3 Pa, or 10 − 2 torr) most likely caused the surfaces of the specimens to be severely depleted of chromium (section 5.2.5 in this chapter), this approach nevertheless appears promising, because the low inert gas pressure in the vacuum furnace can readily be increased. However, chromium depletion would not show up in the electrochemical testing, because the testing surfaces had been polished to a 600-grit finish. Nakamura et al. (Ref 81) added 0.2% B and 0.2% P separately to SUS 304. When tested for intergranular corrosion in boiling HNO_3, the corrosion performance with boron addition was similar to wrought SUS 304, while the phosphorus addition produced localized corrosion because of the presence of phosphides in the grain boundaries. When tested for general corrosion in 5% boiling H_2SO_4, the corrosion resistance of both materials closely followed the density of the specimens.

A higher-alloyed liquid-phase-sintered austenitic stainless steel (23Cr18Ni3.5Mo0.25B) was recently standardized by ASTM International as standard B 853 (Ref 82). This stainless steel is based on Reen's U.S. patent of 1977 (Ref 75). Sintering was in hydrogen or under vacuum at a temperature of approximately 1260 °C (2300 °F) to a density of 7.7 to 7.8 g/cm^3. Distortion-free parts with improved crevice and pitting corrosion resistances over lower-alloyed steels were produced, with a typical ultimate tensile strength of 590 MPa (85,000 psi), a 0.2% offset yield strength of 260 MPa (37,500 psi), an elongation in 2.54 cm (1 in.) of 19%, and a Rockwell B hardness of 83 to 91. For improved dimensional control, Reen also used the double press/double sinter method, that is, first pressing to a green density of approximately 6.65 g/cm^3, followed by sintering below the liquid-phase-forming temperature (for example, at 1204 °C, or 2200 °F), then repressing and sintering above that temperature. Reen explains its superior performance (to boron-containing 316L) in terms of the higher chromium and molybdenum levels. Despite their partial depletion by secondary phase formation, there were still enough of these critical elements in the matrix to provide superior resistance against crevice and pitting corrosion. The latter was measured in a 5% neutral salt spray test and by anodic polarization in a 3% salt solution.

Preusse et al. (Ref 83) employed Cu_3P and Fe_3P additions to 316L. With an optimal amount of approximately 8% Cu_3P and vacuum sintering at 1250 °C (2282 °F), they obtained only marginal corrosion-resistance improvements in H_2SO_4 and observed serious pitting in 3.5% NaCl. The chromium-enriched eutectic phase was identified as an iron/chromium phosphide intermetallic. In a subsequent investigation (Ref 84) the authors report improved resistance in aqueous chloride solutions.

For austenitic stainless steels, silicon additions of up to 5% (Ref 85) produced activated liquid-phase sintering and a duplex structure with good corrosion resistance.

Copper infiltration of sintered stainless steels improves density, hardness, mechanical properties, and corrosion resistance in 0.5% H_2SO_4 (Ref 86). The authors, however, have observed problems with galvanic corrosion in a neutral chloride-containing environment as a result of the different nobilities of copper and austenitic stainless steel.

REFERENCES

1. F.V. Lenel, *Powder Metallurgy—Principles and Applications*, Metal Powder Industries Federation, Princeton, NJ, 1980, p 184
2. R.M. German, *Powder Metallurgy Science,* 2nd ed., Metal Powder Industries Federation, Princeton, NJ, 1994, p 290
3. W. Schatt, *Powder Metallurgy Processing and Materials*, EPMA, Shrewsbury, U.K., 1997
4. G. Dautzenberg and H. Gesell, Production Technique and Properties of Austenitic Cr-Ni Stainless Steel Powders, *Powder Metall. Int.*, Vol 8 (No. 1), 1976, p 14–17
5. E. Klar, Corrosion of Powder Metallurgy Materials, *Corrosion,* Vol 13, *Metals Handbook,* 9th ed., ASM International, 1987
6. T. Tunberg, L. Nyborg, and C.X. Liu, Enhanced Vacuum Sintering of Water Atomized Stainless Steel Powder by Carbon Addition, *Advances in Powder Metallurgy and Particulate Materials,* Vol 3, J. Capus and R. German, Eds., MPIF, 1992, p 383–396
7. F.M.F. Jones, "The Effect of Processing Variables on the Properties of Type 316L Powder Compacts," *Progress in Powder Metallurgy*, Vol 30, Metal Powder Industries Federation, 1970, p 25–50
8. M.Y. Nazmy, W. Karner, and A.A. Al-Gwaiz, *J. Met.*, No. 6, 1978, p 14
9. G. Lei and R.M. German, Corrosion of Sintered Stainless Steels in a Sodium Chloride Solution, *Modern Developments in Powder Metallurgy*, Vol 16, E.N. Aqua and C.I. Whitman, Ed., MPIF, Princeton, NJ, 1984, p 261–275
10. P. Peled and D. Itzhak, *Corros. Sci.*, Vol 30, 1990, p 59
11. T. Raghu, S.N. Malhotra, and P. Ramakrishnan, *Br. Corros. J.*, Vol 23, 1988, p 109
12. D. Itzhak and E. Aghion, *Corros. Sci.*, Vol 23, 1983, p 1085
13. E. Klar and P.K. Samal, Effect of Density and Sintering Variables on the Corrosion

Resistance of Austenitic Stainless Steels, *Advances in Powder Metallurgy and Particulate Materials,* Vol 11, J. Porter and M. Phillips, Eds., MPIF, 1995, p 11–3 to 11–17

14. T. Raghu, S.N. Malhotra, and P. Ramakrishnan, *Corrosion,* Vol 45 (No. 9), 1989, p 698

15. M.G. Fontana and N.D. Greene, *Corrosion Engineering,* 2nd. ed., McGraw-Hill Book Co., 1978, p 42–43

16. D. Ro and E. Klar, Corrosive Behavior of P/M Austenitic Stainless Steels *Modern Developments in Powder Metallurgy,* Vol 13, H. Hausner, H. Antes, and G. Smith, Eds., MPIF, 1980, p 247–287

17. G. Lei, R.M. German, and H.S. Nayar, Corrosion Control in Sintered Austenitic Stainless Steels, *Progress in Powder Metallurgy,* Vol 39, H.S. Nayar, S.M. Kaufman, and K.E. Meiners, Ed., Metal Powder Industries Federation, 1983, p 391–410

18. E. Maahn and T. Mathiesen, "Corrosion Properties of Sintered Stainless Steel," presented at U.K. Corrosion '91 (Manchester), NACE, 1991

19. E. Maahn, S.K. Jensen, R.M. Larsen, and T. Mathiesen, Factors Affecting the Corrosion Resistance of Sintered Stainless Steel, *Advances in Powder Metallurgy and Particulate Materials,* Vol 7, C. Lall and A. Neupaver, Eds., MPIF, 1994, p 253–271

20. T. Mathiesen and E. Maahn, "Effect of Pore Morphology on the Corrosion Behavior of Sintered 316L Stainless Steel," *Advances in Powder Metallurgy and Particulate Materials*, Vol 11, J. Porter and M. Phillips, Eds., MPIF

21. A. Ashurst and E. Klar, Mercury Porosimetry, *Powder Metallurgy,* Vol 7, *Metals Handbook,* 9th ed., American Society for Metals, 1984, p 266–271

22. E. Klar, Relationship Between Pore Characterization and Compacting Properties of Copper Powders, *J. Mater.,* Vol 7 (No. 3), 1972, p 418–424

23. C. Molins, J.A. Bas, and J. Planes, P/M Stainless Steel: Types and Their Characteristics and Applications, *Advances in Powder Metallurgy and Particulate Materials,* Vol 5, J. Capus and R. German, Eds., MPIF, 1992, p 345–357

24. W. Schatt and K.-P. Wieters, *Powder Metallurgy Processing and Materials,* EPMA, 1997, p 151

25. P. Beiss, "Processing of Sintered Stainless Steel Parts," Powder Metallurgy Group Meeting 1991, Powder Materials in Transportation, York, U.K.

26. R.M. Larsen, Ph.D. dissertation, Technical University of Denmark, 1994 (in Danish)

27. E. Klar and P.K. Samal, Optimization of Vacuum Sintering Parameters for Improved Corrosion Resistance of P/M Stainless Steels, *Advances in Powder Metallurgy and Particulate Materials*, Vol 7, C. Lall and A. Neupaver, Eds., MPIF, 1994, p 239–251

28. M.A. Pao and E. Klar, Corrosion Phenomena in Regular and Tin-Modified P/M Stainless Steels, *Progress in Powder Metallurgy,* Vol 39, H. Nayar, S. Kaufman, K. Meiners, Eds., MPIF, 1984, p 431–444

29. R.M. Larsen and K.A. Thorsen, "Removal of Oxygen and Carbon During Sintering of Austenitic Stainless Steels," presented at PM World Congress (Kyoto, Japan), Japan Society for Powder Powder Metallurgy, 1993

30. R.M. Larsen and K.A. Thorsen, Equilibria and Kinetics of Gas-Metal Reactions During Sintering of Austenitic Stainless Steel, *Powder Metal.,* Vol 37 (No. 1), 1994, p 1–12

31. P.K. Samal and E. Klar, Effect of Sintering Atmosphere on Corrosion Resistance and Mechanical Properties of Austenitic Stainless Steels—Part I, *Advances in Powder Metallurgy and Particulate Materials,* R. McKotch and R. Webb, Eds., MPIF 1997, p 14-55 to 14-65

32. P.K. Samal, J.B. Terrell, and E. Klar, "Effect of Sintering Atmosphere on the Corrosion Resistance and Mechanical Properties of Austenitic Stainless Steels—Part II," *Advances in Powder Metallurgy and Particulate Materials,* W. Eisen and S. Kassam, Eds., MPIF, 2001

33. E. Klar and E.K. Weaver, Process for Production of Metal Powders Having High Green Strength, U.S. Patent 3,888,657, 1975

34. G.H. Lei and R.M. German, Corrosion of Sintered Stainless Steels in a Sodium Chloride Solution, *Modern Developments in Powder Metallurgy*, E. Aqua and C. Whitman, Eds., MPIF, 1984

35. R.L. Sands, G.F. Bidmead, and D.A. Oliver, The Corrosion Resistance of Sintered Stainless Steels, *Modern Developments in Powder Metallurgy,* Vol 2, H.H. Hausner, Ed., Plenum Press, 1966, p 73–85

36. *Consultants Corner, In. J. Powder Metall.,* Vol 39 (No. 4), 2003, p 22

37. T. Mathiesan and E. Maahn, Corrosion Behavior of Sintered Stainless Steels in Chloride Containing Enviroments, 12th Scandinavian Corrosion Congress (Helsinki), 1992, p 1–9

38. G. Lei, R.M. German, and H.S. Nayar, Influence of Sintering Variables on the Corrosion Resistance of 316L Stainless Steel, *Powder Metall. Int.,* Vol 15 (No. 2), 1983, p 70–76

39. E.E. Stansbury and R.A. Buchanan, *Fundamentals of Electrochemical Corrosion,* ASM International, 2000

40. M.G. Fontana and N.D. Greene, *Corrosion Engineering,* McGraw-Hill Book Co., 1978

41. E. Klar and P.K. Samal, Sintering of Stainless Steel, *Powder Metal Technologies and Applications,* Vol 7, *ASM Handbook,* 1998, p 476–482

42. A.J. Sedriks, *Corrosion of Stainless Steels,* John Wiley & Sons, 1996, p 20

43. D. Peckner and I.M. Bernstein, *Handbook of Stainless Steels,* McGraw-Hill Book Company, New York, 1977, p 16–80

44. A.J. Sedriks, Effect of Alloy Composition and Microstructure on the Passivity of Stainless Steels, *Corrosion,* Vol 42 (No. 7), July 1986, p 376

45. P.K. Samal and J.B. Terrell, On the Intergranular Corrosion of P/M 316L Stainless Steel, *Advances in Powder Metallurgy and Particulate Materials,* Vol 7., (PM World Congress) V. Arnhold, C.-L. Chu, W.F. Jandeska, Jr., and H.I. Sanderow, Compilers, 2002, p 7–89 to 7–101

46. E.E. Stansbury and R.A. Buchanan, *Fundamentals of Electrochemical Corrosion,* ASM International, 2000, p 349

47. P.K. Samal, E. Klar, and S.A. Nasser, On the Corrosion Resistance of Sintered Ferritic Stainless Steels, *Advances in Powder Metallurgy and Particulate Materials,* R. McKotch and R. Webb, Eds., MPIF, 1997, p 16–99 to 16–112

48. Y.M. Kolotyrkin, V.M. Knyazheva, N.S. Neiman, and V.P. Pancheshnaya, in *Proc. Fifth Int. Cong. Met. Corros.,* N. Sato, Ed., National Association of Corrosion Engineers, 1974, p 232

49. R.J. Hodges, Intergranular Corrosion in High Purity Ferritic Stainless Steels: Isothermal Time-Temperature-Sensitization Measurements, *Corrosion,* Vol 27 (No. 4), April 1971

50. K.H. Moyer, The Burn-Off Characteristics of Common Lubricants in 316L Powder Compacts, *Int. J. Powder Metall.,* Vol 7 (No. 3), 1971, p 33–43

51. D. Saha and D. Apelian, Control Strategy for De-Lubrication of P/M Compacts, *Int. J. Powder Metall.,* Vol 38 (No. 3), 2002, p 71–78

52. H. Zitter and L. Habel, Zur Löslichkeit des Sickstoffs in Reineisen und austenitischen Chrom-Nickel-Stählen, (On the Solubility of Nitrogen in Pure Iron and Austenitic Chromium-Nickel Steels), *Arch. Eisenhüttenwes.,* Vol 44 (No. 3), 1973, p 181–188

53. T. Masumoto and Y. Imai, *J. Jpn. Inst. Met.,* Vol 33, 1969, p 1364

54. N. Dautzenberg, "Eigenschaften von Sinterstählen aus Wasserverdüsten and Fertiglegierten Pulvern," "Properties of Sintered Steels from Water Atomized Elemental and Alloyed Powders"), Paper. 6.18, *Second European Symposium on Powder Metallurgy,* Vol II, EPMA, 1968

55. A. Sieverts, G. Zapf, and H. Moritz, *Z. Phys. Chem., Abt. A,* Vol 183, 1938, p 19–37

56. H. Miura and H. Ogawa, Austenitizing and Hot Compaction of High-Nitrogen Containing Cr-Ni and Cr-Mn Steel Powders Mechanically Alloyed, *Proc. of 2000 Powder Metallurgy World Congress* (Kyoto, Japan), Professional Engineering Publishing Limited, U.K., 2000

57. M. Svilar and H.D. Ambs, P/M Martensitic Stainless Steels: Processing and Properties, *Advances in Powder Metallurgy,* E. Andreotti, P. McGeehan, Eds., Vol 2, MPIF, 1990, p 259–272

58. A. Johannson, Report IM-2913, Institute for Metals Research, Sweden 1991

59. A.J. Sedriks, Effects of Alloy Composition and Microstructure on the Passivity of Stainless Steels, *CORROSION/86,* National Association of Corrosion Engineers, 1986

60. *Corrosion,* Vol 13, *Metals Handbook,* 9th ed., ASM International, 1987, p 581

61. R.M. Larsen and K.A. Thorsen, Influence of Sintering Atmosphere on Corrosion Resistance and Mechanical Properties of Sintered Stainless Steel, *PTM '93, Proc. Int., Conf.,* March 23–26, 1993 (Dresden, Germany), Verlag DGM – Informationsgesellschaft, Germany

62. G. Grützner, Sensitivity of Nitrogen-Alloyed Austenitic Chromium-Nickel Steels to Intergranular Corrosion Caused by Chromium Nitride Precipitation, *Stahl Eisen,* Vol 93, 1973, p 9

63. K. Frisk, A. Johanson, and C. Lindberg, Nitrogen Pickup During Sintering of Stainless Steel, *Advances in Powder Metallurgy and Particulate Materials,* J. Capus and R. German, Eds., Vol 3, Metal Powder Industries Federation, Princeton, NJ, 1992, p 167–179

64. R.H. Shay, T.L. Ellison, and K. Berger, Control of Nitrogen Absorption and Surface Oxidation of Austenitic Stainless Steels in H-N Atmospheres, *Progress in Powder Metallurgy*, Vol 39, H.S. Nayar, S.M. Kaufman, and E.E. Meiners, Ed., Metal Powder Industries Federation, Princeton, NJ, 1983, p 411–430

65. T. Mathiesen, "Corrosion Properties of Sintered Stainless Steels," Ph.D. thesis, Technical University of Denmark, 1993 (in Danish)

66. H.S. Nayar, R.M. German, and W.R. Johnson, The Effect of Sintering on the Corrosion Resistance of 316L Stainless Steel, *Modern Developments in Powder Metallurgy,* Vol 15, H.H. Hausner and P.W. Taubenblat, Ed., MPIF, Princeton, NJ, 1981

67. T. Mathiesen and E. Maahn, "Evaluation of Sensitization Phenomena in Sintered Stainless Steel," Powder Metallurgy World Congress (Paris, France), 1994

68. G. Grützner, "Über die interkristalline Korrosion stickstofflegierter 18/10 Chrom-Nickel-Stähle," ("On the Intergranular Corrosion of Nitrogen Alloyed 18/10 Chromium-Nickel Steels"), Ph.D. thesis, Technical University, Aachen, 1971

69. J.J. Eckenrod and C.W. Kovach, "Properties of Austenitic Stainless Steels and Their Weld Metals," in STP 679, ASTM, 1979, p 17

70. K.H. Moyer and W.R. Jones, How Argon Can Assist in Providing Clean Burn-Off of Lubricants and Binders, *Advances in Powder Metallurgy and Particulate Materials*, H. Ferguson, D. Whychell, Sr., Eds., MPIF, Princeton, NJ, 2000, Part 5, p 5-33/5-42

71. E. Klar, M. Svilar, C. Lall, and H. Tews, Corrosion Resistance of Austenitic Stainless Steels Sintered in Commercial Furnaces, *Advances in Powder Metallurgy and Particulate Materials,* J. Capus and R. German, Eds., Vol 5, MPIF, Princeton, NJ, 1992, p 411–426

72. P. Beiss, Control of Protective Atmosphere during Sintering, *Second European Symposium on Powder Injection Molding,* European Powder Metallurgy Association, 2000, p 147–157

73. R. Tandon and J. Johnson, Liquid-Phase Sintering, *Powder Metal Technologies and Applications,* Vol 7, *ASM Handbook*, ASM International, 1998, p 567

74. W. Karner, M.Y. Nazmy, and R. Arfai, *Werkst. Korros.,* Vol 31, 1980, p 446

75. O. Reen, U.S. Patent 4,032,336, 1977

76. B. Becker, J.D. Bolton, H. Preusse, and M. Youseffi, Application of Computer Modelling of Phase Equilibrium Diagrams to Stainless Steel Alloy Development, *Proc. of the 1998 Powder Metallurgy World Congress and Exhibition*, (Granada, Spain), Vol 3, European Powder Metallurgy Association, 1998, p 513–518

77. P.K. Samal and J.B. Terrell, Effect of Boron Addition on the Corrosion Resistance of P/M 316L Stainless Steel, *P/M Sci. Technol. Briefs,* Vol 3 (No. 3), 2001, p 18–22

78. A. Molinari, J. Kazior, F. Marchetti, R. Canteri, I. Cristofolini, and A. Tiziani, Sintering Mechanisms of Boron Alloyed AISI 316L Stainless Steel, *Powder Metall.*, Vol 37 (No. 2), 1994, p 115–122

79. A.M. Alper, *Phase Diagrams: Materials Science and Technology,* Academic Press, New York and London, 1970

80. A. Sharon and D. Itzhak, Corrosion Resistance of Sintered Stainless Steel Containing Nickel Based Additives, *Powder Metall.*, Vol 37, (No. 1), 1994, p 67–71

81. M. Nakamura, K. Kamada, H. Horie, and S. Hiratsuka, Corrosion Resistance of High Strength P/M Austenite Type Stainless Steels, *J. Jpn. Soc. Powder Metall.,* Vol 40, (No. 4), 1993

82. "Standard Specification for Powder Metallurgy (P/M) Boron Stainless Steel Structural Components," B 853–94, ASTM

83. H. Preusse, J.D. Bolton, and B.S. Becker, "Enhanced Sintering of 316L Stainless Steel with Additions of Cu_3P," 1995 European Conference on Advanced P/M Materials, Oct 23–25, 1995 (Birmingham) EPMA, Shrewsberry, UK

84. J.D. Bolton and H. Preusse, Corrosion Behaviour of 316L Stainless Steels Sintered to High Density through the Effects of a Phosphide Liquid Phase Sintering Aid, *Proc. of the 1998 Powder Metallurgy World Congress and Exhibition* (Grenoda, Spain), Vol 3, EPMA, 1998, p 401–406

85. W.F. Wang and Y.L. Su, *Powder Metall.*, Vol 29, 1986, p 177

86. M. Rosso and O. Morandi, Studies of Infiltration Applied to P/M Stainless Steel, *Proc. 1998 Powder Metallurgy World Congress and Exhibition* (Granada, Spain), Vol 3, European Powder Metallurgy Association, p 435–440

CHAPTER 6

Alloying Elements, Optimal Sintering, and Surface Modification in PM Stainless Steels

ECONOMIC PRODUCTION of sintered stainless steel parts often relies on sintered densities of 80 to 90% of theoretical density, with internal surfaces/porosity that can have negative effects on corrosion resistance as well as the crevice-sensitive region in a neutral saline environment. One basic question is how these negative effects can be ameliorated or eliminated without changing the density per se. Approaches include increasing a material alloy content and modifying its surface composition.

6.1 Alloying Elements

Increasing the alloy content of a sintered stainless steel, provided it is optimally sintered, will more or less follow the relationships that are valid for wrought and cast stainless steels. Figure 6.1 (Ref 1) illustrates schematically the effects of a number of elements used in stainless steels on the various electrochemical corrosion characteristics.

It is noted that silicon, with its various positive and negative effects during powder production (Chapter 3, "Manufacture and Characteristics of Stainless Steel Powders") and sintering (Chapter 5, "Sintering and Corrosion Resistance"), widens the passive range in the final part when present in solid solution rather than as an oxide.

Chromium, above its minimum content of approximately 12%, improves the passivation properties of a steel because it widens the range of potential and pH over which it forms a stable protective oxide layer.

So-called Pourbaix diagrams, derived from thermodynamic data, provide qualitative guidelines for the corrosion behavior of stainless steels. They show the extension of stable passivation and immunity in an aqueous environment for various combinations of electrochemical potential and pH. Figure 6.2 illustrates such diagrams for iron, chromium, and nickel.

By superimposing diagrams of individual elements, the combined effect of the constituents are obtained. Thus, in Fig. 6.2, chromium addition to iron extends the passive region of iron to

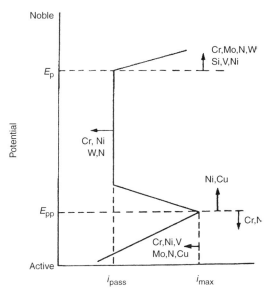

Fig. 6.1 Effects of alloying elements in stainless steels on the anodic polarization behavior. Source: Ref 1. © NACE International 1986

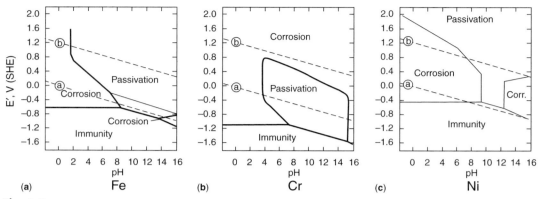

Fig. 6.2 Pourbaix diagrams for (a) iron, (b) chromium, and (c) nickel. Source: Ref 2.

lower pH values for intermediate values of E; the addition of nickel further extends the regions of immunity and passivation. Osozawa and Engell (Ref 3) show such relationships for various chrome-nickel steels.

In section 5.2.4 in Chapter 5, "Sintering and Corrosion Resistance," for wrought stainless steels, an empirical index of equivalent pitting corrosion (electro chemical potentiokinetic reactivation, or EPR is also called pitting resistance equivalent number, PREN), weighs the resistance contribution of chromium, molybdenum, and nitrogen to crevice and pitting corrosion, in accordance with EPR = Cr + 3Mo + 16N. Maahn et al. (Ref 4) have attempted to apply this index to some of their data on sintered austenitic stainless steels (Table 6.1) but found only a weak positive relationship between EPR and corrosion resistance as measured electrochemically (E_{stp}, E_{pit}) and by means of salt spray testing (NSS1 and NSS2) in a 5% NaCl solution.

The lack of such and other relationships with wrought stainless steels has led several investigators to conclude that underperformance in

sintered stainless steels was predominantly caused by crevice corrosion, due to the presence of pores. As mentioned earlier, however, underperformance in many cases can be attributed to metallurgical defects, detectable by metallographic analysis, arising from incorrect or suboptimal sintering. The corrosion data of optimally sintered austenitic stainless steels in Fig. 9.14 in Chapter 9, that is, the data marked "O" for optimized, do indeed give a very good relationship between EPR and corrosion resistance as measured in a 5% NaCl solution by immersion (Fig. 6.3), whereas the nonoptimized data of that figure (marked "N/O") fail to give such a relationship.

A similar positive relationship between pitting resistance equivalent and corrosion resistance applies to ferritic stainless steels (i.e., 410L, 434L, and 434L MOD) (Chapter 2, "Metallurgy and Alloy Compositions"), again, provided that sintering was performed under optimal conditions.

The corrosion resistance of SS-100, an austenitic stainless steel (20Cr, 18Ni, 5Mo), was

Table 6.1   Corrosion properties of sintered stainless steels produced from prealloyed powders(–100 mesh) with different alloy compositions

Sintering, °C/min	Type	Pitting resistance equivalent	E_{pit}[a], mV SCE	E_{stp}[b], mV SCE	NSS1, h	NSS2
1120/30/H₂	316L	25	500	225	600	8
	317L	30	725	725	>1500	9
	18-18-6[c]	37	275	275	48	4
	SS-100[d]	37	575	400	>1500	9
	17-25-8[e]	42	550	425	>1500	9
1250/120/H₂	316L	25	500	150	96	7
	317L	30	500	350	14	6
	18-18-6	37	450	275	50	5
	SS-100	37	>800	>800	>1500	10
	17-25-8	42	675	450	355	8

(a) 0.1% Cl⁻, pH 5, 30 °C (86 °F), 5 mV/min. (b) 0.1% Cl⁻, pH 5, 30 °C (86 °F), 25 mV/8 h. (c) 18.3% Cr, 18.3% Ni, 5.6% Mo, 1.7% Cu, 1.3% Sn, 0.78% Si, 0.23% Mn, bal Fe. (d) 20.0% Cr, 17.0% Ni, 5.0% Mo, 0.75% Si, <0.15% Mn bal Fe. (e) 16.3% Cr, 24.3% Ni, 7.7% Mo, 0.81% Si, 0.25% Si, bal Fe

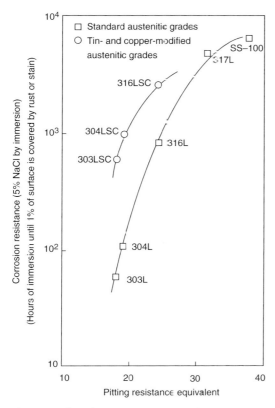

Fig. 6.3 Effect of pitting resistance equivalent on corrosion resistance (5% NaCl by immersion) of standard (□)and tin- and copper- (○) modified austenitic stainless steels (unpublished data)

found to be similar to wrought 316L when tested in 5% NaCl. The lack of Maahn et al.'s data to more clearly show the beneficial effects of increasing alloy content on EPR may be the result of suboptimal sintering. It may also be the result of the use of electrochemical testing that has shown its usefulness for wrought stainless steels but may require modification and/or complementary analysis when applied to sintered stainless steels, which may contain several metallurgical defects.

While not all applications of sintered stainless steels require optimal sintering to achieve a certain corrosion resistance, it is generally more economical to employ optimal sintering in combination with a stainless steel that has a lower content of costly constituents than to employ suboptimal sintering of a costlier stainless steel.

Alloy modification through the addition of fine powders, for example, chromium or molybdenum, is generally ineffective, even when sintering is done at relatively high temperatures, due to the lack of complete homogenization (Ref 5).

Other Alloying Elements. For sintered stainless steels, a number of alloying elements have been investigated, with the goal of modifying electrochemical corrosion characteristics and improving corrosion resistance in general. In most of these investigations, however, sintering conditions were not defined in sufficient detail to exclude suboptimal sintering. The effects of any alloying additions may not be the same as reported if sintering conditions were optimized. Among the investigated additives, copper (up to 9%) was found to improve corrosion resistance in H_2SO_4 and in chloride-containing environments. However, the copper addition also caused the sintered density to decrease (Ref 6). Peled and Itzhak (Ref 7) found the corrosion properties of sintered 316L in H_2SO_4 to improve by the addition of noble elements (copper, nickel, palladium, silver, gold, platinum). Gold, palladium, and platinum added to hot-pressed and sintered 316 stainless steel were observed to increase the oxide film thickness of the passive layer (Ref 8).

6.2 Optimal Sintering

Because of the importance of optimal processing and sintering, and the many processing combinations that can yield optimal results, the processing requirements for optimal results are summarized in Table 6.2, and the critical sintering requirements for optimal sintering of several austenitic stainless steels in a hydrogen atmosphere are given in Table 6.3.

For practical reasons, the amount of residual oxygen has been omitted in the definition of optimal sintering, that is, the difficult-to-reduce oxides originating from the water-atomizing process and consisting predominantly of SiO_2. It is clear, however, from the previous sections that a low oxygen content is beneficial to compacting properties as well as to corrosion resistance. Of particular interest is to what extent the negative effect of the crevice-sensitive density region (section 5.2.2 in Chapter 5, "Sintering and Corrosion Resistance") can be ameliorated through the control of residual oxygen, as well as to what extent corrosion resistance at low sintered densities is improved by low residual oxygen content. Also, exactly how do oxides and their morphologies, up to oxygen contents of 2000 to 3000 ppm, affect corrosion-resistance properties in full or nearly full-dense stainless steels?

Table 6.2 Processing requirements for optimal sintering of stainless steels

- No contamination with galvanic corrosion-causing materials (i.e., iron, low-alloy steels, 410L in austenitic stainless steels), unless the contaminants dissolve and homogenize during sintering (section 3.4 in Chapter 3). Both stainless steel powder and stainless steel green parts should pass the ferroxyl test (section 3.4 in Chapter 3).
- Sintering in hydrogen must be performed at a dewpoint that is reducing to $SiO_{2/316L}$ (Fig. 5.14 and section 5.2.3 in Chapter 5). Otherwise, excessive oxidation will result in a major deterioration of corrosion-resistance characteristics (as well as mechanical properties).
- Cooling after sintering in hydrogen must be fast enough to avoid reoxidation of the surface (section 5.2.3 in Chapter 5), which causes pitting.
- Carbon contents must be low enough (<0.02 to 0.03% for slowly cooled austenitic stainless steels) to prevent sensitization due to the precipitation of chromium carbides during cooling. For higher-carbon contents, cooling rates must be high enough to prevent sensitization due to the precipitation of chromium carbides (section 5.2.3 in Chapter 5), or, for carbon contents >0.03% and slow cooling rates, the stainless steel is stabilized with niobium to prevent sensitization.
- Sintering in nitrogen-containing atmospheres must be done at temperatures high enough to prevent Cr_2N formation during sintering (Fig. 5.44 and section 5.2.4 in Chapter 5). Otherwise, the steel will sensitize and suffer from intergranular corrosion.
- When sintering in a nitrogen-containing atmosphere, the cooling rate must be high enough to prevent sensitization and intergranular corrosion due to the precipitation of Cr_2N during cooling. Sintering in dissociated ammonia requires cooling rates from 200 °C/min (360 °F/min) to over 450 °C/min (810 °F/min), depending on dewpoint (section 5.2.4 in Chapter 5). Critical cooling rates for sintering in $90\%H_2$-$10\%N_2$ are lower (section 5.2.4).
- Sintering in vacuum must be performed under conditions that avoid chromium depletion of the surface through vaporization (section 5.2.5 in Chapter 5). Chromium losses impair the passivation characteristics of a steel.
- Sintered densities in the range from 6.7 to 7.2 g/cm^3 are to be avoided for low-alloy-content austenitics used or tested in a neutral chloride environment, because of their crevice sensitivity in that density range (section 5.2.2 in Chapter 5). Higher alloy content (for example, molybdenum, nitrogen) and low residual oxygen content reduce this problem, but more data are needed for reliable and specific recommendations.

Table 6.3 Approximate critical sintering parameters for optimal sintering of austenitic stainless steels in hydrogen

	Upper critical cooling temperature[a] at dewpoint of:				Critical cooling rate[b] at dewpoint of:			
	−40 °C (−40 °F)		−45 °C (−49 °F)		−40 °C (−40 °F)		−45 °C (−49 °F)	
Alloy	°C	°F	°C	°F	°C/min	°F/min	°C/min	°F/min
316L	1075	1967	1020	1868	400	720	250	450
316LSC	Similar to 316L				280	504	135	243
SS-100	Similar to 316L				175	315	45	81

(a) Derived from Fig. 5.32 in Chapter 5
(b) Derived from Fig. 5.33 in Chapter 5

For sintering in hydrogen and in vacuum, higher sintering temperatures and lower dewpoints bring about more complete reduction of oxides. For sintering in nitrogen-containing atmospheres, both strength and corrosion resistance benefit from dissolved nitrogen, provided that sintering is controlled to prevent sensitization by Cr_2N formation. Sintering atmospheres containing only 5 to 10% N, in combination with high sintering temperatures (>1205 °C or >2200 °F), minimize this danger. For tin-modified stainless steels, however, experimental data suggest that lower sintering temperatures of 1149 to 1177 °C (2100 to 2150 °F) are sufficient for achieving optimal corrosion resistance.

Relatively small improvements in dewpoint significantly lower the critical cooling rates necessary to avoid surface oxidation during cooling. Improving alloy content, either by increasing chromium and molybdenum (316L => SS-100) or by surface modification (316L => 316LSC), significantly lowers the critical cooling rates.

The approximate critical cooling rates in Table 6.3 were derived from Fig. 5.33 in

Chapter 5, and the critical cooling temperatures are from Fig. 5.32. Sintering times of up to 45 min were used in the derivation of the aforementioned data. The residual oxygen contents (not to be confused with oxygen due to reoxidation on cooling, which, when present, is usually much less) ranged from approximately 1400 to 2000 ppm. The general improvement of corrosion-resistance properties with increasing sintering intensity (i.e., time and temperature of sintering) suggests that lower residual oxygen contents in combination with optimal sintering will further improve corrosion-resistance properties of sintered stainless steels.

Due to a lack of quantitative data, particularly on the effects of dewpoint and cooling rate in nitrogen-containing sintering atmospheres upon nitrogen absorption and surface reoxidation during cooling and of chromium losses in vacuum sintering and their effects on corrosion resistance, it is not possible at present to make recommendations for these sintering atmospheres similar to those shown in Tables 6.2 and 6.3 for sintering in hydrogen. For example, the pitting resistance equivalents of the austenitic stainless steels of Fig. 9.14 in Chapter 9 that were sintered in dissociated ammonia are all several points higher than their hydrogen-sintered equivalents. Yet, their corrosion resistances are all suboptimal, undoubtedly because of the precipitation of Cr_2N during cooling, which masks and overshadows the beneficial effect of dissolved nitrogen. Only the highly alloyed and more forgiving SS-100 alloy approaches the optimal corrosion resistance of its hydrogen-sintered equivalent. The marked effect of the cooling rate, in agreement with the data in section 5.2.4 in Chapter 5, suggests that its optimal value in dissociated ammonia should be above that of the H_2-sintered material but also requires a faster cooling rate than that used in Fig. 9.14 in Chapter 9.

6.3 Surface-Modified Stainless Steels

The large surface areas associated with the porosity of sintered stainless steels are generally viewed as detrimental to mechanical and corrosion-resistance properties. There are, however, cases where these surfaces have been exploited to improve certain properties. The enrichment of water-atomized stainless steel powders with SiO_2 (section 3.1.3 in Chapter 3) is an example, the origin and significance of which has been recognized only recently. In this case, the function of silicon (via its oxidation) is to render the particle shape of the water-atomized powder irregular, so as to generate a compactible powder and to prevent excessive oxidation during atomization. More recently, other elements have been investigated with the goal of modifying the hydrogen overvoltage, the nature and composition of the passive film, and other electrochemical properties of stainless steels for the benefit of improved corrosion resistance. The best-known example is the tin enrichment of the surfaces of both austenitic stainless steel powders and parts (Ref 9, 10). When tin is added in amounts of 1 to 2% to the stainless steel powder prior to pressing, in powder form or, preferably, as an alloying constituent prior to water atomization, it appears enriched on the surface of the water-atomized powder as well as on the surfaces of sintered parts. In wrought or cast stainless steels, tin additions are kept below 0.3% because of the segregation of tin to the grain boundaries at higher concentrations as well as its negative effect on some mechanical properties. Auger surface analysis

data (Chapter 3 "Manufacture and Characteristics of Stainless Steel Powders") have shown the surface enrichment in stainless steel powders and parts to amount to 20%, with significant benefits to corrosion resistance and machinability properties. This type of surface modification is very attractive because of its low cost in comparison to other methods, such as chrome plating, ion implantation, laser processing, and so on. As pointed out earlier, compositional changes on the surface of a material, for a few atomic layers, are normal for any alloy. For water-atomized stainless steels, however, such changes can be much larger and extend into the bulk of the material for several hundreds of nanometers. It is particularly the elements with high oxygen affinity that tend to aggregate on the surfaces in the form of their oxides. In the case of tin, whose ΔG for oxide formation is not particularly large, it may also be atomic size and/or solubility characteristics that play a role. Figure 6.4 (Ref 11) shows the Auger composition-depth profile of a tin-modified (i.e., 1.5% Sn-containing) 316L stainless steel part sintered

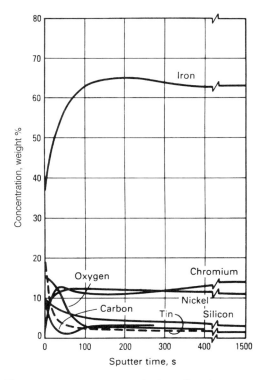

Fig. 6.4 Auger composition-depth profile of 1.5% Sn-containing 316L sintered in dissociated ammonia (–40 °C or –40 °F, dewpoint) at 1177 °C (2150 °F). Source: Ref 11. Reprinted with permission from MPIF, Metal Powder Industries Federation, Princeton, NJ

in dissociated ammonia. The most striking differences with a regular 316L part processed identically are the different profiles, near the surface, for nitrogen, chromium, and tin.

With regular 316L, nitrogen and chromium are enriched (nitrogen has a maximum on the surface and decreases toward the interior; chromium forms a peak near the surface); with modified 316L, tin is enriched on the surface, nitrogen is absent from the surface, and the chromium peak near the surface is only weakly present. Having described the problems with the presence of chromium and nitrogen on the surfaces of stainless steel parts, it comes as no surprise that these problems are greatly decreased with the tin-modified material. The beneficial effect of tin has been confirmed by several investigators (Ref 12–17). Tin may also form stable acid-resistant passive films in a crevice and may cause cathodic surface poisoning. However, for sintering in a nitrogen-containing atmosphere, its major beneficial effect is believed to lie in its formation of a barrier to nitrogen diffusion. This decreases the rate at which nitrogen is absorbed on the surface of the sintered part as it enters the cooling zone of the furnace. The 316LSC, a tin-copper-modified version of 316L, is significantly superior to plain 316L, regardless of whether sintering is performed in hydrogen, dissociated ammonia, or vacuum. For sintering in dissociated ammonia, good corrosion resistances are obtained even at the lower sintering temperatures of approximately 1149 °C (2100 °F). Also, cooling rate requirements for the avoidance of reoxidation (Fig. 5.33 in Chapter 5) are less severe than for plain 316L. These characteristics suggest that tin may lower the diffusion rate of both nitrogen and oxygen and may contribute to the formation of an improved passive film.

Another benefit of the presence of tin in sintered powder metallurgy (PM) stainless steels is its effect on machinability. Stainless steels are difficult to machine. So-called free-machining grades, for example, 303L, contain a small amount of sulfur; MnS powder is added to 304L stainless steel powder prior to compaction. However, in PM 303L, these additives exert a negative influence on corrosion-resistance properties. It is therefore of considerable practical interest that tin-modified 304L stainless steel was found to offer, in addition to its improved corrosion resistance, machinability performance similar to 303L. The latter is obtained with

the combined addition of 1% Sn plus 4% Cu (Ref 16).

In Fig. 6.3, the tin- (plus copper-) modified grades of 303L, 304L, and 316L form a second, improved corrosion-resistance curve versus EPR, which, for the lower-alloyed grades of 303LSC and 304LSC, shows an improvement in corrosion resistance by a whole order of magnitude but which, with increasing amount of alloying, appears to level off.

REFERENCES

1. A.J. Sedriks, Effects of Alloy Composition and Microstructure on the Passivity of Stainless Steels, *Corrosion,* NACE, Vol 42 (No.7), 1986, p 376–389
2. M. Pourbaix, *Atlas of Electrochemical Equilibria,* Pergamon Press, New York, 1966
3. K. Osozawa and H.-J. Engell, *Corros. Sci.,* Vol 6, 1966, p 389
4. E. Maahn, S.K. Jensen, R.M. Larsen, and T. Mathiesen, Factors Affecting the Corrosion Resistance of Sintered Stainless Steels, *Advances in Powder Metallurgy and Particulate Materials,* ed. C. Lall, A. Neupaver, Vol 7, MPIF, 1994, p 253–271
5. S.K. Jensen and E. Maahn, Microstructure of Sintered Stainless Steel Based on Mixed Powders, *PTM '93, Proc. Int. Conf.,* March 1993 (Dresden, Germany), Verlag DGM-Informationsgesellschaff, Germany
6. L. Fedrizzi, A. Molinari, F. Deflorian, L. Ciaghi, and P.L. Bonora, *Corrosion,* Vol 46, 1990, p 672
7. P. Peled and D. Itzhak, The Corrosion Behavior of Double Pressed, Double Sintered Stainless Steel Containing Noble Alloying Elements, *Corros. Sci.,* Vol 30 (No.1), 1990, p 59–65
8. P. Peled and D. Itzhak, The Surface Composition of Sintered Stainless Steel Containing Noble Alloying Elements Exposed to a H_2SO_4 Environment, *Corro. Sci.,* Vol 32 (No.1), 1991, p 83–90
9. T. Hisada, Japanese Patent 79-29285, 1977
10. E. Klar and M. Pao, U.S. Patent 4,420,336, 1983
11. D. Ro and E. Klar, Corrosive Behavior of P/M Austenitic Stainless Steels, *Modern*

Developments in Powder Metallurgy, ed. H. Hausner, H. Antes, G. Smith, Vol 13, MPIF, 1980, p 247–287

12. M.A. Pao and E. Klar, On the Corrosion Resistance of P/M Austenitic Stainless Steels, *Proceedings of the International Powder Metallurgy Conference* (Florence, Italy), Associazone Italiano di Metallurgia, 1982

13. S.K. Chatterjee, M.E. Warwick, and D.J. Maykuth, The Effect of Tin, Copper, Nickel, and Molybdenum on the Mechanical Properties and Corrosion Resistance of Sintered Stainless Steel (AISI 304L), *Modern Developments in Powder Metallurgy,* Vol 16, E.N. Aqua and C.I. Whitman, Ed., Metal Powder Industries Federation, 1984, p 277–293

14. M.A. Pao and E. Klar, Corrosion Phenomena in Regular and Tin-Modified P/M Stainless Steels, *Proceedings of the 1983 National Powder Metallurgy Conference, Progress in Powder Metallurgy* (New Orleans, LA), ed. H. Nayar, S. Kaufman, K. Meiners, Vol 39, MPIF, p 431–444

15. G. Lei, R.M. German, and H.S. Nayar, Corrosion Control in Sintered Austenitic Stainless Steels, Proceedings of the 1983 National Powder Metallurgy Conference, *Progress in Powder Metallurgy,* ed. H. Nayar, S. Kaufman, K. Meiners, Vol 39, MPIF

16. K. Kusaka, T. Kato, and T. Hisada, Influence of S, Cu, and Sn Additions on the Properties of AISI 304L Type Sintered Stainless Steel, *Modern Developments in Powder Metallurgy,* Vol 16, E.N. Aqua and C.I. Whitman, Ed., Metal Powder Industries Federation, 1984, p 247–259

17. D. Itzhak and S. Harush, The Effect of Sn Addition on the Corrosion Behavior of Sintered Stainless Steel, *Corros. Sci.,* Vol 25 (No.10), 1985, p 883–888

CHAPTER 7

Mechanical Properties

STAINLESS STEELS, wrought or powder metallurgy (PM), are primarily selected for their corrosion resistance and physical properties. Nevertheless, for a large majority of applications, it is essential that they offer reasonably good mechanical strength and ductility. The high-corrosion-resistant austenitic family and the low-cost ferritic family of alloys exhibit modest levels of mechanical strength that are satisfactory for many applications. Applications requiring higher strengths and wear resistance command the use of either a martensitic or a precipitation-hardening grade of stainless steel. While alloys from the former family are significantly lower in cost, the use of these alloys is limited because of their very low ductilities. Precipitation-hardening stainless steels offer unique combinations of high strength and ductility as well as good corrosion resistance. However, their chemistry and processing, which includes heat treatment, must be precisely controlled in order to develop optimal properties. In order to take full advantage of their high strength, PM versions of these alloys should essentially be processed to their full or near-full densities. Duplex stainless steels, which have a mixed microstructure of austenite and ferrite, exhibit properties that are some combinations of the properties of these two alloy families, both in terms of corrosion resistance and mechanical properties.

7.1 Strengthening Mechanisms in Stainless Steels

As discussed in Chapter 2, "Metallurgy and Alloy Compositions," the primary mechanisms of strengthening in the five families of stainless steel differ significantly from each other.

Ferritic and austenitic stainless steels do not benefit from strengthening by a second phase or by the austenite- (γ) to-martensite transformation, which is the most popular strengthening mechanism for carbon and low-alloy steels. Solid-solution strengthening is the primary mode of strengthening in these two alloy families. The major alloying element, chromium, is a mild solid-solution strengthener. Figure 7.1 shows the effects of various alloying elements on the Brinell hardness of ferrite (body-centered cubic iron) (Ref 1) that essentially result from the changes in the lattice parameter due to the presence of the solute atoms in the matrix. Molybdenum has a somewhat stronger effect as a solid-solution strengthener, compared to chromium. Medium-chromium ferritic stainless steels, such as 430L, 434L, and 434L-Modified, exhibit somewhat higher yield strengths over the low-chromium ferritic grades (409L, 409LE, and 410L), due to their higher levels of chromium and molybdenum.

In austenitic stainless steels, solid-solution strengthening also results from the change in the lattice parameter due to the presence of solute atoms. The presence of a solute atom in the matrix of a face-centered cubic (fcc) alloy also produces a secondary effect in the form of a change in its stacking fault energy. Table 7.1 lists the effect of various solute elements on the lattice parameter as well as on the yield and tensile strengths of annealed austenitic stainless steel type 302 (Ref 2). Nickel contents, in the range present in standard austenitic alloys (9 to 14%), not only have no positive effect on the yield strength but also lead to a slight lowering of the tensile strength. Nickel does, however, enhance the elevated-temperature yield and tensile strengths, as does molybdenum.

Austenitic and ferritic grades of stainless steel containing similar levels of chromium (and

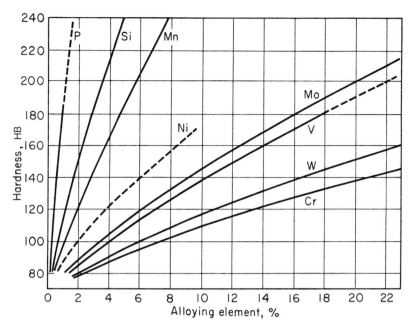

Fig. 7.1 Solid-solution hardening effect of various alloying elements in ferrite. Source: Ref 1

molybdenum) exhibit similar yield strengths. However, austenitic stainless steels exhibit a somewhat higher tensile strength compared to the ferritic grades, due to their higher rate of strain hardening (the slope of the plastic portion of the stress-strain curve is flatter for low-strain-hardening ferritic materials). Austenitic stainless steels, being fcc in structure, are more ductile than the ferritic stainless steels.

While it is not uncommon to alloy wrought austenitic stainless steels with several thousand parts per million of nitrogen to improve their strength, it requires caution to do so with PM austenitic stainless steels (section 5.2.4 in Chapter 5, "Sintering and Corrosion Resistance"). Only with very careful processing is it possible to add even relatively small amounts of nitrogen to PM austenitic stainless steels without the associated harmful chromium nitride precipitation. Nitrogen addition is not an option with ferritic stainless steels, due to the much lower limit of solubility of nitrogen in the ferrite matrix and, in the case of low-chromium ferritic alloys, their tendency to form martensite even with very low concentrations of nitrogen.

Martensitic stainless steels derive their strength from their severely strained and distorted body-centered tetragonal crystal structure, in which movement of dislocations becomes very difficult. In these alloys, the total amount of martensite-forming interstitials, carbon and

nitrogen, determines the relative amount of the martensitic grains formed and the resulting increase in strength of the alloy. In contrast to many low-alloy and standard martensitic steels, the martensitic grades of stainless steels are less sensitive to cooling rate (from austenitizing temperature), due to the strong hardenability effect of chromium. Martensitic stainless steels suffer from low ductility and toughness, the problem being more severe in low-density materials. This may necessitate tempering or annealing in some situations. Grade 410, containing approximately 0.1% C, is the most popular martensitic grade because it can be annealed without forming chromium carbides. Grade 420, which typically contains 0.30% C, does offer higher hardness, but when the material is annealed, the excess carbon leads to the formation of chromium carbides ($Cr_{23}C_6$). This results in a significant loss of chromium (~6 to 7%) from the alloy matrix, with attendant loss of corrosion resistance. This problem can be overcome by selecting a higher-chromium-containing grade, such as 440A, 440B, or 440C, that has sufficient chromium (~17%) to retain corrosion resistance even after forming chromium carbides. These alloys contain 0.6 to 1.0% C, and at these high carbon levels, carbides are still present in the material in the hardened condition. The presence of chromium carbides in a matrix of tempered martensite makes these

Table 7.1 Effect of alloying elements on the lattice parameter and strength of austenite in alloys approximating AISI type 302 stainless steel, in annealed condition

Solute	Type	Change in lattice parameter per at.%, nm	Strength coefficient for yield strength(a)	Strength coefficient for ultimate tensile strength(b)
C	Interstitial, austenite stabilizer	+0.00060	23	35
N	Interstitial, austenite stabilizer	+0.00084	32	55
Si	Substitutional, ferrite stabilizer	−0.00050	1.3	1.2
Nb	Substitutional, ferrite stabilizer	NA	2.6	5.0
Ti	Substitutional, ferrite stabilizer	NA	1.7	3.0
V	Substitutional, ferrite stabilizer	+0.00015	1.2	0.0
Mo	Substitutional, ferrite stabilizer	+0.00033	0.9	0.0
Cr	Substitutional, ferrite stabilizer	NA	0.2	0.0
W	Substitutional, ferrite stabilizer	+0.00033	0.3	0.0
Ni	Substitutional, austenite stabilizer	−0.00002	0.0	−0.1
Mn	Substitutional, austenite stabilizer	+0.00002	0.0	0.0
Cu	Substitutional, austenite stabilizer	+0.00023	0.0	0.0
Co	Substitutional, austenite stabilizer	−0.00004	0.0	NA

(a) Yield strength (in tons/in.^{2}) = 4.1 + Σ (element coefficient) (wt% element) + 0.16 (% ferrite) + 0.46 $d^{-1/2}$, where d is the mean grain diameter in millimeters. (b) Ultimate tensile strength (in tons/in.^{2}) = 29 + Σ (element coefficient) (wt% element) + 0.14 (% ferrite) + 0.82 $t^{-1/2}$, where t is the twin mean free path in millimeters. Source: Ref 2

materials comparable to tool steels with a hardness in the range of 55 to 60 HRC and yield strengths in the 650 to 1900 MPa (95 to 275 ksi) range.

Precipitation-hardening grades of stainless steel derive their strength from the presence of hard, submicroscopic precipitates that form in the matrix of the alloy upon heat treatment. Additional strengthening occurs in the martensitic and semiaustenitic versions of these alloys, due to the martensitic structure of the matrix. The precipitation-hardening mode of strengthening combines high strength with high

toughness and moderate levels of ductility. Thus, these grades are preferred over martensitic grades, from the mechanical properties point of view.

7.2 Factors Affecting Mechanical Properties of PM Stainless Steels

7.2.1 Porosity

Porosity in conventionally processed low-alloy PM steels accounts to a large extent for the differences in mechanical strength between wrought and PM components. However, for PM stainless steels, as seen in the following section, lack of control of the amounts of interstitials (carbon, nitrogen, oxygen) can overshadow the effects of porosity. Conventionally processed PM stainless steels typically reach sintered densities in the range of 6.6 to 7.3 g/cm^3, which translates to pore volumes in the range of 16 to 7%. Theoretical analyses of the effect of porosity on ductility and static mechanical properties, such as yield strength and tensile strength, have been made by a number of investigators. Figure 7.2 schematically shows the relative effect of porosity on a number of static and dynamic mechanical properties (Ref 3).

The adverse effect of porosity is generally more severe on the dynamic mechanical properties, such as fatigue and impact strength, and more so for brittle materials than for ductile materials, when compared to static mechanical properties.

In addition to the overall pore volume, the morphology of the porosity also influences mechanical properties. This includes the size distribution as well as the shape of pores. For the same total pore volume, many small pores are less detrimental than a few large ones.

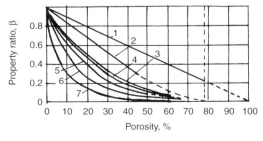

Fig. 7.2 Ratio of properties in porous iron as a function of porosity: 1, density; 2, electrical conductivity; 3, Young's modulus; 4, tensile strength; 5, fatigue limit for rotary bending; 6, elongation; 7, toughness. Source: Ref 3

Similarly, well-rounded porosity, which typically develops in high-temperature sintering, is less damaging than sharp, angular porosity. The detrimental effect of sharp, angular pores is much more pronounced in dynamic than in static mechanical properties. Sharp, angular pore surfaces become points of stress concentration and potential sites for crack initiation.

The loss in strength due to porosity is more pronounced than what may be estimated from the relative density, because of the stress-concentration effects at the pores. Prediction of the strength of a porous material is further complicated by the fact that the effect of stress concentration can vary from one material to another based on ductility. This is because ductile materials can better diffuse stress concentration by undergoing plastic deformation at points of stress (Ref 4).

Various investigators have proposed equations correlating yield strength of PM materials (σ) with their density (ρ).

German (Ref 5) has proposed:

$$\sigma = \sigma_0 K(\rho / \rho_T)^m \qquad \text{(Eq 7.1)}$$

where σ_0 is the wrought strength of the same alloy, ρ_T is the theoretical density, K is a geometric and processing constant similar to the

stress-concentration factor, and m gives the exponential dependence on density.

Salak and Miscovic (Ref 6) tried to correlate the results of a large number of experiments using sintered iron and found a formula proposed by Ryshkievich (Ref 4) to fit their results the best. Ryshkievich's formula is given by:

$$\sigma = \sigma_0 \, e^{-k\rho} \qquad \text{(Eq 7.2)}$$

where k is a coefficient depending on processing and testing conditions. It had a value of 0.043 for sintered iron, with $\sigma_0 = 443$ MPa (64.2 ksi).

Both exponential and linear dependence of strength on porosity are reported in the literature. Experimental data sometimes show a linear dependence of strength versus density, as is seen in the data by Kutsch et al. (Ref 7) for 316L stainless steel sintered in pure hydrogen and a hydrogen-nitrogen mixed atmosphere (Fig. 7.3a, b). It should be noted, however, that curve fitting of experimental data over a narrow density range may lead to the appearance of a linear relationship. Experimental data taken over extended density ranges almost always exhibit an exponential dependence on density.

Kutsch et al. (Ref 7) also noted that Young's modulus did not depend on the sintering atmosphere.

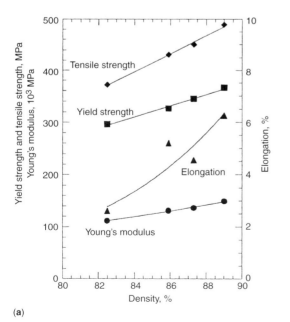

(a)

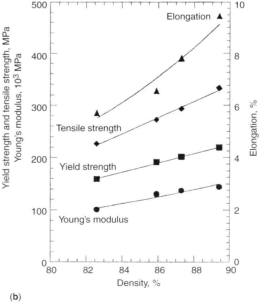

(b)

Fig. 7.3 Mechanical properties of (a) 70% N_2-30% H_2 and (b) 100% hydrogen-sintered 316L as a function of sintered density. Sintering was carried out at 1280 °C (2336 °F) for 20 min. Source: Ref 7

7.2.2 Sintering Atmosphere and Interstitial Content

Although interstitials, to the extent that they form separate phases, should be avoided or minimized from a corrosion-resistance point of view (Chapter 5, "Sintering and Corrosion Resistance"), they are nevertheless a fact of life and account to a large extent for the discrepancies in the mechanical properties data reported in the literature for PM stainless steels. Powder metallurgy processing is prone to wide variations in the interstitial contents (oxygen, carbon, and nitrogen). This is mainly the result of the sintering parameters employed, including the composition and dewpoint of the sintering atmosphere, lubricant, effectiveness of lubricant removal, sintering temperature, time, and cooling rate. If, however, the starting powders have widely differing oxygen contents, the quality of the powder will also be an important factor.

A high residual oxygen content after sintering will adversely affect ductility and mechanical strength. In an early study, Dautzenberg and Gesell (Ref 8) found the tensile strength of a hydrogen-sintered austenitic stainless steel to decrease significantly (from 390 to 300 MPa, or 56.6 to 43.5 ksi) when oxygen content increased from 2000 to 5000 ppm (section 3 1.3 in Chapter 3, "Manufacture and Characteristics of Stainless Steel Powders"). Tunberg et al. (Ref 9) compared the properties of vacuum-sintered 304L stainless steels, with and without graphite addition, using three sintering temperatures. They noted significant increases in tensile strength, ductility, and impact strength as a result of improved oxide reduction. Yield strength was found to be unaffected. The reason for this is probably that the improvement in strength was negated by an increase in grain size. Table 7.2 and Fig. 7.4(a–d) summarize these results. It may be noted that at the lowest sintering temperature employed (1120 °C, or 2048 °F), carbothermal reduction was less effective,

and hence, a higher residual carbon content led to increased yield strength.

Although the carbon content of L-grade PM stainless steels is specified to be less than or equal to 0.03%, it is not uncommon to find sintered parts containing higher amounts of carbon, in many cases as high as 0.07%. High carbon content will result in high strength and low ductility.

Samal et al. (Ref 10) observed that 434L compacted at 690 to 772 MPa (50 to 56 tsi) and sintered at a temperature below 1260 °C (2300 °F) contained higher levels of residual carbon and oxygen when compared with 434L compacted at 483 and 552 MPa (35 and 40 tsi) and sintered at a temperature above 1260 °C (2300 °F). Although all sets of samples had sintered densities in the range of 7.27 to 7.30 g/cm^3, the lower-temperature sintered samples had significantly lower ductility (elongation of 3 versus 16%) and impact strength (42 versus 138 J, or 31 versus 102 ft · lbf) compared to the higher-temperature sintered samples. In this study, alloy 409L, because of its niobium content, was affected to a much lesser extent by the variations in process parameters.

In comparison to oxygen and carbon, nitrogen is usually found to vary more widely in PM stainless steels, due to the fact that much sintering or cooling from sintering is still being carried out in dissociated ammonia. In the early years of sintering stainless steel, dissociated ammonia was often the atmosphere of choice because of its lower cost compared to hydrogen. The amount of nitrogen that a stainless steel absorbs during sintering is dependent on the nitrogen content of the atmosphere (based on Sievert's law), sintering temperature, and cooling rate as well as the chromium content of the steel (section 5.2.4 in Chapter 5, "Sintering and Corrosion Resistance"). Formation of chromium nitride, which commonly occurs during cooling in a nitrogen-bearing atmosphere, is highly undesirable from the corrosion-resistance and ductility points of view. However, if nitrogen is

Table 7.2 Interstitial contents of vacuum-sintered 304L, with and without graphite addition

Material	Admixed carbon, wt %	Sintering temperature		Sintered density, g/cm^3	Interstitial, wt%		
		°C	°F		C	N	O
304L	0	1120	2048	6.59	0.02	0.0236	0.274
304L	0	1200	2192	6.66	0.007	0.0105	0.265
304L	0	1250	2282	6.7	0.010	0.0061	0.271
304L+C	0.19	1120	2048	6.6	0.093	0.0201	0.165
304L+C	0.19	1200	2192	6.66	0.030	0.0071	0.090
304L+C	0.19	1250	2282	6.7	0.014	0.0042	0.048

Source: Ref 9

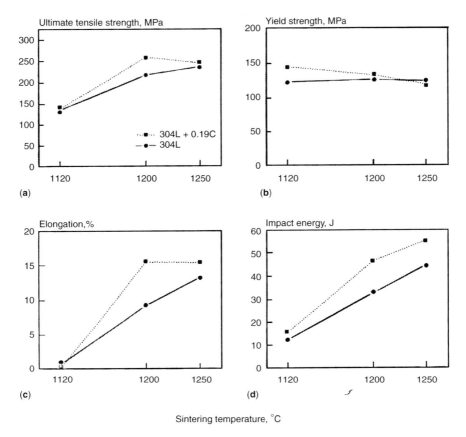

Fig. 7.4 (a–d) Mechanical properties versus sintering temperature of vacuum-sintered 304L, with and without carbon addition. Source: Ref 9

present in solution in the stainless steel matrix, it has a marked beneficial effect on its yield and tensile strengths as well as its corrosion resistance. How much of the total nitrogen is present in solution and how much is present in the form of chromium nitride depends on the composition of the sintering/cooling atmosphere, the sintering temperature, and the cooling rate (Ref 11, 12). Figure 7.3(a, b) compares the mechanical properties of PM 316L, sintered in a mixture of 70% N_2 and 30% H_2, with those sintered in pure hydrogen. Figure 7.5, based on data published by Dautzenberg (Ref 13), shows the effect of nitrogen content on the tensile strength and ductility of 304L austenitic stainless steel. Under most conditions of commercial sintering and cooling, it is not possible to keep more than approximately 1500 ppm of nitrogen in solution, with the excess nitrogen being present as chromium nitride. With careful processing, however, it is possible to minimize nitrogen absorption to less than 1500 ppm in a PM austenitic stainless steel, thus increasing

strength without risking formation of chromium nitride and attendant loss of corrosion resistance. In this regard, a good compromise can be reached by sintering in a 10% N_2-90% H_2 atmosphere, followed by rapid cooling. Samal et al. (Ref 12) determined the yield strength of 90% H_2-10% N_2 atmosphere-sintered 316L to be 17% higher than that of a hydrogen-sintered 316L, combined with improved corrosion resistance, provided that the cooling rate was 222 °C/min (400 °F/min) or higher from the sintering temperature of 1316 to 538 °C (2400 to 1000 °F).

Smith et al. (Ref 14) found the tensile strength of austenitic grades 304L and 316L to increase by 50% when the sintering atmosphere was changed from pure hydrogen to a 90-10 hydrogen-nitrogen mixture. Further increase in the concentration of nitrogen, up to 80%, did not lead to any additional increase in strength. It appears that the larger amounts of nitrogen absorbed during sintering in greater than 10% N_2-bearing atmospheres resulted mainly in the formation of chromium nitrides. Ductility and

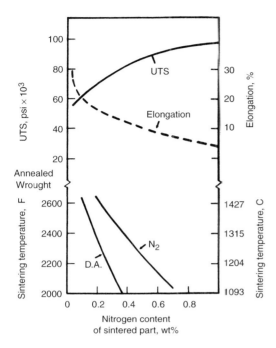

Fig. 7.5 Effect of temperature on the equilibrium nitrogen content of 304L in dissociated ammonia (D.A.) and nitrogen (bottom); and ultimate tensile strength (UTS) and elongation of 304L sintered in pure hydrogen at 1204 °C (2200 °F) and then nitrided in dissociated ammonia long enough to reach equilibrium nitrogen content (top). 316L has a similar response. Source: Ref 8

impact energy decreased steadily as the nitrogen content of the atmosphere was increased, starting with the 10% N_2-bearing atmosphere.

7.2.3 Sintering Temperature and Time

Mechanical properties of PM stainless steels are strongly influenced by the sintering temperature and, to a lesser degree, the sintering time. Figure 5.2 (Chapter 5, "Sintering and Corrosion Resistance") illustrates schematically the relative importance of sintering temperature for various properties of PM materials. At lower sintering temperatures, such as 1149 °C (2100 °F), reduction of surface oxides, interparticle bonding, and pore rounding progress at relatively slower rates. This results in relatively lower ductility and lower impact and tensile strengths. For a large majority of PM stainless steel applications, a sintering temperature in the neighborhood of 1232 °C (2250 °F) can provide good sintering, resulting in satisfactory ductility and impact and tensile strengths. Sintering in the neighborhood of 1316 °C (2400 °F) can provide further improvement in ductility and impact and

tensile strengths as well as high sintered densities. With the higher sintered densities and greater degree of pore rounding, the dynamic mechanical properties, such as fatigue, can be enhanced (Ref 15). Elevated-temperature sintering, such as at 1316 °C (2400 °F), also permits shorter sintering times, which in turn improves furnace throughput and process economy. At these high sintering temperatures, pore rounding is maximized. It must be also noted that pores present in PM materials tend to pin down grain boundaries and, as a result, resist any excessive or abnormal grain growth, except when the sintering time is unusually long. Any inclusions present in a material, including carbides and oxides, may also impede grain growth. At very high sintering temperatures, however, such as above 1343 °C (2450 °F), grain boundaries begin to detach themselves from the pores. Hence, unless the sintering time is kept reasonably short, the rate of grain growth can become rapid at these sintering temperatures, especially for ferritic stainless steels, and this can partially offset strengthening achieved from increased sintered density. Grain coarsening particularly lowers the yield strength of the sintered material. This effect is more frequently observed in ferritic stainless steels because of the higher rates of atomic mobility in these alloys. It is not uncommon to find a high-temperature-sintered PM stainless steel with a lower yield strength compared to its low-temperature-sintered counterpart having the same sintered density. An example of this effect is shown in Fig. 7.6 for hydrogen-sintered 304L. This effect is more common at high sintered densities because of the reduced dispersion of pores.

Inclusions present in the material in the form of stable carbides and oxides also act as barriers

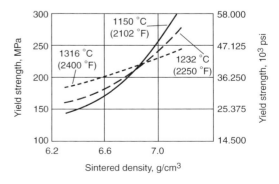

Fig. 7.6 Effect of sintering temperature on the room-temperature yield strength of hydrogen-sintered 304L (unpublished data)

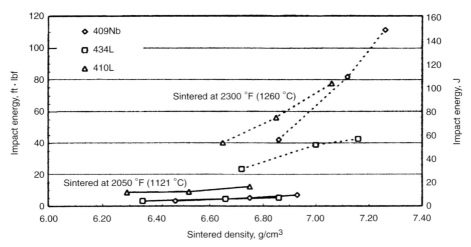

Fig. 7.7 Impact strength of three ferritic stainless steels as a function of sintering temperature and sintered density. Sintering atmosphere was hydrogen, and sintering time was 30 min. Source: Ref 16. Reprinted with permission from MPIF, Metal Powder Industries Federation, Princeton, NJ

to grain growth. Alloys that are relatively free from grain-pinning precipitates, such as 410L, 434L, and 304L, are therefore more prone to rapid grain growth during sintering at high temperatures. This is evidently the reason behind the fact that high-temperature-sintered 409L usually exhibits higher yield strength when compared with a similarly processed 410L at room and elevated temperatures. The significant effect of sintering temperature on the dynamic mechanical properties is illustrated in Fig. 7.7, taken from Rawlings et al. (Ref 16).

The effect of sintering time on mechanical properties is much less pronounced, except for sintering time periods of less than approximately 20 min. The actual time period during which the parts are at or close to the sintering temperature should be taken into account when considering the effects of sintering temperature and time on properties. Ambs et al. (Ref 17) found the mechanical strength of PM 316L sintered at 1232 °C (2250 °F) in dissociated ammonia to increase by 8% when sintering time was increased from 30 to 60 min. The increase in ductility was more significant, with tensile elongation increasing from 10 to 15%.

7.2.4 Thermal History and Cold Work

Unlike wrought stainless steels, the near-net shape PM products are rarely subjected to any cold work. Rather, they are most commonly supplied in the as-sintered condition, in which case the final thermal treatment is the sintering

process. In some instances, sintered parts are re-pressed (or sized) for the purpose of meeting dimensional tolerances. Re-pressing imparts some degree of cold work to the final product. When comparing the mechanical properties of PM materials with those of their wrought counterparts, one must take into consideration the thermal history and degree of cold work. Wrought stainless steel components are commonly formed by stamping, cold forming, extrusion, or machining of cold-rolled or cold-drawn stocks. As a result, they exhibit significantly higher strengths, compared to their strength in the fully annealed state.

Differences in the thermomechanical history between a wrought and a PM stainless steel also contribute to differences in their grain sizes. Depending on sintering temperature, time, atmosphere, interstitial content, and pore structure, PM stainless steels can have a wide range of grain sizes. Similarly, depending on their thermomechanical history, wrought stainless steels can have widely varying grain sizes. Differences in grain size between PM and wrought materials can contribute to differences in their mechanical properties.

Overall, the mechanical properties of PM stainless steels are strongly influenced by their sintered density, interstitial contents, sintering temperature, and, to a somewhat lesser extent, sintering time and grain size.

The presence of oxides (or other nonmetallic inclusions) is highly detrimental to the dynamic mechanical properties, especially for materials with high sintered densities. The effect of such inclusions is less pronounced in static mechanical properties. The presence of nonmetallic inclusions can significantly decrease the tensile elongation of high-density materials.

7.3 Mechanical Property Standards

Because sintered density, nitrogen content, and sintering temperature exert such a significant influence on the mechanical properties of PM stainless steels, it is customary to specify these parameters when reporting mechanical properties of sintered parts. Generally, vacuum- and hydrogen-sintered materials show similar properties. Both contain typically less than 150 ppm of nitrogen, and this level of nitrogen content has a negligible effect on the mechanical properties. It must be kept in mind, however, that if a hydrogen-sintering furnace is operated with nitrogen gas curtains, it is possible to pick up much higher amounts of nitrogen due to back-diffusion of nitrogen into the cooling zone. Similarly, in vacuum sintering, significant amounts of nitrogen can be picked up during cooling if nitrogen is used for backfill. Lack of specification of the amount of interstitials (i.e., oxides and Cr_2N) is the main reason for the large scatter in the dynamic mechanical properties reported in the literature.

The Metal Powder Industries Federation (MPIF), through its Materials Standards Committee, has taken the lead in the standardization of mechanical properties of PM stainless steels. It has developed and published as-sintered mechanical property data for the standard grades of PM stainless steels as a function of sintered density for selected sintering temperatures and atmospheres. Based on the sintering practice, each grade of stainless steel is designated as one of four material classes: N1, N2, L, and H. In addition, martensitic 410, processed with carbon addition, sintered in dissociated ammonia, and then heat treated, is designated as 410-HT. A detailed description of the classification system is covered in section 2.2 in Chapter 2, "Metallurgy and Alloy Composition." Briefly, each class of material is specified as follows:

- N1: Dissociated ammonia sintering at 1149 °C (2100 °F)—for material classes 303N1, 304N1, 316N1, and 410-HT
- N2: Dissociated ammonia sintering at 1288 °C (2350 °F)—for material classes 303N2, 304N2, 316N2, 430N2, and 434N2

- L: Vacuum (partial) sintering at 1288 °C (2350 °F)—for material classes 303L, 304L, 316L, 410L, 430L, and 434L
- H: Hydrogen sintering at 1149 °C (2100 °F)— for material classes 304H and 316H

Within a given class, a number of material codes are assigned each code, specifying a minimum strength that can be obtained by sintering to a target density. Thus, a material code represents an alloy, a set of designated sintering parameters, and a nominal sintered density. For each material code, MPIF standard 35 (volume entitled *Materials Standards for PM Structural Parts*) specifies minimum values of yield strength and elongation and simply lists the typical values for ultimate tensile strength, hardness, and impact energy. The two-digit suffix of the code signifies the minimum yield strength (in 10^3 psi). This system of material designation gives the parts fabricator some flexibility in selecting different process routes, as long as minimum specified yield strength and elongation values are met. The MPIF published standard properties are held as important benchmark properties for PM part design and use. Appendixes 1 and 2 list, respectively, the properties of 300- and 400-series PM stainless steels as published in MPIF standard 35, 2007 edition. In recent years, ASTM International has also started to include mechanical properties of PM stainless steels in its standards. Data published by ASTM International are very similar to those published by MPIF, because both organizations share the same data bank.

7.4 Room-Temperature Mechanical Properties

7.4.1 Static Mechanical Properties

Prior to the late 1980s, the focus of most technical publications on PM stainless steels was on the identification of factors affecting corrosion resistance. Little emphasis was placed on the determination of mechanical properties. Mechanical properties data provided by some powder producers, in the form of product literature, made up the major source for such information. In addition, much of the data published prior to the mid-1990s was expressed as a function of green density (or compaction pressure) rather than as a function of sintered density. In 1992, Sanderow and Prucher, working under grants from MPIF and the U.S. Navy, carried out a systematic study of the mechanical as well as corrosion properties of PM stainless steels. The mechanical property data from the study were published beginning in 1994 (Ref 18) and formed the basis for MPIF standard 35 (Appendixes 1 and 2). At approximately the same time, SCM Metal Products Inc. (currently North American Hoganas) published an extensive product guide covering mechanical properties of various grades of PM stainless steel expressed as functions of sintered density, sintering temperature, and sintering atmosphere.

Also in the mid-1990s, the 400-series stainless steels were first considered for use as exhaust flanges in U.S.-made automobiles. This being a structural application, it warranted development of mechanical property data based on a wide range of processing routes and covering a number of PM stainless steels, most being ferritic grades.

In an attempt to make the reader aware of the wide range of process parameters used in industry and the extent of variations in mechanical properties that can be expected to result from these partly controlled and partly uncontrolled or unspecified process variables (i.e., amount of interstitials, dewpoint of sintering atmosphere, cooling rate, oxygen content of starting powder, etc.), a compilation of published data is presented in Table 7.3, covering the austenitic grades, and in Table 7.4, covering the ferritic grades.

The PM martensitic stainless steels produced by conventional pressing and sintering fall into two categories: high-interstitial (carbon- and/or nitrogen-rich) alloys and nickel-containing low-interstitial alloys. Both types of alloys typically contain 10 to 13% Cr. With most processing, both of these do contain significant amounts of ferrite as well. The structure and characteristics of the martensites that form in these two types of alloys differ significantly from each other. It is possible to develop a high-interstitial-based martensitic material by sintering a standard low-carbon, low-chromium ferritic material ($\leq$13% Cr) in a nitrogen-rich atmosphere. The depth of the martensitic layer and its hardness can vary widely, depending on green density, composition and temperature of the sintering atmosphere, and the cooling rate. Materials containing 0.10 to 0.20% C and/or nitrogen are designated 410, while those containing 0.25 to 0.35% C and/or nitrogen are designated as the 420 stainless steel. Svilar and Ambs (Ref 26)

observed that nitrogen and carbon have a similar effect in terms of hardening and microstructure in 410 and 420 PM martensitic steels. Generally, ductility and toughness of the interstitial-based martensitic materials in the as-sintered condition are quite low, rendering these less suitable for load-bearing applications. Heat treatment (in the form of a low-temperature tempering) is beneficial in improving toughness. Table 7.5 lists the room-temperature mechanical properties of martensitic grades of PM stainless steels, including those of nickel-modified, low-interstitial martensitic material, namely 409LNi. Its microstructure comprises 50% martensite and 50% ferrite. Table 7.6 lists the properties of interstitial-based martensitic materials in their heat-treated condition.

A much tougher PM-based martensitic material can be produced via liquid-phase sintering, with its composition and properties approximating those of wrought 440C stainless steel. These alloys often contain a small amount of boron, in addition to high carbon content, in order to facilitate liquid-phase formation. A suitable heat treatment assures optimal yield strength and ductility. A number of martensitic stainless steels, including 440C, are also processed via the metal injection molding route, with mechanical properties comparable to those of their wrought counterparts.

The PM precipitation-hardening stainless steels are commonly processed via metal injection molding in order to realize the full benefit of their high strength. Table 7.7 (Ref 28) lists the properties of a conventionally processed 17-4PH alloy in the heat treated condition.

7.4.2 Fatigue Properties

There have been a significant number of studies relating to performance of PM materials under cyclic loading. However, only a small number of these studies have included PM stainless steels. Sintered density is the most important material variable influencing fatigue strength as well as other dynamic mechanical properties, such as impact energy. Other material variables of importance include pore structure and cleanliness of microstructure, that is, freedom from oxides, carbides, nitrides, and nonmetallic inclusions. Sintered components having fine and rounded pores exhibit superior dynamic properties, compared to those having coarse and angular pores.

Table 7.3 Room-temperature mechanical properties of austenitic grades

Alloy	Sintered density, g/cm³	Sintering temperature °C	°F	Sintering atmosphere(a)	Ultimate tensile strength MPa	ksi	Yield strength MPa	ksi	Elongation, %	Hardness, HRB	Impact energy J	ft-lbf	Reference
303L	6.36	1121	2050	DA	295.2	42.8	252.4	36.6	1.5	59	7.9	5.8	18
303L	6.51	1316	2400	DA	376.6	54.6	270	39	8.1	60	34.6	25.5	18
303L	6.61	1288	2350	Vacuum	274.5	39.8	145.5	21.1	16.9	25	57.4	42.3	18
303L	6.90	1316	2400	DA	470	68.1	311.7	45.2	12.6	72	54.2	40	18
303L	6.92	1288	2350	Vacuum	333.8	48.4	165.5	24	19.9	40	80.4	59.3	18
303L	6.71	1149	2100	DA	317.3	46	245.5	35.6	7	60	...	...	19
303L	6.82	1316	2400	H₂	325.0	47.1	144.0	20.9	25.1	29	...	...	20
303LSC	6.87	1316	2400	H₂	324.0	47.0	157.0	22.7	21.5	31	...	...	20
304L	6.45	1121	2050	DA	345.5	50.1	280.7	40.7	2.8	60	9.9	7.3	18
304L	6.42	1316	2400	DA	436	63.2	297.3	43.1	10.7	60	40.7	30	18
304L	6.85	1316	2400	DA	525	76.2	338.6	49.1	15.8	72	80	59	18
304L	6.50	1288	2350	Vacuum	309.7	44.9	160.7	23.3	19.8	30	70.1	51.7	18
304L	6.92	1288	2350	Vacuum	393.1	57	184.8	26.8	26.9	52	127.0	94.3	18
304L	7.01	1316	2400	DA	413.8	60	299.3	43.4	...	...	...	...	21
304L	7.06	1316	2400	H₂	403.5	58.5	198	28.7	25	...	...	...	21
304L	6.61	1149	2100	DA	350.4	50.8	272.4	39.5	5.5	63	...	...	19
304L	6.93	1250	2280	H₂	328.3	47.6	...	...	16.5	...	78.6	58	14
304L	6.82	1250	2280	90H₂/10N₂	471.8	68.4	148.0	21.6	8	30	40.7	30	14
304L	6.78	1316	2400	H₂	338.0	49.0	148.0	21.6	27	30	...	...	20
304LSC	6.76	1316	2400	H₂	311.0	45.1	152.0	22.0	22	25	...	...	20
316L	6.44	1121	2050	DA	327.6	47.5	271	39.3	2	61	9.8	7.2	18
316L	6.50	1316	2400	DA	426.9	61.9	298	43.2	9.9	58	37.7	27.8	18
316L	6.81	1316	2400	DA	482.8	70	314.5	45.6	14.8	66	65.5	48.3	18
316L	6.53	1288	2350	Vacuum	281.4	40.8	155.2	22.5	17.3	26	53.3	39.3	18
316L	6.91	1288	2350	Vacuum	404.9	58.7	215.9	31.3	21.6	50	88.1	65	18
316L	6.60	1316	2400	H₂	333.1	48.3	160.7	23.3	22	...	...	...	12
316L	6.57	1316	2400	DA	374.5	54.3	192.4	27.9	17.5	...	...	...	12
316L	6.58	1316	2400	90H₂/10N₂	351.7	51	189.7	27.5	19	...	...	...	12
316L	6.61	1177	2150	DA	344.9	50	296.6	43	4	60	...	...	19
316L	6.97	1177	2150	DA	386.9	56.1	322	46.7	6.5	68	...	...	19
316L	6.72	1149	2100	DA	355.9	51.6	275.9	40	5.5	62	...	...	19
316L	7.07	1316	2400	H₂	375.9	54.5	200	29	21	...	...	...	21
316L	7.11	1316	2400	DA	509.7	73.9	310.4	45	12	...	...	...	21
316L	6.93	1250	2280	H₂	328.3	47.6	...	...	19	40	67.8	50	14
316L	6.88	1250	2280	90H₂/10N₂	472.4	68.5	...	...	10	64	51.5	38	14

Table 7.4 Mechanical properties of powder metallurgy ferritic stainless steels

Grade	Sintered density, g/cm³	Sintering temperature °C	°F	Sintering atmosphere	Ultimate tensile strength MPa	ksi	Yield strength MPa	ksi	Elongation, %	Hardness, HRB	Impact energy J	ft·lbf	Reference
409L	7.17	1260	2300	H₂	374.5	54.3	220.0	31.9	21.0	NA	170	126	21
409L	7.1	1304	2380	H₂	358.6	52.0	193.1	28.0	18.0	57	115	85	22
409L	7.25	1304	2380	H₂	379.3	55.0	220.7	32.0	25.0	60	169	125	22
409L	7.1	1316	2400	H₂	358.6	52.0	189.0	27.4	NA	NA	88	65	23
409L	7.26	1366	2490	H₂	373.0	54.1	214.0	31.0	17.0	NA	163	120	10
409L	7.27	1321	2410	H₂	366.0	53.1	212.0	30.7	16.0	NA	146	108	10
409L	7.25	1271	2320	H₂	357.0	51.8	209.0	30.3	17.0	NA	136	100	10
409L	7.26	1238	2260	H₂	377.0	54.7	208.0	30.2	9.0	NA	104	77	10
409L	7.3	1316	2400	H₂	372.0	53.9	211.0	30.6	32.0	57	NA	NA	20
409L wrought	7.75	...	...	...	408.3	59.2	234.5	34.0	...	...	...	...	21
410L	6.94	1288	2350	Vacuum	343.5	49.8	198.0	28.7	19.8	50	83.1	61.3	18
410L	7.19	1260	2300	H₂	389.7	56.5	319.3	46.3	18.0	NA	NA	NA	21
410L	7.1	1304	2380	H₂	344.8	50.0	206.9	30.0	20.0	50	115	85	22
410L	7.25	1304	2380	H₂	358.6	52.0	220.7	32.0	25.0	55	169	125	22
410L	7.1	1316	2400	H₂	379.3	55.0	186.2	27.0	...	...	81	60	23
410L	6.96	1250	2280	H₂	300.0	43.5	...	...	17.5	39	98	72	14
430L	7.08	1121	2050	DA	413.1	59.9	230.4	33.4	7.0	64	34.8	25.7	18
430L	6.89	1288	2350	Vacuum	341.4	49.5	212.4	30.8	18.0	40	84	62	18
430L	7.17	1288	2350	Vacuum	383.5	55.6	239.3	34.7	24.2	62	146	108	18
430L	6.93	1250	2280	H₂	300.0	43.5	...	...	14.2	55	65	48	14
430L	6.88	1250	2280	90H₂/10N₂	345.5	50.1	...	...	7.5	62	41	30	14
434L	7.09	1316	2400	DA	428.3	62.1	246.9	35.8	10.3	68	22.1	16.3	18
434L	7.25	1316	2400	DA	460.7	66.8	257.9	37.4	17.2	73	24.8	18.3	18
434L	7.06	1288	2350	Vacuum	377.3	54.7	251.1	36.4	18.7	57	102	75	18
434L	7	1200	2190	H₂	358.6	52.0	206.9	30.0	10.0	NA	NA	NA	24
434L	7.2	1290	2355	H₂	400.0	58.0	234.5	34.0	16.0	NA	NA	NA	24
434L	7.2	1316	2400	H₂	386.2	56.0	220.7	32.0	NA	NA	108	80	25
434L	7.11	1260	2300	H₂	404.8	58.7	264.8	38.4	22.0	NA	130	96	21
434L	7.28	1360	2480	H₂	402.0	58.3	246.0	35.6	16.0	...	137	101	10
434L	7.29	1316	2400	H₂	405.0	58.7	248.0	36.0	16.0	...	146	108	10
434L	7.29	1260	2300	H₂	477.0	69.2	277.0	40.2	7.0	...	129	95	10
434L	7.29	1227	2240	H₂	512.0	74.3	329.0	47.7	3.0	...	42	31	10
434L	7.1	1304	2380	H₂	372.4	54.0	220.7	32.0	18.0	60	108	80	22
434L wrought	7.75	...	...	...	379.3	55.0	262.1	38.0	50.0	...	...	...	24

(a) DA, dissociated ammonia

Table 7.5 Room-temperature mechanical properties of powder metallurgy stainless steels in the as-sintered condition

Base alloy	C addition, wt%	Sintering temperature °C	°F	Sintered density, g/cm³	Sintering atmosphere(a)	%N₂	Ultimate tensile strength MPa	ksi	Yield strength MPa	ksi	Elongation, %	Hardness, HRC	Reference
Fe-12Cr	0	1135	2075	6.5	DA	0.3	469	68	NA	NA	0.5	23	26
Fe-12Cr	0.15	1135	2075	6.5	DA	0.26	552	80	NA	NA	0.5	24	26
Fe-12Cr	0.3	1135	2075	6.5	DA	0.34	538	78	NA	NA	0.5	27	26
Fe-12Cr	0	1232	2250	6.8	DA	0.16	655	95	579	84	1	30	26
Fe-12Cr	0.15	1232	2250	6.8	DA	0.17	910	132	827	120	0.5	30	26
Fe-12Cr	0.3	1232	2250	6.8	DA	0.16	848	123	848	123	0.5	31	26
Fe-12Cr	0	1232	2250	6.9	H₂	<0.01	221	32	228	33	5	NA	Not martensitic 26
Fe-12Cr	0.15	1232	2250	6.9	H₂	<0.01	690	100	552	80	1.5	27	26
Fe-12Cr	0.3	1232	2250	6.9	H₂	<0.01	889	129	827	120	1	28	26
Fe-12Cr	0	1250	2282	6.9	90H₂/10N₂	NA	552	80	NA	NA	2	88 HRB	14
Fe-12Cr	0	1121	2050	6.57	DA	<0.01	550.9	79.9	NA	NA	0.5	30	18
Fe-11Cr-1.2Ni	0	1330	2425	7.3	H₂	<0.01	600	87	490	71	8.5	87 HRB	37

(a) DA, dissociated ammonia

For PM steels, Schatt and Wieters (Ref 29) have postulated a linear relationship between density and fatigue strength up to a density of 7.50 g/cm³, and thereafter, the increase in fatigue strength is noted to be asymptotic.

The role of porosity in fatigue crack initiation and propagation has been studied by a number of researchers. According to Lindsted et al. (Ref 30), the presence of pores leads to rapid initial strain hardening as the plastic zones

Table 7.6 Mechanical properties of powder metallurgy martensitic alloys in the heat-treated condition

Base alloy	C addition	Sintering temperature °C	°F	Sintered density, g/cm³	Sintering atmosphere(a)	%N₂	Ultimate tensile strength MPa	ksi	Yield strength MPa	ksi	Elongation, %	Hardness, HRC	Particle hardness	Reference
Fe-12Cr	0.15	1232	2250	6.9	H₂	<0.01	827	120	724	105	1.5	30	609 VPN, 55 HRC	26
Fe-12Cr	0.15	1232	2250	6.8	DA	0.17	896	130	827	120	0.5	32	...	26
Fe-12Cr	0.30	1232	2250	6.9	H₂	<0.01	965	140	896	130	0.8	33	628 VPN, 56 HRC	26

Heat treatment: Vacuum heat treated at 1010 °C (1850 °F) for 1 h; oil quenched, followed by 315 °C (600 °F) air temperature. (a) DA, dissociated ammonia

Table 7.7 Room-temperature mechanical properties of conventionally processed 17-4PH alloy, heat treated to H900 condition in nitrogen

Alloy	Density, g/cm³	Sintering temperature °C	°F	Sintering atmosphere	Ultimate tensile strength MPa	ksi	Yield strength MPa	ksi	Elongation, %
17-4PH	7.45	1260	2300	100% H₂	1172	170	724	105	7
17-4PH	7.30	1260	2300	100% H₂	1030	150	655	95	4

Source: Ref 28

around pores become strain hardened at relatively low strain levels, which is then followed by a more gradual plastic zone growth. This is in contrast to pore-free materials, which undergo strain hardening much more homogeneously. When a crack is induced near a pore, it grows rapidly through the plastic zone created by the pore and then slows down when it reaches the less strained matrix. In this context, the rate of work hardening of the alloy is an important factor in determining crack initiation. Austenitic stainless steels work harden more rapidly compared to ferritic stainless steels and thus are expected to undergo fatigue crack initiation sooner than the ferritic stainless steels. The second most important role of porosity is the stagnation of crack growth due to pore-crack interaction. When the crack reaches a pore, stress concentration at the crack tip is released, and a new blunted crack tip is created at the other side of the pore. The crack may stop growing until there is sufficient stress concentration on the other side of the pore. This interpretation suggests a dynamic and pulsating crack growth behavior. Lindstedt et al. (Ref 30) examined the fatigue mechanism in PM 316L austenitic stainless steels by comparing single-press/single-sinter, double-press/double-sinter, and hot isostatic pressed (HIPed) materials having final densities of 6.9, 7.2, and 8.0 g/cm³, respectively. They confirmed the pore-crack-linking phenomenon from the mean surface crack size measurements in the single-press/single-sinter samples.

In the case of most wrought steels, the fatigue endurance ratio (FER) (fatigue endurance limit ÷ tensile strength) falls predictably in the range of 0.38 to 0.50. For PM steels, however, the FER can vary widely. O'Brian (Ref 31) determined the FER of sintered steels to vary from 0.16 to 0.47. His study did not include stainless steels.

Sanderow et al. (Ref 15) carried out a comprehensive study of fatigue behavior of both 300- and 400-series stainless steels as a function of sintering temperature, sintering atmosphere, and sintered density. Fatigue data from this study were used for developing the MPIF standard 35 material specifications (Appendixes 1 and 2). This study showed a larger scatter in the data for austenitic grades compared to ferritic grades. Also, high-nitrogen-containing austenitic materials showed generally lower FER values (typically 0.32) compared to the low-nitrogen-containing (vacuum-sintered) materials (typically 0.40). Grade 304L showed higher fatigue strength compared to similarly processed 316L and 303L. Work done by Genest et al. (Ref 32), using a hydrogen-nitrogen sintering atmosphere, also showed a relatively low FER of 0.29 for 316L of a sintered density of 6.95 g/cm³. Their samples contained approximately 0.35% N₂, the same as found in N1 and N2 sintered austenitic stainless steels in the Sanderow et al. (Ref 15) study. Figure 7.8 shows the stress-number of cycles (S-N) curve for 316L determined by Genest et al. Figures 7.9 and 7.10 show the S-N curves for austenitic and ferritic grades as determined in the Sanderow et al. study.

In the Sanderow et al. study, ferritic stainless steels showed relatively higher and more consistent FERs, with an average value of 0.46. Figure 7.11 compares the fatigue and tensile strength data for standard 400-series materials, including those for 409LE and 409LNi developed by Shah

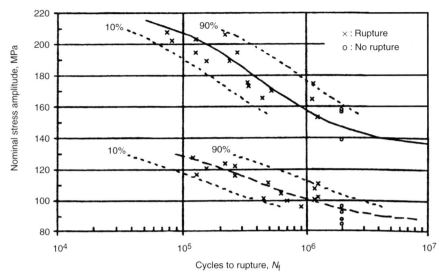

Fig. 7.8 Effect of sintered density on the fatigue strength of 316L sintered in 93% H_2 + 7% N_2 atmosphere at 1290 °C (2354 °F). Sintered densities were 6.31 (dashed line) and 6.95 (solid line) g/cm³ with a stress ratio R = 0.06 and test frequency at 30 Hz. Source: Ref 32. Reprinted with permission from MPIF, Metal Powder Industries Federation, Princeton, NJ

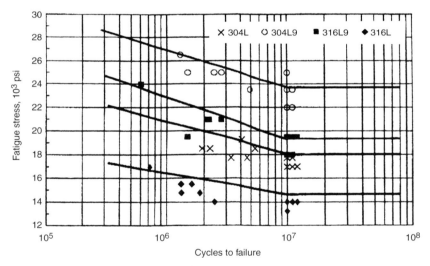

Fig. 7.9 Fatigue curves for vacuum-sintered 304L and 316L as a function of sintered density. Sintered densities of 304L and 316L were 6.51 and 6.54 g/cm³, respectively. Sintered densities of 304L9 and 316L9 were 6.90 and 6.89 g/cm³, respectively. Sintering temperature was 1288 °C (2350 °F). Source: Ref 15. Reprinted with permission from MPIF, Metal Powder Industries Federation, Princeton, NJ

et al. (Ref 33) based on high-temperature hydrogen sintering. An examination of these data indicates that the fatigue strengths of the ferritic grades are governed largely by their sintered densities, and the composition of the alloy has only a minor influence. Sintering in nitrogen-bearing atmospheres lowers FER by a small extent.

It appears that prediction of minimum fatigue strength from tensile strength using a value of

0.38 for FER is only feasible for PM 400-series stainless steels having sintered densities at or above 7.0 g/cm³. Such prediction is not feasible for PM austenitic stainless steels. The lower FER of austenitic stainless steels may be attributed to their high rate of strain hardening, as discussed earlier. Sintering of austenitic stainless steels in a nitrogen-bearing atmosphere further lowers FER.

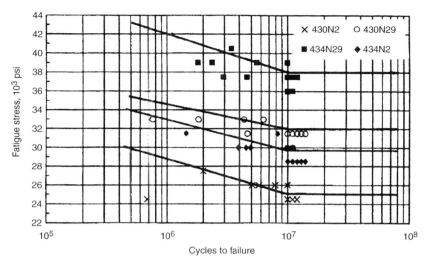

Fig. 7.10 Fatigue curves for two dissociated-ammonia-sintered ferritic stainless steels; parenthetical. Sintered densities of 403N2 and 434N2 were 7.04 and 7.07 g/cm³, respectively. Sintered densities of 430N29 and 434N29 were 7.27 and 7.24 g/cm³, respectively. Sintering temperature was 1316 °C 2400 °F). Source: Ref 15. Reprinted with permission from MPIF, Metal Powder Industries Federation, Princeton, NJ

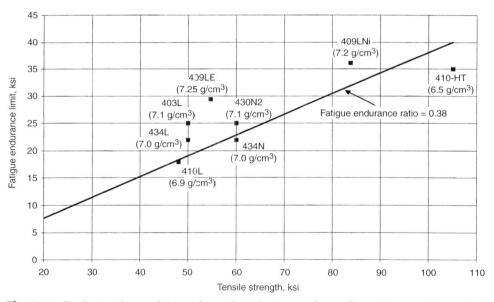

Fig. 7.11 Tensile strength versus fatigue endurance limit of various powder metallurgy 400-series stainless steels. Source: Ref 33. Reprinted with permission from SAE Paper 03M-315 ©2003 SAE International

7.5 Elevated-Temperature Mechanical Properties

7.5.1 Static Mechanical Properties

In 1983, Grinder and Zhiqiang (Ref 34) published elevated-temperature tensile strength data on full-dense PM 304L and 316L, made via HIP/extrusion as a part of a study on the effect of oxide inclusions on the hot workability of full-dense PM austenitic stainless steels. With the exception of these, there had been practically no published data on the elevated-temperature mechanical properties of PM stainless steels until approximately 1997. This reflects the absence of applications for PM stainless steels involving service temperatures much higher than room temperature in the earlier years. In the mid-1990s, with the introduction of PM stainless steel components for automotive exhaust applications, interest grew in the development of elevated-temperature mechanical properties of conventional PM stainless steels. The service temperatures of these components—flanges and hot exhaust gas outlet bosses—were estimated to be fairly high, in the range of 650 to 870 °C (1200 to 1600 °F). In addition to the requirement that these structural components retain their integrity at these high temperatures, it was considered that their elevated-temperature yield strength would be indicative of their resistance to stress relaxation. In the exhaust flange application, the retention of the clamping force (for leak tightness) of the flange is dependent on its resistance to stress relaxation; therefore, it became necessary to develop elevated-temperature mechanical properties data of the PM-based candidate materials for this application. Most of the data developed have been based on high-temperature hydrogen sintering, with sintered densities ranging from 7.10 to 7.35 g/cm^3; most materials were ferritic stainless steels. Only a limited amount of data have been published on austenitic grades of PM stainless steels, because these were not considered the most suitable materials for this application. Table 7.8 lists data on 304L and 316L alloys, along with that for their wrought counterparts.

There have been a fair number of publications covering mechanical properties of PM 400-series stainless steels, including both ferritic and martensitic grades (Table 7.9). Most of these data are in remarkably good agreement, despite differences in the processing parameters. Published data also include properties of modified 400-series alloys, such as niobium-doped 434L.

It is also noted that, at elevated temperatures, sintered PM 400 ferritic stainless steels exhibit higher yield and tensile strengths compared to their wrought counterparts. This is attributed mainly to the relatively larger grain size of high-temperature-sintered ferritic stainless steels compared to that in the wrought ferritic stainless steels. Deformation at elevated temperatures is via grain-boundary sliding, and thus, larger grains are beneficial for elevated-temperature strength. A contributing factor could be the relatively cleaner grain boundary in high-temperature-sintered PM materials, in terms of compounds comprising sulfur, phosphorus, nitrogen, and carbon.

7.5.2 Creep and Stress-Rupture Properties

There have been only a limited number of studies directed toward creep and stress-rupture behavior of PM stainless steels. A majority of these were undertaken as a part of the exhaust flange materials development program. In this application, the PM part is used as a structural component, with intermittent exposure to elevated temperatures, and thus, the performance of the flange as a leaktight clamp is strongly influenced by its creep behavior. Relatively high sintered densities were employed in some of these studies.

Hubbard et al. (Ref 21) determined stress-rupture lives of high-temperature (above 1260 °C, or 2300 °F) hydrogen-sintered PM 304L, 316L, 409L, and 434L having densities in the

Table 7.8 Elevated-temperature mechanical properties of austenitic powder metallurgy stainless steels

Alloy	Sintered density, g/cm^3	Sintered parameters	Test temperature(a) °C	°F	Ultimate tensile strength MPa	ksi	Yield strength MPa	ksi	Elongation, %	Reference
304L	7.01	1260 °C (2300 °F), H$_2$	RT	RT	405	58.7	195	28.3	25	21
			260	500	281	40.7	140	20.3	9	
			538	1000	250	36.2	110	16	10	
			816	1500	110	16	95	13.8	5	
316L	6.96	Vacuum	RT	RT	301	43.7	228.3	33.1	1.8	35
			649	1200	158.6	23	122	17.7	2.9	
			870	1600	67	9.7	53.1	7.7	4.2	
316L	7.59	Vacuum + hot isostatic pressing	RT	RT	541.5	78.5	211.7	30.7	NA	35
			649	1200	282	40.9	97.2	14.1	21	
			870	1600	113.8	16.5	73.8	10.7	21.7	
316L wrought	8.0	. . .	RT	RT	565.5	82	269	39	. . .	36
			482	900	503.5	73	172.4	25	. . .	
			649	1200	386.2	56	144.8	21	. . .	
			870	1600	172.5	25	103.5	15	. . .	

(a) RT, room temperature

Table 7.9   Elevated-temperature mechanical properties of 400-series powder metallurgy stainless steels

Alloy	Sintered density, g/cm³	Sintering parameters	Test temperature(a) °C	°F	Ultimate tensile strength MPa	ksi	Yield strength MPa	ksi	Elongation, %	Reference
409LE	7.3	1330 °C (2425 °F), H₂	RT	RT	393	57	222	32.2	24	37
			482	900	267	38.7	142	20.6	18	
			649	1200	226	32.8	111	16.1	12	
			760	1400	69	10	55	8	43	
			870	1600	31	4.5	26	3.8	93	
409L	7.15	1370 °C (2500 °F), H₂	RT	RT	357	51.8	251	36.4	NA	22
			649	1200	194	28.1	126	18.3	. . .	
			760	1400	65	9.4	50	7.2	. . .	
409L	7.2	1288 °C (2350 °F), H₂	RT	RT	379	55	205	28.7	. . .	25
			649	1200	269	39	124	18	. . .	
			760	1400	66	9.6	41	5.9	. . .	
409LE	7.25	1288 °C (2350 °F) H₂	RT	RT	359	52	221	32	15	37
			482	900	283	41	159	23	20	
			566	1050	269	39	131	19	16	
			649	1200	234	34	90	13	11	
			760	1400	90	13	62	9	32	
			870	1600	28	4	21	3	65	
409L wrought	7.75	. . .	RT	RT	427	61.9	241	35	. . .	16
			649	1200	157	22.8	86	12.5	. . .	
			760	1400	42	6.1	30	4.4	. . .	
			870	1600	25	3.6	16	2.3	. . .	
409L wrought	7.75	. . .	RT	RT	407	59	234	34	. . .	36
			649	1200	155	22.5	85	12.3	. . .	
			760	1400	40	5.8	30	4.4	. . .	
			870	1600	21	3	17	2.5	. . .	
434L	7.2	1288 °C (2350 °F), H₂	RT	RT	407	59	234	34	. . .	25
			649	1200	303.5	44	138	20	. . .	
			870	1600	62	9	48	7	. . .	
434L	7.27	1315 °C (2400 °F), H₂	RT	RT	405	58.7	248	36	16	27
			649	1200	169	24.5	118	17.1	29	
			871	1600	36.5	5.3	20.7	3	74	
434L	7.11	>1260 °C (>2300 °F), H₂	RT	RT	410.4	59.5	262	38	. . .	21
			649	1200	301	43.6	138	20	. . .	
			870	1600	33	4.8	30.3	4.4	. . .	
434L wrought	7.75	. . .	RT	RT	510	74	331	48	. . .	21
			649	1200	269	39	179	26	. . .	
			870	1600	33	4.8	33	4.8	. . .	
434Nb	7.2	1288 °C (2350 °F), H₂	RT	RT	365	53	241	35	. . .	25
			649	1200	186	27	138	20	. . .	
			870	1600	55	8	41	6	. . .	
410L	7.25	1288 °C (2350 °F) H₂	RT	RT	372	54	238	35	27	27
			482	900	324	47	159	23	26	
			566	1050	234	34	124	18	18	
			649	1200	117	17	62	9	35	
			760	1400	56	8	28	4	47	
			870	1600	21	3	14	2	83	
409L+1.2%Ni	7.3	1330 °C (2425 °F), H₂	RT	RT	600	87	489	70.8	8.5	27
			482	900	534.5	77.5	377	54.7	13	
			649	1200	277	40.1	238	34.5	12	
			760	1400	79	11.5	65.5	9.5	57	
			870	1600	69	10	47	6.8	65	
409L+1.0%Ni	7.17	>1260 °C (>2300 °F), H₂	RT	RT	559	81	455	66	. . .	21
			549	1200	350	50.8	320.4	46.5	. . .	
			870	1600	58.5	8.5	56.5	8.2	. . .	
409L+1.0%Ni	7.1	1370 °C (2500 °F), H₂	RT	RT	537.5	78	434	63	. . .	22
			549	1200	357	51.8	318	46.2	. . .	
			870	1600	67.5	9.8	48.2	7	. . .	

(a) RT, room temperature

range of 7.04 to 7.15 g/cm³. Stress-rupture tests were conducted at 676 °C (1250 °F), using ten different stress levels. Strength levels for 100 and 1000 h rupture life are listed in Table 7.10. The best performance was exhibited by 316L, followed closely by 304L and

409L. The worst performance was exhibited by PM 434L.

Clase and Sanderow (Ref 35) compared 100 h stress-rupture performance of PM 316L, 409L, and 434L with their wrought counterparts at a temperature of 677 °C (1250 °F). The PM samples

Table 7.10   Stress-rupture data from the Hubbard et al. study

Alloy	Density, g/cm³	%C	%N	%O	100 h rupture strength		1000 h rupture strength	
					MPa	ksi	MPa	ksi
409L	7.15	0.061	0.003	0.095	56	8.12	50	7.25
434L	7.11	0.027	0.003	0.111	28	4.06	NA	NA
304L	7.04	0.027	0.017	0.061	62	8.99	40	5.8
316L	7.11	0.020	0.010	0.052	74	10.73	56	8.12

Source: Ref 21

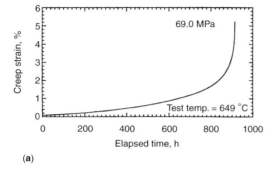

(a)

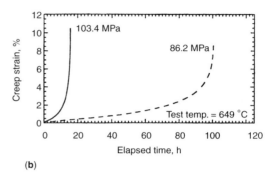

(b)

Fig. 7.12   Creep strain vs. elapsed time for high-temperature (1330 °C, or 2425 °F) hydrogen-sintered 409LE having a density of 7.35 g/cm³ at 649 °C (1200 °F). (a) Stressed to 69.0 MPa, and (b) stressed to 86.2 and 103.4 MPa. Source: Ref 37

were processed in two different ways: low-temperature (1120 °C, or 2050 °F) vacuum sintering to 6.96 g/cm³ density, and high-temperature (1250 °C, or 2280 °F) vacuum sintering followed by HIP to 7.55 g/cm³ density. The 100 h rupture strength of 316L increased with density, from 62 MPa (9 ksi) for 6.96 g/cm³ to 138 MPa (20 ksi) for HIPed samples with 7.55 g/cm³ density. However, even at this high density, the 100 h rupture strength was significantly lower than that of wrought 316L (186 MPa, or 27 ksi). Performance of PM 409L was found to be unaffected by density; at both densities, it showed higher rupture strengths compared to wrought 409L. The low-density data of the Clase and Sanderow study agree well with data obtained in the Hubbard et al. study.

Samal et al. (Ref 37) determined creep rates and rupture lives of high-temperature (1330 °C, or 2425 °F) hydrogen-sintered 409LE having a density of 7.35 g/cm³. These tests were carried out at 649 °C (1200 °F). The results are shown in Fig. 7.12 and in Table 7.11, along with other published data on PM and wrought 409L.

Although it is not possible to make direct comparisons, these data appear to be somewhat superior to those reported by Hubbard et al. This may be due to the fact that the Hubbard et al.

material had a lower sintered density and higher carbon content (0.61 versus 0.014%) compared to the Samal et al. study data.

Based on these limited published data, it appears that PM 409LE has a superior creep resistance when compared to wrought 409L. The dominant mechanism for creep in ferritic stainless steels is grain-boundary migration/diffusion of vacancies, rather than via bulk diffusion. Hence, a large grain size results in greater resistance to creep. Sun et al. (Ref 40) have shown that in wrought ferritic 430 stainless steel, an increase in grain size from 30 to 100 μm results in a tenfold decrease in creep rate at 730 °C (1346 °F). The average grain size of 409LE samples in the Samal et al. (Ref 37) study was 120 μm, compared to a typical grain size of 20 μm for wrought 409L. Additionally, the high-temperature hydrogen-sintered PM stainless steels are considered to be relatively free of impurities and deleterious grain-boundary precipitates, compared to the wrought materials, as a result of their low interstitial content and overall low impurity levels. Because of this, the rate of diffusion/migration of vacancies along the grain boundaries is expected to be slower for the PM materials. This may be partially responsible for the higher rupture strength.

Table 7.11 Creep and stress-rupture data on powder metallurgy (PM) ferritic stainless steels

Material and density	Test temperature °C	°F	Stress MPa	ksi	Rupture life, h	Steady-state creep rate, %/h	Ref
PM 409LE 7.35 g/cm^3	649	1200	69.0	10.0	910	0.0014	37
PM 409LE 7.35 g/cm^3	649	1200	86.2	12.5	101	0.028	37
PM 409LE 7.35 g/cm^3	649	1200	86.2	12.5	99	0.028	37
PM 409LE 7.35 g/cm^3	649	1200	103.4	15.0	15.7	0.10	37
PM 409LE 7.35 g/cm^3	649	1200	103.4	15.0	16.0	0.10	37
PM 409L 7.30 g/cm^3	677	1250	57	8.3	100	. . .	35
Wrought 409L	677	1250	30	4.4	100	. . .	35
PM 409L 7.15 g/cm^3	677	1250	68	9.9	30	. . .	21
PM 409L 7.15 g/cm^3	677	1250	60	8.7	174	. . .	21
PM 409L 7.15 g/cm^3	677	1250	55	8.0	900	. . .	21
Wrought 409L	704	1300	28	4.1	100	. . .	38
Wrought 409L	704	1300	22	3.2	1000	. . .	38
Wrought 430L	649	1200	30	4.4	1000	. . .	39

Table 7.12 Mechanical properties of metal injection molded (MIM) stainless steels

Grade	Density, g/cm^3	Ultimate tensile strength MPa	ksi	Yield strength MPa	ksi	Elongation, %	Hardness	Ref
316L, as-sintered	7.88	500	73	. . .	. . .	67	44 HRB	43
316, as-sintered	7.80–7.90	500–550	73–80	170–200	25–29	60–80	55–65 HRB	44
17-4 PH, as-sintered	7.60	900	130	. . .	. . .	3	28 HRC	43
17-4 PH, heat treated	7.60	1225	178	. . .	. . .	2	40 HRC	43
17-4 PH, heat treated (H-900)	7.60–7.75	1250–1350	181–196	1100–1200	160–174	4–8	37–43 HRC	44
17-4 PH, H1025, MIM	. . .	158	22.9	148	21.5	10	. . .	45
17-4 PH, H1025, wrought	. . .	159	23	164	23.8	13	. . .	45
410	. . .	380	55	. . .	. . .	. . .	. . .	. . .
444	. . .	450	65	. . .	. . .	. . .	. . .	. . .
262	. . .	570	83	. . .	. . .	. . .	. . .	. . .
420J	. . .	1000	145	. . .	. . .	. . .	. . .	. . .
310S	. . .	520	75	. . .	. . .	. . .	. . .	. . .
316L	. . .	530	77	. . .	. . .	. . .	. . .	. . .

Another possible factor behind the superior creep resistance of PM 409L may be its niobium content. Wrought 409L is commonly stabilized with titanium. Swindeman et al. (Ref 41) found the creep rate of wrought 304 to increase significantly when its niobium content was increased from 50 to 500 ppm. The authors tentatively attributed this effect to the presence of fine niobium carbide precipitates in the matrix. It is quite possible that niobium-containing PM 409L may be benefiting from the presence of NbC precipitates, similar to the way the niobium-containing wrought 304L does.

7.6 Mechanical Properties of Metal Injection Molded Stainless Steels

Metal injection molded (MIM)-processed stainless steels, due to their high sintered densities, exhibit mechanical properties that are comparable to those of their wrought counterparts. Typically, the sintered densities of MIM-processed stainless steels range from 95 to 97%

theoretical. Further densification by HIP has been shown to further enhance mechanical properties. Some variations in the mechanical properties can be expected from differences in the residual carbon content in MIM-processed materials. This is more commonly experienced with MIM-processed 17-4PH, because residual carbon can lead to nonuniform formation of martensite in the material.

In the case of 17-4PH, the processing route employed (i.e., sintering temperature, sintering atmosphere, and cooling rate) can significantly affect the mechanical properties of the as-sintered material. However, heat treatment (solutionizing and aging) is found to minimize the differences in properties (Ref 42).

Table 7.12 lists the typical room-temperature mechanical properties of MIM stainless steels. Table 7.13 lists standard properties of MIM stainless steels as specified by MPIF and the Japanese Standards Association.

In one study, the elevated-temperature tensile strength of MIM 316L was found to decline steadily with increase in the test temperature. Yield strengths were 258, 177, 121, 71, and 62 MPa (37.4, 25.7, 17.5, 10.3, 9.0 ksi) at test

Table 7.13   Metal injection molded (MIM) material property standards

Material designation condition	Standards organization	Minimum values — Ultimate tensile strength MPa	ksi	Yield strength MPa	ksi	Elongation, %	Density, g/cm³	Typical values — Ultimate tensile strength MPa	ksi	Yield strength MPa	ksi	Elongation, %	Impact energy J	ft·lbf	Young's modulus GPa	10⁶ psi	Apparent hardness
MIM-316L(a)	MPIF	450	65	140	20	40	7.6	520	75	175	25	50	190	140	190	28	67 HRB
MIM-430L(a)	MPIF	345	50	205	30	20	7.5	415	60	240	35	25			...	...	65 HRB
MIM-17-4PH(a)	MPIF	795	115	650	94	4	7.5	900	130	730	106	6	190	140	70	10	70 HRC
MIM-17-4PH(b)	MPIF	1070	155	965	140	4	7.5	1165	169	1090	158	6			...	...	33 HRC
SUS316L	JIS	480	69.6	...	...	...	...	...	...	...	...	...	...	...	...	...	...
SUS630	JIS	1310	190	...	...	...	...	...	...	...	...	...	...	...	...	...	...
SUS410L	JIS	360	52	...	...	...	...	...	...	...	...	...	...	...	...	...	...
SUS444	JIS	410	59.5	...	...	...	...	...	...	...	...	...	...	...	...	...	...
SUS262	JIS	410	59.5	...	...	...	...	...	...	...	...	...	...	...	...	...	...
SUS420J	JIS	740	107	...	...	...	...	...	...	...	...	...	...	...	...	...	...
SUS310S	JIS	520	75.4	...	...	...	...	...	...	...	...	...	...	...	...	...	...

(a) As-sintered (b) Treated and aged

temperatures of 180, 300, 500, 700, and 900 °C (356, 572, 932, 1292, and 1652°F), respectively (Ref 46).

Typically, fatigue endurance limits of 500 to 600 MPa (72.5 to 87 ksi) are reported in the literature for MIM 17-4PH in the heat-treated condition.

REFERENCES

1. M. Youseffi, C.S. Wright, and F.M. Jeyacheya, Effects of Silicon Addition and Process Conditions Upon α-Phase Sintering, Sinter Hardening and Mechanical Properties of Fe-1.5 Mo Powder, *Powder Metall.*, Vol 45 (No. 1), 2002, p 53

2. C.J. Novak, Structure and Constitution of Wrought Austenitic Stainless Steels, *Handbook of Stainless Steels,* D. Peckner and I.M. Bernstein, Ed., McGraw-Hill Book Co., New York, 1977, p 4–19

3. B. Kubicki, *Sintered Machine Elements,* Ellis Horwood, New York, 1995, p 52

4. H. Danninger, G. Jangg, B. Weiss, and R. Stickler, Microstructure and Mechanical Properties of Sintered Iron, Part I, *Powder Metall. Int.*, Vol 25 (No. 3), 1993, p 111–117

5. R.M. German, *Powder Metallurgy Science,* 2nd ed., MPIF, Princeton, NJ, 1994, p 381

6. A. Salak, and V. Miscovic, Porosity Dependence of the Mechanical Properties of Sintered Iron Compacts, *Powder Metall. Int.,* Vol 6 (No. 3), 1974, p 129

7. U. Kutsch, P. Beiss, and H.-J. Jager, Effect of Density on Mechanical Properties, Thermal Conductivity, and Machinability of Sintered Stainless Steels, *Proc. Euro PM 97 Conference* (Munich, Germany), EPMA, Shrewsbury, U.K., 1997, p 174–182

8. L. Dautzenberg and H. Gesell, Production Techniques and Properties of Austenitic Cr-Ni Stainless Steel Powders, *Powder Metall. Int.*, Vol 8 (No. 1), 1976, p 14–17

9. T. Tunberg, L. Nyborg, and C.X. Liu, Enhanced Vacuum Sintering of Water Atomized Austenitic Stainless Steel Powders by Carbon Addition Increases Product Properties, *Ind. Heat.*, Nov. 1992, p 37–40

10. P.K. Samal, O. Mars, and I. Hauer, Effect of Processing Parameters on the Room and Elevated Temperature Mechanical Properties of P/M 409L and 434L Stainless Steels, *Advances in Powder Metallurgy and Particulate Materials,* Vol 10, W.B. James and R.A. Chernenkoff, Compilers, MPIF, Princeton, NJ, 2004, p 10-122 to 10-133

11. A. Frisk, A. Johansson, and C. Lindberg, Nitrogen Pick Up During Sintering of Stainless Steel, *Advances in Powder Metallurgy and Particulate Materials,* ed. J. Capav, R. German, Vol 3, MPIF, Princeton, NJ, 1992, p 167–181

12. P.K. Samal, J.B. Terrell, and E. Klar, Effect of Sintering Atmosphere on the Corrosion Resistance and Mechanical Properties of Stainless Steels, Part II, *Advances in Powder*

Metallurgy and Particulate Materials, MPIF, Princeton, NJ, 2001, p 7-111 to 7-120

13. N. Dautzenberg, Eigenschaften von Sinterstählen aus Wasserverdüsten Unlegierten und Fertiglegierten Pulvern (Properties of Sintered Steels from Water Atomized Elemental and Alloyed Powders), *Second European Symposium on Powder Metallurgy*, May 8–10, 1968 (Stuttgart, Germany), Section 6–18, p 1–27

14. D.C. Smith, J. Liu, L.N. Smith, and R.M. German, Impact of Variations in Sintering Atmosphere on Stainless Steel Properties, *Advances in Powder Metallurgy and Particulate Materials*, ed. C. Rose, M. Thibodeau, Vol 3, MPIF, Princeton, NJ, 1999, p 3-45 to 3-54

15. H. Sanderow, J.R. Spirko, and T.G. Friedhoff, Influence of Density, Sintering Conditions and Microstructure on the Fatigue Properties of PM Stainless Steels, *Advances in Powder Metallurgy and Particulate Materials*, ed. C. Rose, M. Thibodeau, Vol 9, MPIF, Princeton, NJ, 1999, p 9-105 to 9-117

16. A.J. Rawlings, H.M. Kopech, and H.G. Rutz, The Effect of Processing and Service Temperature on the Properties of Ferritic PM Stainless Steels, *Advances in Powder Metallurgy and Particulate Materials*, ed. R. McKotch, R. Webb, Vol 9, MPIF, Princeton, NJ, 1998, p 9-19 to 9-36

17. H.D. Ambs and A. Stosuy, The Powder Metallurgy of Stainless Steels, *Handbook of Stainless Steels*, D. Peckner and I.M. Bernstein, Ed., McGraw-Hill Publishing Co., New York, 1977, p 29-16

18. H.I. Sanderow and T. Prucher, Mechanical Properties of PM Stainless Steel: Effect of Composition, Density and Sintering Conditions, *Advances in Powder Metallurgy and Particulate Materials*, ed. M. Phillips, J. Porter, Vol 3, Part 10, MPIF, Princeton, NJ, 1995, p 10-13 to 10-28

19. J.A. Reinshagen and T.J. Brockius, Stainless Steel Based P/M Alloys with Improved Corrosion Resistance, *Advances in Powder Metallurgy and Particulate Materials*, ed. M. Phillips, J. Porter, Vol 3, MPIF, Princeton, NJ, 1995, p 11-19 to 11-30

20. P.K. Samal, O. Mars, and I. Hauer, Means to Improve Machinability of Sintered Stainless Steels, *Advances in Powder Metallurgy and Particulate Materials 2005*, C. Ruas and T.A. Tomlin, Compilers, Vol 7, MPIF, Princeton, NJ, 2005, p 7-66 to 7-78

21. T. Hubbard, K. Couchman, and C. Lall, "Performance of Stainless Steel PM Materials in Elevated Temperature Applications," SAE Paper 970422, SAE International Congress and Expo., Feb 1997 (Detroit, MI)

22. S.O. Shah, J.R. McMillen, P.K. Samal, and E. Klar, "Development of Powder Metal Stainless Steel Materials for Exhaust System Applications," SAE Paper 980314, presented at SAE International Convention, Feb 1998 (Detroit, MI)

23. T.R. Albee, P. dePoutiloff, G.L. Ramsey, and G.E. Regan, "Enhanced Powder Metal Materials for Exhaust System Components," SAE Paper 970281, presented at SAE International Convention, Feb 1997 (Detroit, MI)

24. M.C. Baran, A.E. Segall, B.A. Shaw, H.M. Kopech, and T.E. Haberberger, Evaluation of P/M Ferritic Stainless Steel Alloys for Automotive Exhaust Applications, *Advances in Powder Metallurgy and Particulate Materials*, ed. R.A. McKotch, R. Webb, MPIF, Princeton, NJ, 1997, p 9-45 to 9-59

25. P.F. Lee, S. Saxion, G. Regan, and P. dePoutiloff, "Requirements for P/M Stainless Steel Materials in Order to Meet Future Exhaust System Performance Criteria," SAE Paper 980311, presented at SAE International Convention, Feb 1998 (Detroit, MI)

26. M. Svilar and H.D. Ambs, PM Martensitic Stainless Steels: Processing and Properties, *Advances in Powder Metallurgy*, ed. E. Andreotti, P. McGeehan Vol 2, MPIF, Princeton, NJ, 1990, p 259–272

27. P.K. Samal, J.B. Terrell, and S.O. Shah, Mechanical Properties Improvement of P/M 400 Series Stainless Steels via Nickel Addition, *Advances in Powder Metallurgy and Particulate Materials*, ed. C. Rose, M. Thibodeau, Vol 9, MPIF, Princeton, NJ, 1999, p 9-3 to 9-14

28. J.H. Reinshagen and J.C. Witsberger, Properties of Precipitation Hardening Stainless Steel Processed by Conventional Powder Metallurgy Techniques, *Advances in Powder Metallurgy and Particulate Materials*, ed. C. Lall, A. Neupaver, Vol 7, MPIF, Princeton, NJ, 1994, p 7-313 to 7-324

29. W. Schatt and K.P. Wieters, in *Powder Metallurgy Processing and Materials*, EPMA, Shrewsbury, U.K., 1997, p 209

30. U. Lindstedt, B. Karlsson, and R. Masini, Influence of Porosity on Deformation and Fatigue Behavior of PM Austenitic Stainless Steels, *Int. J. Powder Metall.*, Vol 33 (No. 8), 1997, p 49–60

31. R.C. O'Brian, "Fatigue Properties of PM Materials," SAE Paper 880165, presented at SAE International Congress and Expo., March 1988

32. C. Genest, M. Guillot, E. Beaulieu, and D. Ouellet, High Cycle Fatigue of Sintered 316L Stainless Steel, *Advances in Powder Metallurgy and Particulate Materials,* ed. C. Lall, A. Neupaver, Vol 7, MPIF, Princeton, NJ, 1994, p 325–339

33. S.O. Shah, J.R. McMillen, P.K. Samal, and L.F. Pease, "Mechanical Properties of High Temperature Sintered PM 409LE and 409LNi Stainless Steels Utilized in the Manufacturing of Exhaust Flanges and Oxygen Sensor Bosses," SAE Paper 03M-315, presented at SAE International Conference and Expo., March 2003 (Detroit, MI)

34. O. Grinder and Z. Zhiqiang, Effect of Oxide Inclusions on the Hot Ductility and the Recrystallization of Powder Austenitic Stainless Steels, *Scand. J. Metall.,* Vol 12, 1983, p 67–77

35. S.M. Clase and H.I. Sanderow, "The Effect of Nearly Full Theoretical Density on Critical Performance for Stainless Steel Powder Metal," SAE Paper 980312, SAE International Congress and Expo., Feb 1998 (Detroit, MI)

36. J. Davis, *Stainless Steels,* ASM Speciality Handbook, ASM International, 1994

37. P.K. Samal, J.B. Terrell, and S.O. Shah, Creep and Elevated Temperature Mechanical Properties of PM Stainless Steels, *Proc. PM World Congress* (Kyoto, Japan), 2000

38. "Automotive Exhaust System Materials Comparator," ARMCO Steel Bulletin, ARMCO Inc. 2800-0042, ARMCO Steel, Middletown, OH, 1992, p 11–92

39. W.F. Simmons and H.C. Cross, *Report on Elevated Temperature Properties of Chromium Steels,* STP 228, ASTM, 1958, p 94

40. D. Sun, T. Yamane, and S. Saji, Deformation Mechanism Maps for High Temperature Creep of a 17% Cr Ferritic Stainless Steel, *J. High Temp. Soc. Jpn.,* Vol 20 (No.7), 1994, p 53–57 (in Japanese)

41. R.W. Swindeman, V.K. Sikka, and R.L. Klueh, Residual and Trace Element Effects on the High Temperature Creep Strength of Austenitic Stainless Steels, *Metall. Trans. A,* Vol 14, April 1983, p 581–593

42. J. Mascerahanas and G. Schlieper, High Strength MIM Materials, *Proc. Second European Symposium on Powder Injection Molding,* Oct 18–20, 2000 (Munich)

43. J. Hamill, C. Schade, and N. Myers, Water Atomized Fine Powder Technology, *Powder Metall. Sci. Technol. Briefs,* Vol 3 (No. 3), 2001, p 10–13

44. D.S. Hotter, P/M Breathes Life into Medical Products, *Mach. Des.,* Vol 9, Oct 1997, p 78

45. J.C. LaSalle, B. Sherman, K. Bartone, R. Bellows, D. Lowery, P. Hartfield, and R. Dawson, Microstructure and Mechanical Properties of Aqueous Based Binder Metal Injection Molded 17-4 PH Stainless Steel for Aircraft Engine Components, *Advances in Powder Metallurgy and Particulate Materials,* ed. C. Rose, M. Thibodeau, Vol 2, Part 6, MPIF, Princeton, NJ, 1999, p 6-19 to 6-26

46. R.M. German and A. Bose, *Injection Molding of Metals and Ceramics*, MPIF, Princeton, NJ, p 297

CHAPTER 8

Magnetic and Physical Properties

POWDER METALLURGY (PM) offers many advantages with regard to the production of both soft and hard magnetic materials. Powder metallurgy processing is often the most convenient method for producing unique material combinations that may be either difficult or considered not feasible for production by wrought metallurgical processes. The PM process can lend itself to precise control of chemistry, reduction of harmful impurities, and formation of homogeneous microstructures. Its ability to form near-net shapes is an added plus, especially in the case of brittle and difficult-to-machine materials.

This chapter reviews PM processing with respect to physical and magnetic properties. Section 8.1 of this chapter describes the fundamental relationships governing magnetic properties of materials. Readers familiar with this subject may want to proceed to section 8.2.

8.1 Fundamental Relationships

Before a quantitative analysis of various types of magnetic fields can begin, it is necessary to define some of the customary units of magnetism. A unit pole, or a pole of unit strength, is defined in the centimeter-gram-second (cgs) system as one that exerts a force of 1 dyne on another unit pole, located at a distance of 1 cm. However, a magnetic pole simply creates a field of strength H, around it, and this field H produces a force on the second pole. Experiments show that the resulting force, F, is the product of pole strength, p, and field strength, H:

$$F = p \cdot H \qquad \text{(Eq 8.1)}$$

This relationship is helpful in defining H as a magnetic field of unit strength that exerts a force of 1 dyne on a unit pole and is named 1 oersted (Oe) in cgs units. In other words, a field H of 1 Oe exerts a force of 1 dyne on a unit pole. By analogy, the magnetic field created by a unit pole has an intensity of 1 Oe at a distance of 1 cm from the pole. This field decreases with the inverse square of the distance from the pole.

The strength of a magnetic field, or its field strength, H, can also be quantified by defining it as the number of lines of force passing through a unit area perpendicular to the field. Therefore, another way of quantifying the field strength is by representing each oersted by one line of force per square centimeter. A line of force when defined in this manner is called a Maxwell. In other words, 1 Oe = 1 line of force/cm^2 = 1 Maxwell/cm^2.

A unit pole located at the center of a sphere of radius 1 cm will exert a field intensity of 1 Oe at the surface. Because the surface area of the sphere is 4π, it must have 4π lines of force passing through it. If the pole strength is p, $4\pi p$ lines of force will be produced from it.

In vacuum, the magnetic flux density, B, is directly proportional to the magnetic field intensity, H, and may be represented by the equation:

$$B = \mu_0 H \qquad \text{(Eq 8.2)}$$

where μ_0 is the proportionality constant, defined as the permeability of vacuum (or free space). When a material is placed in this magnetic field, the magnetic flux density, B, in the material becomes:

$$B = \mu_0 \mu_r H \qquad \text{(Eq 8.3)}$$

where, μ_r is the relative permeability of the material. It is the ratio of the flux density in the material to the flux density that would be produced in vacuum under the same magnetic field, H. Consequently, the value of μ_r is a dimensionless number. Unlike μ_0, μ_r is not a constant; it varies with the flux density in the material. As a result, the relationship between B and H is not linear and does not make Eq 8.3 very practical to use.

While the value of μ_r is 1 for vacuum, it is slightly less than 1 for diamagnetic materials and slightly greater than 1 for paramagnetic materials. Ferromagnetic materials have much higher values of relative permeability, and as a result, they are either moderately or strongly magnetic. Iron, cobalt, and nickel are the only elemental metals that are ferromagnetic at room temperature. Gadolinium and some other rare earth metals are ferromagnetic at very low temperatures.

Figure 8.1 (Ref 1) is a graphical representation of Eq 8.2 and 8.3. By subtracting the flux density of vacuum from the total flux density, a quantity is obtained that represents the added flux density due to the material. This quantity may be represented as $\mu_0 M$, where M is defined as the magnetization of the material (M has the same units as H). Hence:

$$B = \mu_0 (H + M) \qquad \text{(Eq 8.4)}$$

Because $\mu_0 M = B$ material $- B$ vacuum:

$$M = (\mu_r - 1)H \qquad \text{(Eq 8.5)}$$

Magnetic susceptibility χ is defined as $\chi = \mu_r - 1 = M/H$. Paramagnetic materials have a very small positive value for magnetic susceptibility, of the order of 10^{-4} to 10^{-5}, and diamagnetic materials have a very small negative value for magnetic susceptibility, of the order of 0 to 10^{-5}. Ferromagnetic materials have high magnetic susceptibilities, in the range of 10^2 to 10^6.

The strong response of ferromagnetic materials to external magnetic fields is due to the presence of what is known as the magnetic domains in the material. The elements iron, cobalt, and nickel have atoms with several unpaired 3-d electrons. In ferromagnetic materials, the electron spins align themselves in a small volume of the material to produce a magnetic domain. Each domain acts like a small magnet. In the demagnetized state, the domains are oriented randomly, producing no net magnetic field. When the material is placed in a magnetic field, the domains orient themselves magnetically (by aligning electron spins) with the external magnetic field, thus becoming magnetized. The thickness of the domain wall d is a material property and can be as wide as 100 nm. Figure 8.2 (Ref 2) is a schematic of the structure of a domain wall (also known as a Bloch wall). In the case of a permanent, or hard, magnet, the domains remain lined up even after removal of an external magnetic field, and the material remains either moderately or strongly magnetic.

The critical characteristics of a ferromagnetic material can be determined from its B-H curve, also known as its hysteresis curve (Fig. 8.3) (Ref 3). When a ferromagnetic material, starting in its demagnetized state, is magnetized by an external magnetic field, H, it produces a magnetic flux density, B. As H is increased from zero, B increases. The increase in B is nonlinear with respect to H. The line OC in Fig. 8.3 represents the initial magnetization of the material. The tangent OA, representing B/H when H is close to zero, is referred to as the initial permeability (μ_i), which, incidentally, is difficult to measure. As H is further increased, both B and μ_r,

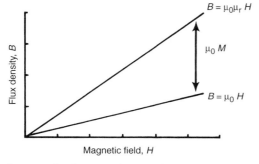

Fig. 8.1 Flux density as a function of magnetic field. Source: Ref 1

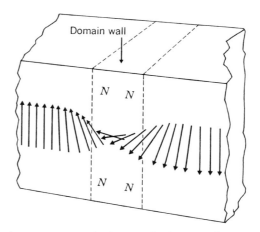

Fig. 8.2 Structure of a domain wall (schematic). All moments lie in the plane of the wall. Source: Ref 2. Reprinted with permission from McGraw-Hill

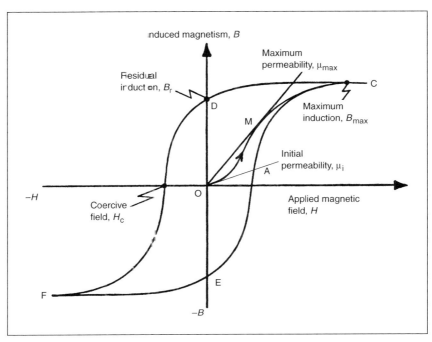

Fig. 8.3 Hysteresis curve of a typical hard magnetic material. Source: Ref 3. Reprinted with permission from MPIF, Metal Powder Industries Federation, Princeton, NJ

the relative permeability, increase but not in a linear fashion. Relative permeability, μ_r (often simply referred to as permeability), is defined as the slope of the tangent drawn from origin O to a particular point on the B-H curve and is not the actual slope of the curve. In a physical sense, application of H leads to the movement of domain walls within the material, resulting in the alignment of more and more domains with the direction of H. At some point in this process, the alignment of domains by the movement of domain walls is more or less complete. After this point, further magnetization occurs mainly via the rotation of remaining domains. The latter process requires greater external force than what was necessary for domain wall movement. The rate of magnetization becomes slower, and as a result, a knee in the curve develops, which is indicated by the letter "M". The line represented by joining points O and M, or the permeability at the knee, is known as the maximum permeability, μ_{max} (which is also referred to as maximum relative permeability). As H is increased beyond point M, further increase in B becomes smaller and smaller, and at point C, practically no more gain in B is realized, regardless of any increase in H. This value of B is called saturation induction, B_{sat}. In an actual test, however, a maximum value of H is selected

($H_{applied}$), and the value of B corresponding to this value of H is determined, which is indicated as B_{max}, such as B_{max} at $H_{applied}$ of 15,000 Oe, and so on. If, after reaching B_{max} (or at any other point on the initial magnetization curve), the magnetizing field H is decreased, then induction B would decrease in an irreversible manner. Line CD in Fig. 8.3 represents a typical reversal from B_{max}. When the value of H equals zero, the corresponding value of B is called the residual induction, B_r (also called remanent induction or remanence). It requires a reversal of the direction of the applied field ($-H$) in order to reduce B to zero, that is, fully demagnetize the material. This value of H is known as the coercive field, H_c. At this point, the domains are sufficiently randomized to yield zero net magnetization. Further increase of H in the negative direction causes B to become more and more negative until a point of saturation is reached, indicated by point F in Fig. 8.3. At this point, all domains are aligned in the direction opposite to the alignment at point C. If the magnetic field is now reversed and increased, B varies with H according to the line FEC. Thus, by varying H suitably to produce a complete loop, $CDFEC$, one cycle of the hysteresis loop has been traversed.

Determination of the hysteresis curve (also called the B-H loop) is conveniently made using a ring or toroid specimen of the material. ASTM standard test method A 773M-01 describes one such method for determining hysteresis curves. An electrical wire is wound around the specimen, covering the entire circumference of the toroid, which serves as the primary coil. A second coil, covering only a segment of the toroid, serves as the secondary coil. The primary coil is energized with a small electric current, which produces a magnetic field H (given by Eq 8.2). The induced magnetic field is detected by the secondary coil and is measured with a ballistic galvanometer or a flux meter. The applied magnetic field, H, is gradually varied by varying the current, i, in the primary coil. The resulting value of magnetic flux induced in the toroid, B, is recorded along with H. Because the relative values of B are much smaller than the corresponding values of H, a larger scale is typically used for recording B, compared to that for H.

A hysteresis curve generated by maximizing the applied field H (i.e., when both tips represent saturation) is called a major loop. It is symmetrical about the origin as the point of inversion; that is, if the right half of the curve is rotated by $180°$, it will be the mirror image of the left half. There can be an infinite number of minor hysteresis loops within the major loop, some of which may not be symmetrical about either coordinate. One way of fully demagnetizing a ferromagnetic material is to continually cycle it in an external field while progressively reducing the strength of the maximum applied field until the loop shrinks further and further and approaches the origin. The only other means of demagnetizing a ferromagnetic material is to heat it above its Curie point.

The main properties of interest obtained from a hysteresis curve are maximum permeability, maximum induction, residual induction, and coercive field. The area enclosed in a hysteresis curve represents the energy expended when the magnet is subjected to a forward and reverse magnetization cycle. This energy is considered redundant work, which is essentially released as heat.

Ferromagnetic materials are divided into two distinct classes: soft and hard magnetic materials. Soft magnetic materials have a low coercive field. As such, these are magnetized and demagnetized easily; they are suitable for applications requiring alternating magnetic flux. These magnets are also expected to have large μ_{max} and large B_{max}, so that a large output of flux is achievable with the smallest possible H. Hysteresis loss is kept low by selecting materials that have narrow and tall hysteresis loops.

Soft magnetic materials are used in applications such as transformers and motor cores, where the primary requirement is large power-handling capacity with low energy losses. In addition to hysteresis loss, a second kind of energy loss stems from the eddy currents that generate in the material due to continual change in flux density. The magnitude of the eddy current is inversely proportional to the electrical resistivity of the material. Hence, a large electrical resistivity is desirable for minimizing eddy current losses.

Hard magnetic materials, on the other hand, are permanent magnets that have high coercive fields. When magnetized, they retain their induced magnetic field for long periods of time, even when exposed to stray magnetic fields. A hard magnet is characterized by its "power of the magnet" (also called maximum energy product of the magnet), which is defined as the maximum value of the product of B and H, or $(BH)_{max}$. This is determined from the area of the second quadrant of the hysteresis loop. While the hysteresis curve of a typical soft magnet is narrow and tall, that of a hard magnet is wide and short (Fig. 8.4) (Ref 4).

8.2 Powder Metallurgy Magnetic Materials

As noted, powder metallurgy offers advantages with regard to the production of both soft and hard magnetic materials in terms of chemistry control and the additional benefit of near-net shaping. The conventional press-and-sinter route is well suited for the manufacture of high-volume, low-cost, near-net shape magnetic components. For more demanding magnetic applications, metal injection molding (MIM) can offer near-full-dense components. Notable examples of PM soft magnetic materials are iron, iron-silicon, iron-phosphorus, nickel-iron, cobalt-iron, and ferritic stainless steels. The mechanical properties of these materials are adequate for most engineering applications. Table 8.1 summarizes the typical magnetic properties of some of these materials. The magnetic properties of the PM soft magnetic materials are generally lower than those of their wrought counterparts, due to their lower density. Nevertheless, with optimal processing, the properties of PM soft magnetic materials can approach those of their wrought counterparts.

Soft magnetic powders are also formulated into polymer composites. In these, the powder, prior to compaction, is coated with a polymer

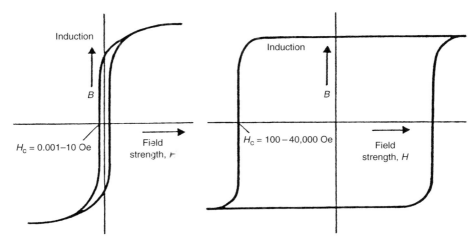

Fig. 8.4 Hysteresis loops of typical soft (left) and hard (right) magnetic materials. Source: Ref 4. Reprinted with permission from MPIF, Metal Powder Industries Federation, Princeton, NJ

Table 8.1 Typical properties of powder metallurgy soft magnetic materials

Alloy	Density, gm/cm³	μ_{max}	B_{max}, kG at 15 Oe	B_r, kG	H_c, Oe	Resistivity(a), $\mu\Omega\cdot$cm	Relative cost
Fe	7.00	2200	10.5	9.3	2.00	10	1.0
	7.20	2800	12.0	10.7	2.00		
Fe-0.45P	7.00	2800	11.5	9.0	1.65	30	1.2
	7.20	3000	12.5	10.0	1.65		
Fe-0.80P	7.00	4000	12.0	10.7	1.40	…	1.2
	7.20	4500	13.0	11.7	1.40		
Fe-3Si	7.00	3500	11.5	9.5	1.20	60	1.4
	7.20	5500	13.5	11.5	1.00		
Fe-50Ni	7.00	8000	9.0	7.5	0.30	45	10.0
	7.50	10,000	2.0	9.0	0.30		
SS 430L	7.25	1900	0.5	8.0	1.80	50	3.5
SS 434L	7.35	1600	9.7	7.7	1.80	50	3.5

(a) Resistivity data are for pore-free material of the alloy

for electrical insulation. The goal is to achieve good electrical insulation with a minimum amount of polymer, as well as to obtain a high density. Spherical or near-spherical powder particles are preferred to irregular-shaped particles, because they better withstand compaction to a high density without breakdown of the polymer insulating layer. Electrical insulation between particles leads to low eddy current losses. However, permeability is significantly reduced due to the presence of air gaps between the particles. The most common applications of these powder composites are in the form of magnetic cores for radio-frequency filters and power transformers. Powders of iron, iron-nickel, and iron-silicon are employed in these applications.

Examples of PM hard magnetic materials are Al-Ni-Co, $Nd_2Fe_{14}B$, $SmCo_5$, and Cr-Co-Fe. These materials are processed to their full theoretical

density. Compared to directionally solidified cast materials and to rolled sheets, PM hard magnetic materials have a drawback because of their lack of texture or crystallographic alignment.

8.2.1 Effect of Density and Morphology

The final density, that is, the sintered density or the re-pressed density for re-pressed parts, has a strong influence on the magnetic performance of PM soft magnetic materials. Porosity not only results in the absence of flux-carrying mass but also generates internal demagnetizing fields. Pores hinder domain wall movement in the same way as nonmagnetic impurities. The effect of porosity is notably stronger in the induction values of PM magnetic materials. Both the maximum induction, B_{max}, and the remanent induction, B_r, are related to density (Ref 3).

Figure 8.5, taken from Baum (Ref 5), shows that for sintered iron, both B_r and B_{max} have a linear relationship with density, while μ_{max} increases exponentially with density.

Moyer et al. (Ref 4) proposed an empirical relationship between density and induction:

$$B_s/B_n = 1 - aP \qquad \text{(Eq 8.6)}$$

where B_s and B_n are the saturation inductions of the porous and pore-free materials, respectively, P is fractional porosity, and a is a constant having a value between 1.5 and 2.0. Adler et al. (Ref 4) have suggested a stronger-than-linear dependence because of the demagnetizing effect of pores. They have proposed:

$$B_s/B_n = (1 - P)^n \qquad \text{(Eq 8.7)}$$

where $n = 1.5$. Both relations are approximately equivalent for porosities less than 15%.

8.2.2 Applications of PM Soft Magnetic Materials

Major users of PM soft magnetic materials are the automotive, computer, office equipment, appliance, and telecommunications industries. In the automotive industry, two types of soft magnetic materials are employed, depending on the operating principles (Ref 7). The first type involves the conversion of an electrical signal into motion. The material is required to exhibit a strong and quick response to an applied field, as well as a low remanence. It is also necessary that the material possesses high permeability, high induction, and a low coercive field. Typical examples are electromagnetic couplings for power steering, solenoid valves for fuel injection, hydraulic control units, and controls for electric locks. The second type of soft magnetic material involves the conversion of motion into an electrical signal. In these applications, a moderate induction and coercive field are required. A rapidly shifting flux density produces a change in the voltage generated. A high permeability produces a higher voltage. The primary example is the sensor rings of an antilock brake system in automobiles.

Selection of a soft magnetic material in a given application is based on a number of factors. The critical magnetic properties that frequently play a role in the decision-making process include permeability, coercive field, saturation induction, and electrical resistivity. Other factors that enter into consideration include mechanical properties, ease of fabrication, and cost. For most soft magnetic applications, the designer has the option of choosing from a number of different materials to satisfy the magnetic performance requirements, especially if component size, shape, and the electronics of the design are flexible.

8.2.3 Powder Metallurgy Stainless Steels

The ferritic and martensitic grades of stainless steel, known as the 400-series alloys, are magnetic. The austenitic grades of stainless steel are

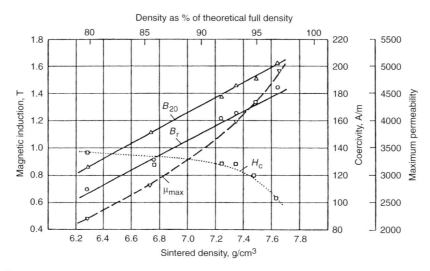

Fig. 8.5 Influence of sintered density on magnetic properties of sintered iron. B_{20}, magnetic induction at H of 2000 A/m[-1] (25.1 Oe); B_r, remanence; H_c, coercive field; μ_{max}, maximum permeability. (One tesla, T = 10[-4] gauss). Source: Ref 5

not magnetic in a practical sense. However, it is not uncommon to find austenitic stainless steel powders or parts to be mildly magnetic. This may arise from a number of reasons. Typically, a water-atomized austenitic powder would contain a small amount of delta ferrite as a result of rapid cooling from the molten state. Thus, a small fraction of the as-atomized powder would be magnetic. However, during sintering of the compacted powder, delta-ferrite phase will transform into the stable austenitic structure. An austenitic stainless steel may be found to be mildly magnetic if its actual composition falls outside the fully austenitic regime in the Schaeffler diagram. For example, excessive chromium nitride formation during sintering in a nitrogen-bearing atmosphere can deplete the matrix of chromium so severely that the composition of the depleted alloy matrix is shifted outside the fully austenitic regime. This condition can be corrected by annealing the magnetized material in either vacuum or hydrogen to expel most of the nitrogen and to homogenize the chromium content. Another condition that could make an austenitic alloy mildly magnetic is excessive cold work. Cold working, such as re-pressing, can transform some of the austenitic grains to martensite, which is magnetic.

Factors Affecting the Magnetic Properties of PM Stainless Steels. Just as mechanical properties and corrosion resistance of PM stainless steels are influenced by processing parameters, so are their magnetic properties. This is reflected in the wide variation among published data on magnetic properties of PM stainless steels. Only with careful selection of raw materials and processing parameters can the magnetic properties of PM stainless steels approach those of their wrought counterparts. The most commonly used PM magnetic stainless steels are the ferritic grades, 409L, 409LE, 410L, 430L, and 434L. The PM martensitic materials, such as 410 and

420, are selected if abrasion resistance is a requirement, in addition to good magnetic response. Grades 430L and 434L are also processed via MIM to near-full theoretical densities. In the mid-1980s, PM ferritic stainless steels found a major new application as sensor rings (or tone wheels) of antilock brake systems in U.S.-made automobiles. Because these components are exposed to road salts, it was deemed essential that they possess adequate corrosion resistance in addition to satisfactory magnetic response. The ring material was also required to possess good ductility, to permit press fitting of the rings onto the shafts of automobiles. The PM-processed 410L and 434L sensor rings did satisfy all of the aforementioned requirements. Also, because this application is based on alternating currents, the low hysteresis and eddy current losses of the PM stainless steels were found to be highly beneficial. This application generated much interest for studying the effects of various processing routes (e.g., vacuum versus atmosphere sintering) and processing variables on the magnetic properties of PM ferritic stainless steels.

Table 8.2 lists typical magnetic properties of 400-series PM magnetic materials produced under optimal sintering conditions. Optimal sintering refers to processing conditions that lead to the achievement of low interstitials, freedom from nonmagnetic inclusions, large grain size, and a relatively high sintered density. Following this, the effects of specific process parameters on the magnetic properties of PM 400-series materials are discussed by drawing information from published data. The purpose is to critically assess the relative effect of each of the variables on the magnetic behavior of PM stainless steels.

For the purpose of discussion, it is convenient to divide these process variables into process-related (independent) and material-related (dependent) variables. Because the goal is to maximize magnetic induction and permeability

Table 8.2 Typical magnetic properties of 400-series stainless steels

Alloy	Density, g/cm³	Sintering conditions	μ_{max}	B_{max}, kG	B_r, kG	H_c, Oe	Reference
Wrought 430FR(a)	7.80	. .	2500	15 at H = 15 Oe	6	2.51	8
409L	6.96	NA	730	7.66 at H = 15 Oe	4.37	3.28	9
410L	7.29	1260 °C (2300 °F), H2, 60 min	2166	11.47 at H = 15 Oe	7.69	1.68	10
410L	7.1	1260 °C (2300 °F), vacuum, 45 min	2200	10.9 at H = 25 Oe	9.4	2.0	10
430L	6.67	1121 °C (2050 °F), H_2, 30 min	1040	7.9 at H = 15 Oe	5.1	2.32	11
434L	6.65	1121 °C (2050 °F), H_2, 30 min	1170	7.9 at H = 15 Oe	4.8	1.9	11

(a) Proprietary alloy of Carpenter Technology, considered as the reference alloy for soft magnetic ferritic stainless steels and to minimize coercivity and remanence, the desired material characteristics are those that permit a high degree of domain wall mobility, along with a low volume of porosity and non-magnetics. There are two kinds of hindrances to domain wall movement: inclusions and residual stresses. The latter can be divided into macro- and microstresses, based on scale. From a magnetic response point of view, an inclusion in a domain is a region that has a different spontaneous magnetization from the surrounding, or none at all (Ref 12). It thus covers microstructural variations of one sort or another, including particles of a second phase, oxides, sulfides, carbides, nitrides, pores, cracks, nonmetallic inclusions, and grain boundaries. Residual microstress is caused by crystal imperfections of various kinds, particularly dislocations. Unlike the residual macrostresses, these are not fully removed by annealing.

Figure 8.6 shows the process- (independent) and material- (dependent) related variables that influence the magnetic behavior. Each of the process or independent variables directly or indirectly influences one or more of the material or dependent variables. As a result, it is the combined effect of many process variables that determines the key characteristics, such as density, purity, and the metallurgical condition of the processed material. Thus, the parts producer has considerable latitude with the selection of

processing parameters for arriving at an optimal combination of material characteristics, from a magnetic performance point of view.

Effect of Powder and Process Variables. The characteristics of the starting powder can influence the magnetic performance of a sintered product. Moyer (Ref 9) compared powders made by a number of different manufacturers and determined that a high oxygen content in the powder does have an adverse effect on the magnetic performance of the sintered product. The oxygen content of water-atomized stainless steel powders can vary quite significantly, typically ranging from 1700 to 3500 ppm. Only a small fraction of this total oxygen is reduced during a commercial sintering operation. Moyer found hydrogen-sintered 410L parts made from a powder that contained 2450 ppm oxygen to exhibit 16% lower maximum permeability and 10% higher coercive field compared to parts made from a powder containing 1700 ppm oxygen. In a study based on vacuum sintering, Shimada et al. (Ref 13) reduced the oxygen content of 410L sintered parts by preblending the powder with various amounts of graphite, ranging from 0.05 to 0.15%. This resulted in an oxygen content of 1860 ppm for no graphite addition to 970 ppm for 0.15% graphite addition. The sample with a graphite addition of 0.05%, whose oxygen content was reduced to 1280 ppm, was found to be the optimal one,

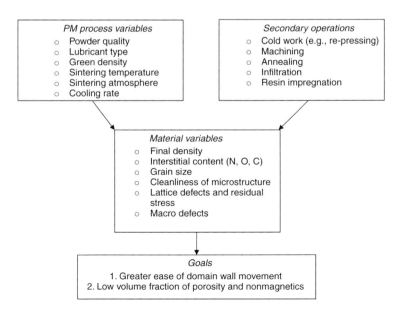

Fig. 8.6 Process and material variables affecting the magnetic behavior of powder metallurgy (PM) ferritic stainless steels, as well as of most magnetic materials

from the magnetic performance point of view. Its maximum permeability was increased by 11% compared to the reference material (no graphite addition). With higher amounts of graphite addition, the oxygen content further decreased, but the residual carbon content increased sufficiently to lower the maximum permeability. With a 0.15% graphite addition, the residual carbon content increased to 0.088%, resulting in a decrease in the maximum permeability by 6% as compared to the reference material.

Compressibility of the starting powder also plays an important role. A powder with higher compressibility will yield higher green density under the same compaction pressure and will result in a higher sintered density.

Several investigators have studied the effect of various types of PM lubricants, as well as their addition amounts, on the magnetic performance of PM ferritic stainless steels. Lubricant effect can be twofold. The type of lubricant, its amount, and the delubrication practice employed will determine the residual carbon level in the sintered product; as such, these variables can influence the magnetic performance. Secondly, the lubricant type and amount can have an effect on the green density achieved under a given compacting pressure, which in turn will influence the sintered density. Frayman et al. (Ref 14) compared the three most commonly used lubricants, namely, lithium stearate, zinc stearate, and Acrawax C, at the 0.5% level. They found that lithium stearate gave slightly higher green and sintered densities in 410L stainless steel compared to the other two. Kopech et al. (Ref 10) compared the effect of various lubricant types on the magnetic performance of hydrogen-sintered 410L Their study included lithium stearate, Acrawax C, zinc stearate, and Kenolube; the amount used was 1% in all cases. Despite yielding a slightly lower sintered density compared to the other three, Kenolube produced higher maximum permeability and maximum induction and a reduced coercive field. These improvements were attributed to lower levels of oxygen, nitrogen, and carbon achieved with Kenolube (Table 8.3).

The effects of compaction pressure and the resulting green density on the magnetic performance have been studied by a number of researchers. As long as an increased green density does not impede the delubrication process or otherwise affect the concentration of interstitials, a higher induction and a higher maximum permeability are anticipated on the basis of the higher sintered density. Frayman et al. (Ref 14) noted an increase of maximum induction (B_{100}) from 10.05 to 11.02 × 10³ gauss in PM 434L when the compaction pressure was increased from 482 to 620 MPa (35 to 45 tsi). Kopech et al. (Ref 10), on the other hand, experienced a decrease in the maximum permeability and an increase in the coercive field when they increased the compacting pressure, despite an increase in sintered density. These unexpected results were explained by increased levels of residual carbon, oxygen, and nitrogen found in the samples made by using the higher compaction pressure.

Sintering temperature has perhaps the most significant impact on the magnetic performance of PM stainless steels. High sintering temperatures not only lead to high sintered densities but also help achieve significantly reduced levels of interstitials (nitrogen, oxygen, carbon). In addition, a high sintering temperature produces a larger grain size and a greater degree of pore rounding, both of which make domain wall movement easier.

The benefits of high-temperature sintering are achieved with both atmosphere sintering and vacuum sintering. Significant improvements in microstructure and reduction of interstitials are observed when the sintering temperature is raised above 1200 °C (2192 °F). Kopech et al. (Ref 10) noted a threefold increase in maximum permeability by increasing the sintering temperature from 1120 to 1200 °C (2048 to 2192 °F) (Table 8.4). Further increase to 1260 °C (2300 °F) had no effect on maximum permeability but improved maximum induction for a given density. Significant improvements were also noted in the

Table 8.3 Effect of lubricant type on magnetic properties

Lubricant type	Sintered density, g/cm³	H (15), Oe	B_{max} (15), kG	B_r (15), kG	μ_{max}	H (30), Oe	B_{max}, kG	B_r (30), kG	%C	Oxygen, ppm	Nitrogen, ppm
Lithium stearate	7.23	1.92	10.86	6.83	1717	1.92	12.62	6.98	0.01	1600	66
Acrawax	7.2	1.75	10.99	7.14	1983	1.75	12.43	7.18	0.01	1400	72
Zinc stearate	7.2	1.79	11.01	7.16	1941	1.81	12.96	7.43	0.01	1800	70
Kenolube	7.13	1.7	11.99	7.39	2111	1.71	12.96	7.6	0.01	1300	54

Compacted at 690 MPa (50 tsi) and sintered at 1260 °C (2300 °F) for 30 min in hydrogen. Source: Ref 10.

Table 8.4 Effect of sintering temperature on interstitial content and magnetic properties of PM 410L

Sintering temperature		Sintered density, g/cm³	H (15), Oe	B_{max} (15), kG	B_r (15), kG	μ_{max}	% C	Oxygen, ppm	Nitrogen, ppm
°C	°F								
1260	2300	6.88	1.6	10.26	6.54	1916	0.005	1100	22
1260	2300	7.09	1.96	10.68	6.18	1630	0.004	1300	25
1200	2192	6.6	1.62	9.36	7.2	1946	0.009	1200	20
1200	2192	6.88	1.89	9.64	6.38	1600	0.008	1800	28
1120	2048	6.16	2.56	5.77	3.22	645	0.045	1600	79
1120	2048	6.55	2.62	6.34	3.49	695	0.029	1800	96

coercive field as a result of the higher sintering temperatures. It decreased by 30% as the sintering temperature was increased from 1120 to 1200 °C (2048 to 2192 °F). Further increase to 1260 °C (2300 °F) resulted in a negligible reduction in coercive field. The authors attributed the dramatic improvement in magnetic properties in going from 1120 to 1200 °C (2048 to 2192 °F) to a significant reduction in the carbon content. However, the data also suggest that reduction of oxygen levels may also have contributed to the improved properties.

Moyer and Jones (Ref 15), in a study based on vacuum sintering, also determined that a sintering temperature of 1120 °C (2048 °F) is ineffective from the viewpoint of reduction of oxygen and carbon contents in the material. They found that sintering at 1260 °C (2300 °F) in a vacuum furnace under a partial pressure of hydrogen significantly improved the magnetic performance of PM 410L and 434L. The PM 410L stainless steel sintered in this manner, with a sintered density of 7.25 g/cm³, had magnetic properties comparable to those of wrought 410L. Similarly, Shah et al. (Ref 16) noted a 70% increase in maximum permeability in hydrogen-sintered PM 410L when the sintering temperature was increased from 1120 to 1260 °C (2048 to 2300 °F). Significant improvements were also noted in maximum induction and coercive field.

Effect of Nitrogen-Containing Sintering Atmospheres. Sintering of stainless steel in atmospheres other than hydrogen, argon, or vacuum can lead to a precipitous reduction in magnetic properties. A number of researchers have investigated the effect of sintering in nitrogen-rich atmospheres, because of the potential cost savings involved. Sintering in atmospheres containing as little as 25% N_2 (balance hydrogen, i.e., dissociated ammonia) can lead to the absorption of several thousand parts per million of nitrogen, much of which may precipitate as chromium nitride (section 5.2.4 in

Chapter 5, "Sintering and Corrosion Resistance"). Slow cooling from the sintering temperature can lead to additional absorption of nitrogen and the precipitation of chromium nitride (Cr_2N). The chromium nitride precipitates will hinder domain wall movement, significantly reducing maximum permeability and induction. The presence of nitrogen in solution, even in very small amounts, makes the alloy significantly harder and increases internal strain, thus increasing the coercive field. Svilar and Ambs (Ref 17) noted a fourfold increase in the coercive field of 410L when sintering in a vacuum furnace with a backfill of (up to several thousand microns of mercury) 25% N_2-75% H_2 compared to a backfill of pure hydrogen. Maximum induction at an applied field of 15 Oe was reduced from 9800 to 382 gauss. Sintering was carried out at 1230 °C (2246 °F). Kopech et al. (Ref 10) compared 75% H_2-25% N_2 sintering to hydrogen sintering and noted a twofold increase in coercive field and a 50% reduction in maximum induction, due to nitrogen pickup. The sintering temperature in their study was 1260 °C (2300 °F). The discrepancy between the results of these two studies may be attributed to differences in the nitrogen and carbon contents of the sintered samples. The nitrogen and carbon contents of the 25% N_2-75% H_2-sintered samples in the Svilar and Ambs (Ref 17) study were 700 and 500 ppm, respectively, whereas those for the Kopech et al. study were 430 and 70 ppm, respectively. The ~400 ppm figure for nitrogen in the Kopech et al. study is suspect, because the equilibrium nitrogen absorption under their process condition is at least 1000 ppm (section 5.2.4 in Chapter 5, "Sintering and Corrosion Resistance"). Thus, the different properties in the two studies are probably due to the variation in the amount of interstitials, as well as their form and distribution.

Except when sintering is carried out in a nitrogen-containing atmosphere, a slower cooling rate is beneficial, from the magnetic performance point of view. Shimada et al. (Ref 13) observed an 8% increase in maximum permeability of sintered 410L as a result of decreasing the cooling rate from 4.8 to 1.9 °C/min (8.6 to 3.4 °F/min). Lower cooling rate reduces lattice strain and strain anisotropy. Work done by Bas et al. (Ref 7) supports this finding.

Effect of Secondary Operations. Cold work leads to an increase in lattice strain and dislocation density. Grain size is also decreased. All these changes are unfavorable to domain wall movement. Shah et al. (Ref 16) compared the magnetic properties of as-sintered and re-pressed 410L sensor rings. The decrease in maximum permeability due to re-pressing was typically over 80%. Annealing restored magnetic performance, to a large extent. Figure 8.7 is a typical example of the hysteresis curves obtained from as-sintered, re-pressed, and annealed 410L sensor rings. Note the higher B_{max} after annealing, due to the higher sample density, as compared to that of the as-sintered sample. Table 8.5 lists the magnetic properties of as-sintered, re-pressed, and annealed materials.

Machining can lead to distortion of the crystal structure and also increase dislocation density, although it would be limited to a thin

layer on the surface. These effects can again be eliminated by annealing.

Resin impregnation is commonly practiced to enhance machinability and also to seal off porosity in a low-density material. Frayman et al. (Ref 14) found no change in magnetic properties as a result of resin impregnation. They also studied the effect of copper infiltration and found that it severely degraded magnetic performance. The effects of resin impregnation, infiltration, as well as re-pressing are summarized in Table 8.6.

Benefits of Metal Injection Molding. Because MIM processing results in high sintered densities, 400-series stainless steels produced via MIM exhibit superior magnetic properties compared to conventionally processed PM stainless steels. In addition, it is feasible to increase the silicon content of stainless steels in MIM processing, which is found to be highly beneficial in enhancing magnetic properties. Suzuki and Ohtsubo (Ref 18) have shown that maximum permeability of MIM-processed stainless steel 430 increases from 1150 to 1550 when silicon content is increased from 0.5 to 2.0%. Silicon addition lowered the coercive field and increased resistivity. In the same study, MIM-processed stainless steel with a chemistry of 18.4% Cr,

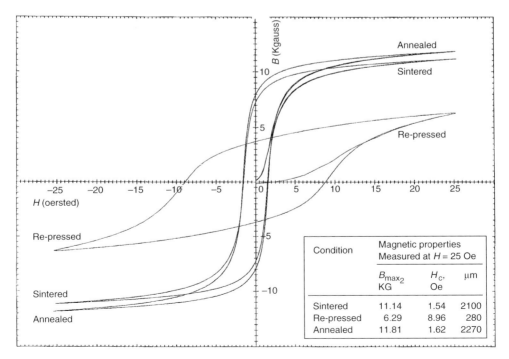

Fig. 8.7 Hysteresis curves of as-sintered, annealed, and re-pressed 410L stainless steel sensor rings. Source: Ref 16. Reprinted with permission from SAE Paper 930449 ©2003 SAE International

Table 8.5   Effect of cold work and annealing on the magnetic properties of PM 410L

Condition	Sintered		Re-pressed		Annealed	
Density, g/cm³	7.07		7.38		7.37	
Applied field, Oe	15	25	15	25	15	25
Maximum induction, B_{max}, kG	10.52	11.14	4.22	6.29	11.15	11.81
Retentivity, B_r, kG	7.16	7.24	2.67	3.71	7.83	7.92
B_r/B_{max}	0.681	0.65	0.633	0.59	0.702	0.671
Coercive field, H_c, Oe	1.53	1.54	7.49	8.96	1.6	1.62
Maximum permeability, μ_{max}	2110	2100	280	280	2260	2270

Notes: 1. Sintered in pusher furnace at 1260 °C (2300 °F), 45 min, H_2, dewpoint –35 °C (–31 °F). 2. Re-pressed at 690 MPa (50 tsi). 3. Annealed in pusher furnace at 900 °C (1652 °F), 45 min, H_2, dewpoint –35 °C (–31 °F)

Table 8.6   Effects of various secondary processing on magnetic performance of powder metallurgy stainless steel

Process	Material	Condition	Sintered density, g/cm³	B_{max} at 15 kG	B_r, kG	H_c, Oe	μ_{max} × 10	Reference
Resin impregnation	SS 409L	As-sintered	6.96	7.66	4.37	3.28	0.73	14
		Resin impregnation	6.96	7.68	4.41	3.22	0.75	
Re-pressing	SS 410L	As-sintered	6.91	7.08	4.3	2.41	0.94	14
		Re-pressed	. . .	3.94	2.52	7.68	0.26	
Re-pressing	SS 410L	As-sintered(a)	6.65	8.32	5.25	2.42	1.24	16
		Re-pressed	7.06	4.17	2.68	7.51	0.28	
Copper infiltration	SS 410L	As-sintered	6.98	8.32	4.74	2.34	1.09	14
		Infiltrated	7.73	0.41	0.07	2.15	0.6	

(a) Sintered at 1120 °C (2048 °F) in H_2

0.9% Si, and 0.5% Mo gave favorable magnetic properties of 1.09 T for B_{max}, 0.66 T for B_r, 88.2 A/m for H_c, and 2205 for μ_{max}.

8.3   Physical Properties

Because published data on the physical properties of PM stainless steels are very limited, one often relies on estimated property values derived from the properties of an equivalent wrought alloy. Theoretical as well as empirical relationships between density and various physical properties have been proposed by a number of researchers. Discrepancies between the estimated values based on various proposed equations are usually found to be small, and thus, in the absence of reliable test data, these equations can serve as useful tools for estimating the physical properties of PM materials. With this in view, some of the physical properties of wrought stainless steels are listed.

8.3.1   Physical Properties of Wrought Stainless Steels

A wide range of physical properties data for wrought stainless steels has been compiled by Lewis (Ref 19), including some measured at elevated temperatures. Lewis noted some discrepancies within individual data sets, which he attributed to differences in the test methods and possible compositional differences in the test materials employed by individual researchers. Data presented here were taken from Lewis as well as from Davis (Ref 20).

Definitions of some of the less commonly used physical properties are as follows. Melting range is defined by the solidus and liquidus temperatures of an alloy.

Specific heat is the quantity of heat required to change by 1 degree the temperature of a body of material of unit mass.

Thermal conductivity is the measure of the rate at which the material transmits heat. If a thermal gradient of 1 degree per unit length is established in the material, then the thermal conductivity is defined as the quantity of heat that is transmitted across a unit cross-sectional area in 1 s. Thermal diffusivity is that property of the material that determines the rate at which a temperature front moves through the material. Thermal diffusivity is determined by the ratio of thermal conductivity to the product of density and specific heat.

Compared to ferritic, austenitic stainless steels show a greater diversion in physical properties from those of common irons and steels. In comparison to plain and low-alloy steels, their moduli of elasticity are slightly lower and their thermal conductivities are substantially lower, whereas the coefficients of thermal expansion and electrical resistivities are significantly higher. The coefficients of thermal expansion of austenitic stainless steels differ significantly from both ferritic stainless steels and plain carbon steels. Because of this, one must exercise caution when designing assemblies or structures using these dissimilar materials for elevated-temperature service.

Table 8.7 lists physical properties of some of the popular grades of wrought stainless steels. Much of these data were obtained by using the standard grades of these alloys, rather than their "L" versions (low carbon). Table 8.8 shows the effect of temperature on some of the properties. The coefficient of thermal expansion differs significantly from one temperature range to another.

8.3.2 Physical Properties of PM Stainless Steels

Specific Heat. Touloukian et al. (Ref 21) demonstrated that heat capacity is not only independent of density but also is unaffected by the presence of chromium nitrides. They found hydrogen-nitrogen-sintered 316L, containing 6 wt% Cr_2N (density of 5.69 g/cm^3), to have essentially the same specific heat as pore-free and nitrogen-free 316L, over a temperature range of 0 to 1000 °C (32 to 1832 °F).

Thermal Diffusivity and Conductivity. Using a laser flash method, Beiss et al. (Ref 22) determined the thermal diffusivity of 316L sintered in hydrogen and in 30% H_2-70% N_2 atmospheres over a temperature range of 0 to 800 °C (32 to 1472 °F). Sintered densities ranged from 5.69 to 7.12 g/cm^3. They found the hydrogen-sintered materials to have typically 6% higher thermal conductivity compared to the nitrogen-sintered material, for the same sintered density and test temperature. Thermal conductivity calculations were made from data on thermal diffusivity, specific heat, and density at various temperatures (density estimates were based on the coefficient of thermal expansion). The results of their study are summarized in Fig. 8.8. The accuracy of their method was confirmed by a close match found between the calculated and the known values of thermal conductivity of pore-free 316L, over the range of test temperatures.

Based on these studies, they proposed a relationship between conductivity and density as follows:

$$\lambda/\lambda_c = (\rho/\rho_c)^m \qquad \text{(Eq 8.8)}$$

where λ and λ_c are thermal conductivities of the porous and pore-free materials, respectively, and ρ and ρ_c are densities of the porous and pore-free materials, respectively. The exponent m was determined to be 2.428, with minimal dependence on temperature.

Table 8.7 Nominal room-temperature physical properties of wrought stainless steels

SI No.	Property	Unit	303	304	316	409	410	430	434
1	Density	g/cm^3	8	8	8	7.75	7.75	7.75	7.75
2	Melting range	°C	1398–1420	394–1440(a)	1405–1445(a)	1483–1532	1427–1532	1427–1510	1427–1510
		°F	2550–2590	2541–2624(a)	2561–2633(a)	2700–2790	2700–2790	2600–2750	2600–2750
3	Modulus of elasticity	GPa	193	193	193	200	206	206	200
		$10^6 \times$ psi	28	28	28	29	30	30	29
4	Specific heat	cal/g·°C	0.39	0.29	0.36	0.36	0.36	0.36	0.36
		B/lb·°F	0.12	0.09	0.11	0.11	0.11	0.11	0.11
5	Thermal conductivity	W/m·K	12.2	13.9	13.6	25.2	25.2	20.7	NA
		Btu/ft·h·°F	7	8	7.8	14.5	14.5	11.9	NA
6	Mean coefficient of expansion	µm/m·°C	17.6	18.2	17.5	11.7	11.2	11.2	11.9
		µin./in.·°F 20–425 °C (70–800 °F)	9.8	10.1	9.7	6.5	6.2	6.2	6.6
7	Thermal diffusivity	m^2/h	. . .	0.014	0.014	. . .	0.021	0.021	. . .
		ft^2/h		0.15	0.15		0.23	0.23	
8	Electrical resistivity	µΩ·cm	72	70	73	61	58	60	60

(a) Melting ranges of "L" versions. Source: Ref 19, 20

Table 8.8 Effect of temperature on physical properties of wrought stainless steels

Property	Units	Alloy	Temperature, °C (°F)						
			−196 (−321)	20 (68)	100 (212)	200 (390)	400 (750)	600 (1110)	1000 (1470)
Density	g/cm³	316	8	8	7.9	7.8	7.7	7.6	7.5
Modulus of	GPa	304	208 (30.2)	193 (28)	191 (27.7)	183 (26.5)	168 (24.3)	148 (21.5)	NA
elasticity	(106 psi)	316	NA	193 (28)	192 (27.8)	185 (26.8)	168.5 (24.4)	151 (21.9)	NA
		410	NA	206 (30)	200 (29)	191 (27.7)	175 (25.4)	158 (22.9)	NA
		430	NA	206 (30)	198 (28.7)	191 (27.7)	165 (23.9)	139 (20.2)	NA
Specific	J (Btu/lb·°F)	304	NA	0.09	NA	NA	0.104	0.115	0.126
heat		410	NA	0.11	NA	0.11	0.113	0.13	0.215
		430	NA	0.11	NA	0.11	0.12	0.14	0.24
Electrical	μΩ·cm	304	55	70	NA	85	98	111	NA
resistivity		316	60	73	NA	87	98	108	NA
		410	NA	58	NA	68	84	103	NA
		430	NA	60	NA	76	91	111	NA

Source: Ref 20

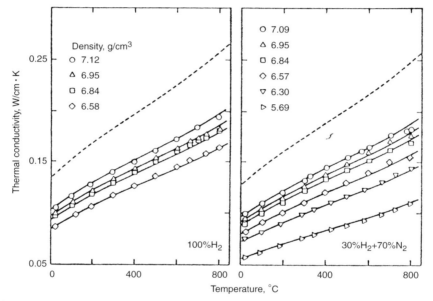

Fig. 8.8 Thermal conductivity of sintered 316L as a function of sintered density for hydrogen (left) and 30% H_2-70% N_2 sintering atmosphere (right). Broken lines represent pore-free 316L

German (Ref 23) has proposed a relationship between porosity and conductivity (thermal and electrical) as:

$$\lambda/\lambda_c = 1 - \varepsilon \cdot \omega \qquad \text{(Eq 8.9)}$$

where ε is the fractional porosity, and ω is a coefficient with a value between 1 and 2 for porosities of less than 30%. Equation 8.9 was found to have good fit with experimental data based on sintered stainless steel. Equation 8.8 shows good agreement with Eq 8.9 for ω values that are closer to 2.

Both electrical resistivity (1/conductivity) and conductivity are also dependent on pore morphology in addition to pore volume (relative density). Spheroidal and equiaxed pores result in higher conductivity (lower resistivity) compared to irregular ones, apparently because of larger interparticle bond areas.

The coefficient of thermal expansion of a material is decreased by a small extent by the presence of porosity. Thermal expansion results from interatomic bonding and atomic vibrations. The presence of pores reduces available mass but does not alter atomic bonding (Ref 23). A model proposed by German gives the relationship as:

$$C_T = C_0 (\rho/\rho_T)^{1/3} \qquad \text{(Eq 8.10)}$$

where C_T is the effective thermal expansion coefficient, C_0 is the bulk thermal expansion coefficient, and ρ/ρ_T is the fractional density.

Elastic Modulus. Like most mechanical properties, Young's modulus of a porous material is highly sensitive to pore structure and the effects of stress concentration (section 7.2.1 in Chapter 7, "Mechanical Properties"). According to Haynes and Eegdiege (Ref 24), elastic modulus decreases sharply with increase of porosity from 3 to approximately 20%; beyond 20% porosity, the rate of decrease becomes much less (Fig. 8.9).

German (Ref 23) has proposed a power law relating Young's modulus to porosity:

$$E = E_0 (\rho/\rho_T)^Y \qquad \text{(Eq 8.11)}$$

where E and E_0 are the elastic moduli of the porous and pore-free materials, respectively, and ρ/ρ_T is the fractional density. The exponent Y has a value of 0.3 to 4. McAdam (Ref 25) has empirically determined the value of Y to be 3.4 for PM plain carbon steels.

Poisson's ratio also varies with fractional density. An empirically determined relation proposed by German (Ref 23) is as follows:

$$\nu = 0.068 \exp\{1.37\rho/\rho_T\} \qquad \text{(Eq 8.12)}$$

where ν is the effective Poisson's ratio, and ρ/ρ_T is the fractional density.

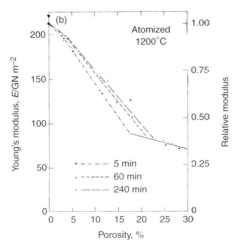

Fig. 8.9 Effect of porosity on Young's modulus of sintered plain carbon steels for three different sintering times (5, 60, and 240 min). Source: Ref 24

REFERENCES

1. G.F. Carter, *Principles of Physical and Chemical Metallurgy*, American Society for Metals, 1979, p 109
2. C.A. Wert and R.A. Thomson, *Physics of Solids,* McGraw-Hill Book Co., New York, 1964, p 379
3. C. Lall, *Soft Magnetism, Fundamentals for Powder Metallurgy and Metal Injection Molding*, MPIF, Princeton, NJ, 1992, p 11
4. E. Adler, G.W. Reppel, W. Rodewald, and H. Warlimont, Matching PM and the Physics of Magnetic Materials, *Int. J. Powder Metall.,* Vol 25 (No. 4) 1989, p 319–335
5. L.W. Baum, Jr., Magnetic Properties of High Density PM Alloys, *Precis. Met.,* Vol 32 (No. 3) 1974, p 47–51
6. K.H. Moyer, M.J. McDermott, M.J. Topoloski, and D.F. Kearney, Magnetic Properties of Iron Alloys, *Magnetic and Electrical PM Technology and Applications, PM Seminar,* MPIF, Princeton, NJ, 1980, p 37
7. J.A. Bas, J. Penafiel, and C. Molins, Jr., PM Continues to Expand Soft Magnetic Role, *Metal Powder Report*, Elsevier Science Ltd., May 1996, p 42–48
8. D.A. De Antonio, Soft Magnetic Ferritic Stainless Steels, *Adv. Mater. Process.,* Oct 2003, p 29–32
9. K. Moyer, Technical Report, Magnatech PM Labs, Aug 3, 1992
10. H.M. Kopech, H.G. Rutz, and P.A. dePoutiloff, Effects of Powder Properties and Processing on Soft Magnetic Performance of 400-Series Stainless Steel Parts, *Advances in Powder Metallurgy and Particulate Materials,* Vol 6, MPIF, Princeton, NJ, 1993, p 217–250
11. F.G. Hanejko, H.G. Rutz, and C.G. Oliver, Effects of Processing and Materials on Soft Magnetic Performance of Powder Metallurgy Parts, *Advances in Powder Metallurgy and Particulate Materials,* Vol 6, J.M. Capus and R.M. German, Ed., 1992 PM World Congress (San Francisco, CA), MPIE, Princeton, NJ, p 376–403
12. B.D. Cullity, *Principles of Physical and Chemical Metallurgy,* American Society for Metals, 1979, p 317–322
13. K. Shimada, H. Hirata, M. Yamaguchi, and H. Ohta, Improvement of Magnetic Properties of Sintered 410L Stainless Steel,

Proc. 1993 PM World Congress, Japan Society of Powder and Powder Metallurgy, p 991–994

14. L.I. Frayman, D.R. Ryan, and J.B. Ryan, Selecting PM Soft Magnetic Materials, *Metal Powder Report,* Elsevier Publications Co., May 1997, p 44–48

15. K.H. Moyer and W.R. Jones, Stainless Steels for Improved Corrosion Resistance, *Advances in Powder Metallurgy and Particulate Materials,* MPIF, Princeton, NJ, 1991, p 145–158

16. S.O. Shah, J.R. McMillen, P.K. Samal, and E. Klar, "Properties of 410L PM Stainless Steel Antilock Brake Sensor Rings," SAE Paper 930449, SAE International Congress and Expo. (Detroit, MI), March 1993

17. M. Svilar and H.D. Ambs, PM Martensitic Stainless Steels, *Advances in Powder Metallurgy,* Vol 2, MPIF, Princeton, NJ, 1990, p 259–272

18. H. Suzuki and H. Ohtsubo, Magnetic Properties of Injection Molded Stainless Steels for Electric Use, *Proc. of 1993 Powder Metallurgy World Congress,* Part 1, Japan Society of Powder Metallurgy (Kyoto, Japan), 1993, p 265–268

19. J.R. Lewis, Physical Properties of Stainless Steels, Chapter 19, *Handbook of Stainless Steels,* D. Peckner and I.M. Bernstein, Ed., McGraw-Hill Book Co., New York, 1977, p 19-1 to 19-36

20. J.R. Davis, *Stainless Steels,* ASM Speciality Handbook, ASM International, 1994

21. Y.S. Touloukian, E.H. Buyco, and P.G. Klemens, Thermophysical Properties of Matter, *Specific Heat — Metallic Elements and Alloys,* Vol 4, IFI/Plenum, New York, Washington, 1990

22. P. Beiss, U. Kutsch, H.-J. Jager, F. Schmitz, and H.R. Maier, Thermal Conductivity of Sintered Stainless Steel 316L, *Proc. 1998 PM World Congress* (Granada), Vol 3, EPMA, Shrewsbury, U.K., 1998, p 425–434

23. R.M. German, *Powder Metallurgy Science,* 2nd ed., MPIF, Princeton, NJ, 1994, p 389.

24. R. Haynes and J.T. Eegdiege, Effect of Porosity and Sintering Conditions on Elastic Constants of Sintered Irons, *Powder Metall.,* Vol 32 (No.1), 1989, p 47–52

25. G.D. McAdam, Some Relations of Powder Characteristics to the Elastic Modulus and Shrinkage of Sintered Ferrous Compacts, *J. Iron Steel Inst. (U.K.),* Vol 168, 1951, p 346–358

CHAPTER 9

Corrosion Testing and Performance

CORROSION DATA of sintered stainless steels should be viewed somewhat differently from those of wrought and cast stainless steels. On one hand, this is due to the lack of corrosion-resistance standards for sintered stainless steels; on the other, it arises from the larger effective surface areas of sintered stainless steels and the chemical reactions taking place during atomizing and sintering. These reactions mainly concern the concentration and distribution of oxides, carbides, and nitrides, as discussed in Chapters 3 and 5. With improved control of these interstitials, corrosion properties improve, and useful applications, at various levels of control, have been identified and are in commercial use.

To date, corrosion resistances of compacted and sintered stainless steel parts with complete control of all interstitials, including low-oxygen contents of 300 ppm or less, have not been published; therefore, it is not surprising that the published corrosion values, particularly those for neutral saline environments, are usually below those of the corresponding wrought stainless steels. With the current knowledge, however, many new applications for sintered stainless steels appear possible, particularly when some of the other advantages of powder metallurgy (PM) processing (Chapter 1, "Introduction") make the PM process more economical. The flexibility of PM processing, including opportunities with alloying and surface modification, should eventually close any existing gaps and probably result in the development of superior materials, as has been the case in other material groups.

This chapter describes corrosion-resistance testing and data of sintered stainless steels. The corrosion data in this section are from published literature references of the past few years, although it should be clear that improvements in corrosion resistance can be realized with process optimization.

9.1 Corrosion-Resistance Testing and Evaluation

Corrosion testing is performed for several reasons:

- To qualify a material for its intended use
- As an acceptance criterion between manufacturer and purchaser/user of stainless steels
- To develop standards of corrosion
- To develop new and superior materials
- For studying corrosion mechanisms and for trouble-shooting corrosion failures
- For monitoring processing variables in the production of stainless steel powders and parts and for general quality control

The majority of the corrosion tests described in textbooks and in the *Annual Book of ASTM Standards* (Ref 1) are accelerated tests and as such should be used with caution. Some of these tests are directly applicable to sintered stainless steels, while others require modification in order to account for the presence and effects of porosity in sintered materials.

For stainless steel powder and parts producers, monitoring of process variables is important because of the many processing variables that can affect the corrosion properties of sintered stainless steels and because of still-lingering misconceptions regarding powder quality and the sintering process.

The so-called ferroxyl test is a quick chemical test that can reveal metallurgical defects of a stainless steel as a result of suboptimal sintering.

Corrosion tests addressing crevice and pitting, intergranular, galvanic, and general corrosion are the most common, whereas information on stress-corrosion cracking, corrosion fatigue, high-temperature corrosion, erosion, and other types of corrosion is just beginning to be developed for sintered stainless steels.

Long-term exposure testing, particularly when combined with metallographic examination and chemical analysis, as discussed in Chapters 3 and 5, permits positive identification of the causes of sensitization.

Electrochemical testing provides information on the passivation properties and resistance to localized corrosion of an alloy as well as on the mechanism of corrosion.

In using corrosion test methods developed for wrought stainless steels, complications arise with some of the tests because of the presence of pores and metallurgical defects that may be present in sintered stainless steels but absent in wrought stainless steels. For poorly or suboptimally sintered stainless steels, it is often necessary to reduce the strength of a testing solution or the length of exposure, or to lower the test temperature, in order to obtain useful corrosion data. Also, the open or interconnected surface area of a sintered part can be many times that of its exterior surface area (Ref 2,3). Maahn and Mathiesen (Ref 4), however, observed that penetration of the testing solution into a sintered stainless steel of a relative density of 86% was only approximately 50% after a 3 week exposure. This signifies that the effective, that is, the actually wetted, surface area was significantly less than that measured by the Brunauer–Emmett-Teller method. They found the high corrosion currents, I_{pass}, of sintered stainless steels to decrease with decreasing porosity (Table 9.1).

Fedrizzi et al. (Ref 5) estimated the internal surface area of a sintered part from mercury porosimetry data and performed comparative electrochemical tests on sintered and wrought 316. The corrosion of the sintered part was found to be over a factor of 10 larger than what would be expected on the basis of its internal surface area alone; this was attributed to the formation of a galvanic couple between the outer free surface of the part and its internal pores. The latter would act as an anode and be subject to corrosion. However, it is not clear to what extent metallurgical weaknesses were responsible

for the inferior results of the sintered material, because the authors provided no information on critical processing variables, such as dewpoint of the sintering atmosphere, dissociated ammonia, and the cooling rate employed in the sintering process.

Because most corrosion tests are accelerated tests, the conditions of testing are much more severe than those usually encountered in the field. Obviously, accelerated tests are in the interest of saving time, because natural corrosion often progresses very slowly. The risk of accelerated testing lies in the uncertainty of applying short-term data to long-term exposure situations. Hence, it is always advisable to simulate actual exposure conditions whenever possible.

At present, only ASTM standard B 895 and MPIF standard 62 (2003) have been adapted for salt solution immersion testing of sintered stainless steels and for 2% sulfuric acid immersion testing of metal injection molded stainless steels, respectively. The majority of corrosion data of sintered stainless steels have so far been conducted in neutral chloride environments because of their practical importance.

Part Preparation. In comparison to wrought materials, the preparation of sintered specimens or parts for corrosion testing is usually minimal. In most cases, the actual parts or specimens are tested in their as-sintered or secondary treated conditions. Because most sintered parts are used without any further processing, it would be counterproductive in most cases to clean the surfaces by mechanical means. This may result in material removal, and the new surfaces may differ in their compositions and structures from the original surfaces. Only the composition and structure of the as-produced original surface is relevant to the corrosion resistance and service life of the part. Even simple degreasing of a porous PM part with an organic solvent should be done with care, because most cleaning liquids will be trapped within the pores and must be removed prior to testing.

It should be mentioned that the primary purpose of so-called passivation treatments—that is,

Table 9.1 Effect of sintering temperature and time on porosity and passive current, I_p

Sintering		Open pores, %	N, %	O, %	C, %	I_{pass}, μA/cm²	E_{pit}, mV SCE	Salt spray, h	FeCl₃ mass loss, %
°C/min	°F/min								
1120/30	2050/30	17.4	380	2230	250	3.4	383	>1500	11.4
1250/30	2280/30	16.3	110	1980	130	3.1	357	>1500	11.1
1120/240	2050/240	15.8	70	1640	130	3.0	508	>1500	11.8
1250/240	2280/240	13.8	20	450	70	1.8	561	260	11.2

Sintering atmosphere: hydrogen; dewpoint: –70 °C (–94 °F). Testing solution: 1000 ppm Cl⁻, pH 5, acetate buffered

pickling of wrought stainless steels in an acid solution (typically a mixture of nitric and hydrofluoric acids)—is to remove free iron and the oxide scales that were formed in the steel mill during heat treatment. A clean stainless steel surface will spontaneously form a passive film in air. It is not necessary to artificially form a passive layer. Nevertheless, the presence of various types of inclusions in stainless steels and their differing responses to mechanical and/or chemical surface treatments can result in significant differences in their corrosion-resistance properties (Ref 6).

In the case of sintered stainless steels, a passivation treatment is expected to remove any surface imperfections, such as contamination with iron or silicon dioxide formed during cooling after sintering. However, metallurgical imperfections present within the bulk of the material, such as precipitates of chromium carbide, chromium nitride, and silicon dioxide that arise from excessive amounts of carbon, nitrogen, and original oxygen (i.e., silicon dioxide formed during water atomization and extending into the bulk for several hundred to several thousand angstroms), are not expected to be removed by passivation. The corrosion resistance of optimally sintered stainless steel parts is not improved by a passivation treatment. Thus, as-sintered parts are used in most cases without further mechanical processing, and, because of the presence of porosity, it is preferable to minimize any metallurgical defects through optimal processing and to omit any cleaning and passivating treatments.

9.1.1 Immersion Testing

Due to its simplicity, immersion testing in acids and neutral salt solutions has been most common for sintered stainless steels. Immersion tests are useful for comparing the corrosion performance of various alloys and for process optimization. Because this type of testing is sensitive to various types of corrosion, it is not possible, in general, to obtain information on the mechanism of corrosion from plain weight losses or from the development of rust. In combination with metallographic and/or chemical analysis, however, such information may be obtained.

Testing in acids usually involves loss of weight, which is used to calculate corrosion rates. For wrought stainless steels, penetration rates are calculated from weight loss data as follows:

$$mm/yr = 87.6 \ W/DAT$$

where D is the density in g/cm^3, W is the weight loss in milligrams, A is the surface area in cm^2, and T is the time in hours. Corrosion rates of less than 0.02 mm/yr are considered outstanding; those from 0.1 to 0.5, 0.5 to 1, and 1 to 5 mm/yr are considered good, fair, and poor, respectively.

While the same procedure can formally be used for sintered, that is, porous, stainless steels, it should be kept in mind that the calculated results may be misleading, because the procedure is not as straightforward as it is for wrought materials:

- There is the problem of entrapment of corrosion products within the pores.
- Expression of weight loss per unit surface area is usually based on the outer geometric surface area of a part, but it actually depends on the effective surface area, that is, on the interior surfaces that participate in the corrosion process. It is sometimes difficult to determine how much of the internal surface area of a specimen takes part in the corrosion process.
- The internal surface area that is accessible to the corrosive medium can vary significantly, depending on the density of a part and its pore structure.

Immersion Tests with Mass Loss. Immersion tests at elevated temperature or in boiling solutions, fashioned after ASTM A 262 and A 763, practice Z, for assessing susceptibility of wrought stainless steels to intergranular corrosion, have also been used for sintered stainless steels. Samal and Terrell (Ref 7) found that while ASTM standard test A 262–01 for intergranular corrosion of wrought stainless steels also gave useful results for sintered stainless steels, weight loss data from immersion tests in 2% (MPIF standard 35) and 10% H_2SO4 (room temperature, 24 h) did not correlate with the damage inflicted on the specimens as a result of intergranular corrosion.

Corrosion rates of sintered stainless steels have been reported for immersion testing in nitric (Ref 8), sulfuric (Ref 9–11), and acetic (Ref 12) acids. Compared to their wrought counterparts, the PM parts exhibited much greater weight losses. In the absence of information on concentration and form of interstitials and actual surface areas (i.e., including pore space), the inferior performance was usually ascribed to either crevice corrosion or the larger surface areas in sintered stainless steels. Frequently, however, excessive corrosion in sintered stainless steels is due to the presence of metallurgical defects arising from inappropriate sintering. Recent data on optimally processed niobium-stabilized 444L (444LNb), for example, showed the sintered material (at a density of 6.6 g/cm^3, when exposed to 2% H$_2$SO$_4$ for 24 h at room temperature) to be fully resistant, while wrought 410 and 430 corroded severely under the same conditions (Ref 13).

Immersion Tests without Mass Loss. Immersion testing without a resultant mass loss includes testing in neutral salt solutions. Again, complementary metallographic analysis is required for gathering information on the mechanism of corrosion.

One of the authors has developed a salt solution immersion test that has now become standardized as ASTM B 895–99 ("Test Methods for Evaluating the Corrosion Resistance of Stainless Steel Powder Metallurgy (P/M) Parts/Specimens by Immersion in a Sodium Chloride Solution"). It is widely used for sintered stainless steels because it is simple, inexpensive, and flexible; PM parts producers can use it in-house for optimizing their sintering processes for stainless steels. Each test is based on five or more pressed-and-sintered replicate specimens that are exposed to a 5% aqueous NaCl solution in individual beakers at room temperature. The specimens are examined visually, at predetermined time intervals, for the onset of staining or rusting and thereafter for an estimation of the percentage of surface area covered by stain or rust, in accordance with the following four rating classes:

- A: Sample free from any corrosion
- B: Up to 1% of surface covered by stain or rust
- C: 1 to 25% of surface covered by stain or rust
- D: >25% of surface covered by stain or rust

A chart containing photographs of a series of specimens for each of the four rating classes is referred to, for maintaining the accuracy of rating (Fig. 9.1). The time intervals between inspections are short at the beginning of the test and are increased gradually as the test progresses. Figure 9.2 shows a typical plot for the test data listed in Table 9.2.

From this plot, the mean lives (in hours) for the specified degrees of surface corrosion are obtained as the intersection points of the respective rating curves with the horizontal line at 50%. If the specimens are not fabricated properly or handled carefully (for example, wide variations in oxygen, nitrogen, or carbon contents; density variations; contamination; etc.), the scatter of the test data will increase.

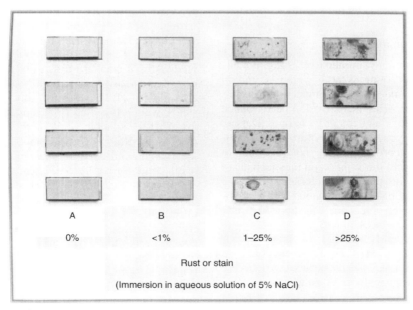

| A | B | C | D |
| 0% | <1% | 1–25% | >25% |

Rust or stain

(Immersion in aqueous solution of 5% NaCl)

Fig. 9.1 Photographic chart of sintered stainless steel transverse-rupture specimens tested in 5% aqueous NaCl by immersion. Extracted, with permission, from ASTM standard B 895–05. Source: Ref 14

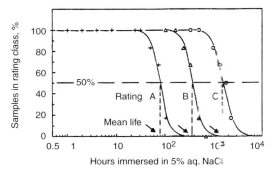

Fig. 9.2 Plot of percentage of replicate specimens with a given rating versus immersion time. Extracted, with permission, from ASTM standard B 895–05. Source: Ref 14

With immersion testing in 5% NaCl, corrosion resistances up to a C-rating (<25% rust) usually have little effect on the mechanical properties of a PM part. At these levels of rusting, weight-loss measurements are unreliable, because some of the rust, as well as salt, may penetrate the pores and cause weight gains instead of weight losses. For applications that can tolerate larger amounts of corrosion, the determination of the amount of surface rust is often insufficient for part qualification. In cases of pitting, determination of number and depth of pits in porous materials can be complex, although their effects on mechanical properties are less severe than in wrought steel. In such cases, it is often useful to remove surface rust (for example, by sandblasting) and to provide photographic evidence of pitting in combination with relevant mechanical properties in accordance with section 7.5 of ASTM G 46, "Practice for Examination and Evaluation of Pitting Corrosion." Electrochemical testing (section 9.1.3 in this chapter) appears to be a better technique for identifying and assessing pitting corrosion in sintered stainless steels.

Yuan et al. (Ref 15) have used a colorimetric method to determine the amount of corrosion of several stainless steels, sintered under various conditions and immersion and spray tested in a neutral salt solution. The method is based on dissolving particulate iron and reducing Fe^{3+} to Fe^{2+}, then determining the amount of Fe^{2+} photometrically.

Correlation with visual assessment of the amount of rust and stain was moderate to good.

9.1.2 Salt Spray Tests

Salt spray testing, in accordance with ASTM B 117, has been widely used in the PM industry in recent years for qualifying new stainless steel parts or PM substitutes for wrought parts. Because many PM parts are subject to atmospheric exposure during their service life, long-term salt spray testing, despite certain shortcomings, appears to be a realistic test. Like immersion testing, it is sensitive to metallurgical defects as well as to porosity. It may also be used for process optimization. The neutral salt spray test is more aggressive than the neutral salt immersion test. The visual rating system described previously may also be used for salt spray testing. A similar rating system is described in ISO 4540 (Ref 16). Mathiesen and coworkers (Ref 17, 18) have provided the most comprehensive account to date on the corrosion resistances of sintered austenitic stainless steels processed under varying conditions. Besides salt spray testing, they used several electrochemical tests on the same stainless steel specimens. In general, there was reasonable agreement between the various methods. Discrepancies were either attributed to differences in corrosion mechanism or to metallurgical defects from suboptimal processing. In Table 9.3, the results of a 5% NaCl salt spray test with 316L specimens of variable density show that corrosion resistance drops rapidly as the sintered density approaches values in the vicinity of 6.6 to 6.9 g/cm^3.

Table 9.2 Example of corrosion rating chart for a set of six replicate specimens of sinstered 316L stainless steel

Hours immersed in 5% aqueous NaCl																		
0.5	1	2	4	8	24	31	50	74	104	168	240	336	496	696	984	1364	1804	2283
A	A	A	A	A	A	A	A	B	B	B	B	C	C	C	C	D	D	D
A	A	A	A	A	A	A	A	A	A	B	B	B	B	C	C	C	D	D
A	A	A	A	A	A	A	B	B	B	B	C	C	C	D	D	D	D	D
A	A	A	A	A	A	A	A	A	B	B	B	B	C	C	C	C	C	D
A	A	A	A	A	A	A	A	A	B	B	B	B	C	C	C	C	C	C
A	A	A	A	A	A	A	A	A	B	B	B	B	C	C	C	C	C	D

A: Sample free from any corrosion; B: up to 1% of surface covered by stain or rust; C: 1 to 25% of surface covered by stain or rust; D: >25% of surface covered by stain or rust

Mathiesen (Ref 18) was unable to develop this relationship with anodic polarization under conditions of rapid scanning nor with the much slower stepwise polarization under the standard conditions of 0.17% NaCl and where the measured initiation potentials were approximately 250 mV (SCE). However, he obtained a strong relationship with stepwise polarization in 0.5% Cl⁻.

Figure 9.3 and Table 9.4 show salt spray corrosion data for several sintered stainless steels processed under various conditions (Ref 15, 19). The sintering conditions, sintered densities, and interstitial analyses of the sintered specimens of Fig. 9.3 are identified in Table 9.5. Although the corrosion-resistance evaluation details in Fig. 9.3 and Table 9.4 differ somewhat, it is apparent that, for identical stainless steels, the corrosion resistances of Fig. 9.3 are, in some cases, over an order of magnitude inferior to those of Table 9.4. The lower corrosion resistances of Fig. 9.3 are due to suboptimal sintering, thus demonstrating again the importance of controlling the sintering conditions.

In some cases, one can recognize the suboptimal conditions directly from the chemical analyses of the various interstitials (Table 9.5) and/or from the sintering conditions employed. (In the authors' experience, gas analyses often exhibit excessive scatter due to sample preparation and/or from the analysis itself. The values denoted by a question mark in parenthesis in Table 9.5 are by the authors and are considered suspect.) Samples sintered in nitrogen-enriched dissociated ammonia (31% N_2), for example, had particularly low corrosion resistances. In accordance with Fig. 5.44 in Chapter 5, 304L, 316L, and SS-100, sintered at 1316 °C (2401 °F), have equilibrium nitrogen contents from approximately 0.24 to 0.28%, signifying that these should also be the approximate nitrogen levels in the sintered specimens. Measured nitrogen contents (Table 9.5), however, ranged from 0.43 to 0.82%, reflecting the large amount of nitrogen absorbed during cooling. The push rate of the pusher furnace was 2.5 cm/min (1 in./min), apparently without any provision for accelerated cooling. The lower-density samples of this series (except for 304N2) show higher nitrogen values, reflecting the greater absorption of nitrogen during cooling due to improved gas diffusion in the more open pore structure. However, with rapid cooling, as required for sintering in dissociated ammonia, nitrogen contents will be close to their high-temperature equilibrium values and independent of sintered density (section 5.2.4 of Chapter 5, "Sintering and Corrosion Resistance").

In comparison to the regular austenitic alloys of Table 9.5, their tin-modified counterparts have nitrogen contents close to their equilibrium values of approximately 0.2% and exhibit less dependence on sintered density. This is because tin, residing on the surfaces of a material, reduces the absorption rate of nitrogen during cooling. This explains, at least in part, the superior performance of these grades of stainless steel, even when processed in a nitrogen-enriched atmosphere with no provision for accelerated cooling.

With suboptimal sintering, ranking of different alloys with respect to corrosion resistance does not necessarily reflect the ranking obtained with optimal or nearly optimal sintering. Table 9.4 shows corrosion data for testing in 5% NaCl by immersion. The data in column C (hours to 1% strain) for 303L, 304L, 316L, and their copper/tin-modified LSC versions (values in parentheses), particularly the upper values of their ranges, show very good agreement with the authors' own optimal data (Fig. 9.14). This suggests that sintering had been performed under close-to-optimal conditions.

Table 9.3 Effect of 316L sintered density on stepwise initiation potential, E_{stp}

Compaction pressure		Sintered density, g/cm³	Open pores, %	O, ppm	i_{peak}[a], μA/cm²	i_{pass}[a], μA/cm²	E_{pit}[a], mV SCE	E_{stp}[b], mV SCE	NSS1, h	NSS2
Mpa	ksi									
295	43	6.34	19.4	340	31	20	475	0	>1500	9
390	57	6.62	15.5	1260	18	19	425	−100	985	7
490	71	6.86	12.3	970	25	15	475	−75	36	4
540	78	6.94	10.8	1900	18	15	500	−200	60	3
590	86	7.02	9.7	1410	21	14	450	−125	28	2
685	99	7.13	7.6	2150	9	7	500	−225	48	2
785	114	7.23	5.7	2040	7	7	475	−200	24	2

(a) 0.1% Cl⁻, pH 5, 30 °C (86 °F), 5 mV/min. (b) 5% NaCl, 30 °C (86 °F), 25 mV/8 h

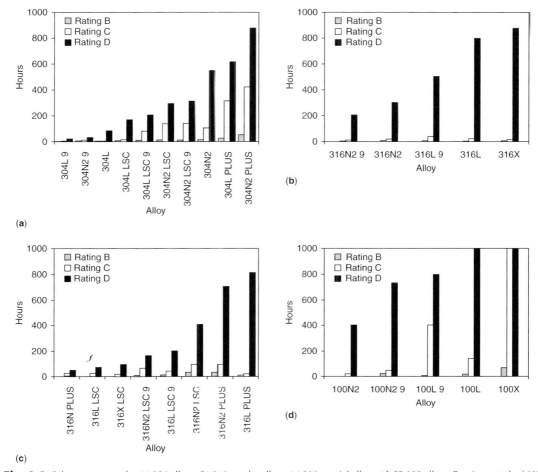

Fig. 9.3 Salt spray test results. (a) 304 alloys. (b) 316 regular alloys. (c) 316 special alloys. (d) SS-100 alloys. B-rating, attack of 1% or less of the surface; C-rating, attack of 1 to 25% of the surface; D-rating, attack of more than 25% of the surface. Source: Ref 15. Reprinted with permission from MPIF, Metal Fowder Industries Federation, Princeton, NJ

While the immersion data reflect the effect of increasing alloy content from 303L to 316L, the 100 h salt spray data for these three alloys do not show this relationship, because the exposure times of the three alloys are identical, 20 to 24 h, for reaching 0.1% of surface stain, depending on sintering conditions.

9.1.3 Electrochemical Tests

Fundamentals. Electrochemical test methods are widely used for wrought and cast stainless steels because they are fast and provide quantitative information on corrosion resistance and corrosion rate, as well as on the mechanism of corrosion. Although the same basic electrochemical criteria are used to characterize sintered materials as are used for wrought metals, the presence of pores and metallurgical defects in sintered materials can complicate the interpretation of electrochemical data. Also, care must be exercised to minimize influences from polishing, cleaning, and degreasing sintered materials.

Most electrochemical data are derived from polarization experiments. Figure 9.4 shows a schematic polarization curve for wrought stainless steel (Ref 20) As the potential is increased from its open-circuit value or its corrosion potential, E_{corr} to what is known as primary passivation potential, E_{pp}, the current density becomes the critical current density, i_{crit}. Between E_{corr} and E_{pp}, the metal is in its active state and undergoes dissolution. At E_{pp}, the passive film begins to form. Thus, the lower the i_{crit}, the easier it is to passivate a material or to remain passive. With the formation

Table 9.4 Salt spray cabinet test results

AISI	Ames	MPIF	Density g/cm³	Hours to 0.1% stain(a)	Weight loss(b), %	Hours to 1% stain(c)	Current density(d) mA/cm²
303	SFN-Cr 18-N11-64	SS-303N1-25	6.4	20	N/A	1.5 (200–500)	25
303	SFAN-Cr 18-N11-66	SS-303N2-35	6.6	20	N/A	N/A	16
303	SFA-Cr 18-N11-70	SS-303L-12 (approx)	7.0	24	N/A	N/A	9 (7)
304	SFN-Cr19-N10-64	SS-304N1-30	6.4	20	N/A	50–100 (500–1200)	27
304	SFAN-Cr19-N10-66	SS-304N2-33	6.6	20	N/A	N/A	18
304	SFA-Cr19-N10-70	SS-304L-13 (approx)	7.0	24	N/A	N/A	11
316	SFN-Cr 17-N12-M1-64	SS-316N1-25	6.4	20	N/A	200–500 (500–1500)	26
316	SFAN-Cr 17-N12-M1-66	SS-316N2-33	6.6	20	N/A	N/A	15
316	SFA-Cr 17-N12-M1-70	SS-316L-15 (approx)	7.0	24	7	N/A	3
410	SFAN-Cr12-67	SS-410-90HT (approx)	6.7	5	N/A	N/A	>40
410	SFA-Cr12-70	...	7.0	5	N/A	N/A	>40
420	SFAN02-Cr12-67	SS410-90HT (approx)	6.7	5	N/A	N/A	N/A
434	SFA-Cr17-M1-70	...	7.0	24	9	N/A	0.2
632	SFA-Cr16-N5-M1-70	...	6.9	22	5	N/A	N/A

Source: Ref 19

Note: N/A = data not available. (a) 5% aqueous NaCl salt spray per ASTM B117. (b) Weight loss after 8 hours in 1% HC1. (c) 5% aqueous NaCl salt immersion in accordance with ASTM G31. Values in parentheses refer to tin-modified grades. (d) Electrolytic corrosion testing in accordance with ASTM B627 measured at IV.

Table 9.5 Sintering conditions and resulting properties of stainless steel samples of Fig. 9.3

Designation	Alloy	Sintering conditions(a)			Sintered density(b), g/cm³	C	N	O
304N2	304 SS	1316 °C	(2400 °F)	DA(c)	6.32	0.002	0.550	0.170
304N2 9	304 SS	1316 °C	(2400 °F)	DA(c)	6.90	0.016	0.560	0.190
304L	304 SS	1288 °C	(2350 °F)	vacuum	6.31	0.002	0.010	0.220
304L 9	304 SS	1288 °C	(2350 °F)	vacuum	6.89	0.004	0.010	0.200
304N2 LSC	304 LSC	1316 °C	(2400 °F)	DA(c)	6.49	0.006	0.220	0.180
304N2 LSC 9	304 LSC	1316 °C	(2400 °F)	DA(c)	6.91	0.010	0.220	0.150
304L LSC	304 LSC	1288 °C	(2350 °F)	vacuum	6.42	0.002	0.008	0.180
304L LSC 9	304 LSC	1288 °C	(2350 °F)	vacuum	6.92	0.006	0.078	0.450
304N2 PLUS	304 PLUS	1316 °C	(2400 °F)	DA(c)	6.40	0.008	0.140	0.170
304L PLUS	304 PLUS	1288 °C	(2350 °F)	vacuum	6.41	0.004	0.044	0.280
316N2	316 SS	1316 °C	(2400 °F)	DA(c)	6.54	<0.001	0.570	0.150
316N2 9	316 SS	1316 °C	(2400 °F)	DA(c)	6.83	0.010	0.430	0.190
316L	316 SS	1288 °C	(2350 °F)	vacuum	6.46	0.007	0.007	0.250
316L 9	316 SS	1288 °C	(2350 °F)	vacuum	6.86	0.006	0.014	0.230
316N2 LSC	316 LSC	1316 °C	(2400 °F)	DA(c)	6.45	0.008	0.180	0.100
316N2 LSC 9	316 LSC	1316 °C	(2400 °F)	DA(c)	6.87	0.010	0.210	0.210
316L LSC	316 LSC	1288 °C	(2350 °F)	vacuum	6.49	0.002	0.052	0.120
316L LSC 9	316 LSC	1288 °C	(2350 °F)	vacuum	6.83	0.005	0.091	0.250
316N2 PLUS	316 PLUS	1316 °C	(2400 °F)	DA(c)	6.42	0.004	0.110	0.120
316L PLUS	316 PLUS	1288 °C	(2350 °F)	vacuum	6.45	0.007	0.045	0.150
316X	316 SS	1288 °C	(2350 °F)	vacuum	6.65	0.005	0.001	0.295
316X LSC	316 LSC	1288 °C	(2350 °F)	vacuum	6.69	0.006	0.002	0.174
316X PLUS	316 PLUS	1288 °C	(2350 °F)	vacuum	6.47	0.007	0.008	0.215
100N2	316 SS	1316 °C	(2400 °F)	DA(c)	6.29	0.006	0.820	0.039
100N2 9	100 SS	1316 °C	(2400 °F)	DA(c)	6.82	0.010	0.610	0.054
100L	100 SS	1288 °C	(2350 °F)	vacuum	6.25	<0.001	0.011	0.140
100L 9	100 SS	1288 °C	(2350 °F)	vacuum	6.71	<0.001	0.011	0.140
100X	100 SS	1288 °C	(2350 °F)	vacuum(d)	6.72	0.007	0.001	0.144

(a) DA, dissociated ammonia. (b) Density was measured with oil impregnation in accordance with ASTM B 328. (c) Actual furnace atmosphere was 91.8 vol% DA + 8.2 vol% N$_2$. (d) Followed by a hold at 1150 °C (2100 °F) for 10 min

of the passive film, the current density decreases rapidly and reaches a value known as passive current density, i_p. The lower the passive current density, the lower the corrosion in the passive region. In the absence of an aggressive species, such as chloride ions, the material remains passive until transpassive dissolution or, for example, oxygen evolution occurs at the transpassive potential, E_t. In the presence of chloride ions, the steel is subject to pitting corrosion, which commences as the potential reaches the pitting potential, E_{pit} (also known as the breakdown potential, E_{bd}). The higher the pitting potential and the lower the corrosion

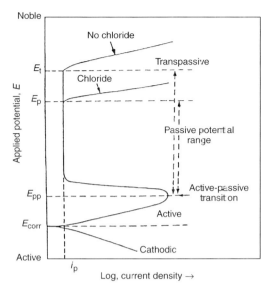

Fig. 9.4 Polarization curve for a stainless steel in a sulfuric acid solution. Source: Ref 20. ©NACE International 1986

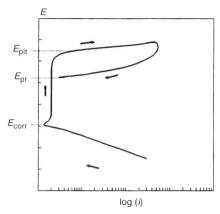

Fig. 9.5 Schematic cyclic polarization curve illustrating hysteresis

potential, that is, the more extended the passive region (or the greater $E_{pit} - E_{corr}$), the better the resistance to pitting corrosion and the better the passivating characteristics. Of the various features of a polarization curve, E_{pit} is probably the most widely studied characteristic, because it is very marked and distinct.

Reversal of the direction of the potential scan from a current density value beyond E_p (Fig. 9.5) produces a hysteresis. The intersection of the forward and reverse scans of this hysteresis is denoted as protection (against pit propagation) potential, E_{pr}, which is defined as the most noble potential where pitting and crevice corrosion will not propagate.

In wrought alloys, both E_{pit} and E_{pr} have been found to depend on the potential scan rate in a cyclic potentiodynamic polarization test. This is related to the induction time required for pitting and repassivation. To overcome such problems, potentiostatic and galvanostatic methods for localized corrosion have been developed.

The effect of various alloying elements on the anodic polarization behavior of stainless steels has already been shown in Fig. 6.1 in Chapter 6, "Alloying Elements, Optimal Sintering, and Surface Modification in PM Stainless Steels."

Metallurgical Defects and Porosity. As mentioned earlier, sintered stainless steels may possess metallurgical weaknesses or defects in addition to their porosity. The metallurgical defects arise from the presence of excessive amounts of the interstitials nitrogen, carbon, and

oxygen and/or from the formation of intermetallic phases. The interaction of such defects among themselves combined with porosity can make the interpretation of corrosion data rather difficult. Comparing materials of different origin and/or different processing may yield unexpected or inconsistent results. It is best to compare materials on the basis of parametric studies, keeping all variables fixed except that under study. However, even under such conditions, if the material to be studied is not in its optimally sintered condition, the results from parametric studies can still be complex and inconclusive. Many studies quoted in the literature omit a full description of the sintering conditions employed, although, in some cases, the approximate conditions can be gleaned indirectly from other data.

To date, Mathiesen (Ref 4, 18, 21, 22) has given perhaps the most comprehensive account of electrochemical testing methods for sintered stainless steels. He has investigated the effects of polarization scan rate, activation period, buffering of the electrolyte, neutral chloride and acidic environments, sintered density, and several other sintering parameters. Occasionally, however, the simultaneous presence of several metallurgical defects and porosity obscured the expected relationships.

Short-Term Exposure Tests. It is convenient to distinguish between short-term and long-term exposure tests. Short-term exposure tests include cyclic polarization, anodic polarization, and the electrochemical potentiokinetic reactivation (EPR) method.

Cyclic polarization provides information on both corrosion characteristics and the corrosion mechanism (Fig. 9.6). The ease of passivation of a material and its corrosion rate in the passive region are described by I_{peak} and i_{pass} respectively. Its susceptibility to pitting corrosion is described by E_{pit} (Ref 4, 18, 21, 22). Mathiesen observed variable results with cyclic polarization that starts from the corrosion potential. Because of this, he standardized polarization experiments by exposing a specimen for 10 min to an active potential of –650 mV saturated calomel electrode (SCE) prior to the potential scan. Figure 9.7 shows examples of cyclic polarization curves of various specimens of sintered 316L, performed with prior activation.

Figure 5.6 in Chapter 5 illustrates that decreasing the pore size increases the width of the

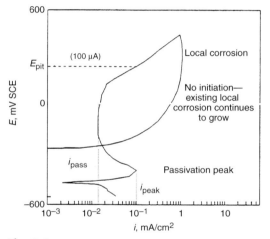

Fig. 9.6 Cyclic polarization curve. Source: Ref 4, 18, 21, 22

hysteresis, as defined by the potential difference E_b (or E_p, E_{bd}) minus E_{pr}. The relationship is similar to Fig. 5.5, which shows the effect of sintered density on the corrosion resistance of 316L, as measured by immersion in 5% NaCl. At low densities, corresponding to larger pore sizes, the detrimental effect of pores is small and gradual. At higher densities, it becomes very pronounced, producing very low corrosion resistances or very large hystereses.

Anodic Polarization. Mathiesen found the passive current density to decrease significantly with decreasing scan rate, whereas the initiation potential was only slightly affected. In most of the tests, he used a scan rate of 5 mV/min as a compromise between practicality and reproducibility.

Anodic polarization test results for sintered 316L stainless steels (Fig. 9.8) show that both oxygen and carbon lower the pitting potential. Samples marked with an asterisk (F08) had extremely low E_{pit} values due to the presence of 7180 ppm of nitrogen from sintering in dissociated ammonia and an insufficient cooling rate.

EPR Method. Electrochemical potentiokinetic reactivation permits the evaluation of sensitization in stainless steels. The single-loop EPR method is more complex and requires, in addition to the electrochemical test, the determination of the material grain-boundary area. In the more rapid double-loop EPR method, polarization starts at the free corrosion potential (at a rate of 6 V/h), increasing to 300 mV (SCE), followed by a reversal to the starting potential (Fig. 9.9) (Ref 4, 18, 21, 22).

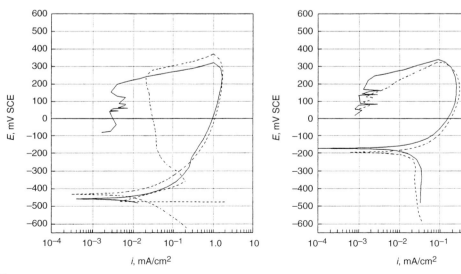

Fig. 9.7 Cyclic polarization curves of sintered 316L specimens. Polarization is started at the free corrosion potential

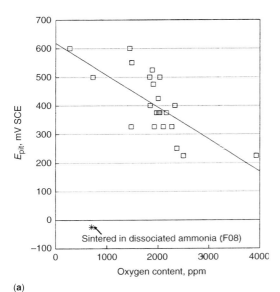

(a)

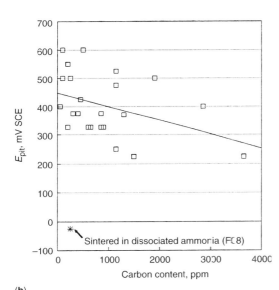

(b)

Fig. 9.8 Pitting potential versus (a) oxygen and (b) carbon content for 316L

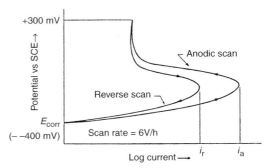

Fig. 9.9 Schematic of double–loop electrochemical poten tiokinetic reactivation technique. Source: Ref 4, 18, 21, 22

Susceptibility to sensitization is determined from the ratio of the maximum current on the reverse scan, i_r, to that measured on the forward scan, i_a. This current ratio also has been correlated with the so-called step, dual, and ditch structures observed in oxalic-acid-etched sensitized wrought stainless steels (ASTM A 262) (Fig. 9.10) (Ref 23).

Current ratios between 0.0001 to 0.001 correspond to step structures (free of grain-boundary attack), ratios of 0.001 to 0.05 to dual structures (carbides or nitrides partially in grain boundaries), and ratios of 0.05 to 0.3 to ditch structures (carbides or nitrides throughout grain boundaries). Figure 9.11 shows examples of this technique for 316L specimens sensitized by excessive nitrogen content and the presence of an intermetallic phase from liquid-phase sintering.

Baran et al. (Ref 24) proposed a modified, basically milder version of ASTM A 763, practice Z, test procedure for identifying intergranular corrosion due to sinter sensitization, that is the sensitization occurring during slow cooling of sintered parts, particularly from 800 to 425 °C (1470 to 800 °F).

Long–Term Exposure Tests. In order to determine the complex effect of pores, it is necessary to use more realistic, long time-exposure tests, which permit the time-consuming development of localized corrosion within pores. Open-circuit potential versus time curves and stepwise polarization are two such methods that permit a characterization of time-dependent corrosion phenomena. The former method provides information on the nature of corrosion, the latter on the passive behavior and susceptibility to pitting and crevice corrosion (Ref 25–27).

Open-Circuit Potential. The open-circuit or corrosion potential of an alloy is usually determined prior to the initiation of a polarization experiment. Figure 9.12 shows potential- time curves of wrought and PM 316L stainless steels in 5% NaCl solution at room temperature (Ref 28).

The change of the potential of wrought 316L toward more positive values indicates that the passive oxide film is self-healing (Ref 29) and/or increasing in thickness. In contrast, the potential of a PM 316L often shifts toward more

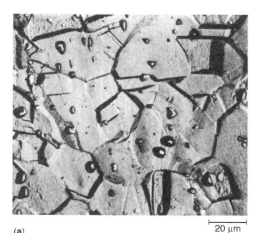

(a)   20 µm

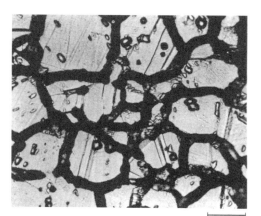

(b)   20 µm

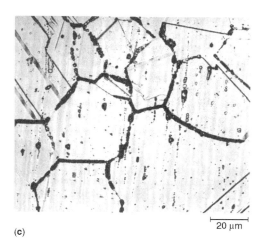

(c)   20 µm

Fig. 9.10 Oxalic acid etches at original magnification 500x.
Etched 1 A/cm² for 1.5 min. (a) Step structure. (b)
Ditch structure. (c) Dual structure. Source: Ref 23

negative values. If the density of the sintered part falls within the crevice-sensitive region (section 5.2.2 in Chapter 5), this activation can be interpreted as destruction of the passive film on the interior pore surfaces as a result of the limited access of oxygen and the acidification of the stagnant solution within the pores (Ref 30). Or, if a part was sintered suboptimally, other defects may interfere with the formation of a continuous passive film (Ref 4).

Whereas the wrought stainless steel remains passive during the entire testing period, the sintered materials, although passive during most of the time, exhibit characteristic, short-lived activation peaks. These instabilities may arise from the penetration of the testing solution into the pore space of a sintered part.

The higher corrosion currents of the sintered materials arise from their large internal surfaces; they decrease with increasing sintered density.

Polarization in the vicinity of the free corrosion potential exhibits a very nonlinear behavior for sintered stainless steels.

Stepwise Polarization. In contrast to normal polarization, stepwise polarization, E_{stp}, is performed over an extended time period. This permits the time-dependent development of corrosion within crevices and pores. Thus, in contrast to normal, accelerated anodic polarization, stepwise polarization can reveal the crevice effect of pores. Table 9.6 shows corrosion data of 316L specimens pressed to various densities and sintered at 1250 °C (2282 °F) in hydrogen. With increasing density, E_{stp} in 0.5 Cl⁻ decreases (Ref 31).

The detrimental effect of small pores, causing increased crevice corrosion at densities of approximately 6.7 g/cm³ and higher, does not show up in a lower E_{pit} because of the fast scanning normally employed in potentiodynamic polarization tests. By reducing the scanning rate from 5 mV/min to 25 mV/8 h, steady-state conditions are assured and allow for the induction time needed for the onset of crevice corrosion. The stepwise potential in Table 9.6 indeed registers lower potential values that decrease with increasing density.

9.1.4   Ferric Chloride and Ferroxyl Tests

Ferric Chloride Test. The use of ferric chloride solution for testing wrought stainless steels for pitting and crevice corrosion is described in ASTM G 48. It has been used for sintered stainless steels to qualitatively identify these types of corrosion.

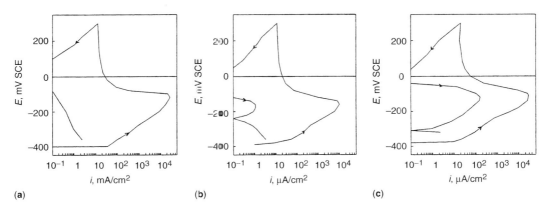

Fig. 9.11 Polarization curves for 316L powder metallurgy steels obtained by the electrochemical potentiokinetic reactivation double-loop technique in 0.5 M H_2SO_4 + 0.01 M KSCN (30 °C, or 86 °F). (a) Steel without sensitization. (b) Sensitized steel with 1850 ppm nitrogen. (c) Liquid-phase-sintered steel with addition of boron

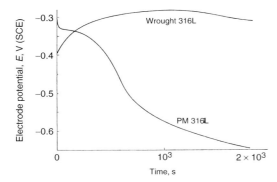

Fig. 9.12 Potential-time curves for wrought and sintered 316L stainless steel in aerated 5% NaCl solution. Source: Ref 28. Reprinted with permission from MPIF, Metal Powder Industries Federation, Princeton, NJ

Ferroxyl Test. This test is based on the use of hexacyanoferrate (II/III) solution with additions of variable amounts of NaCl (Ref 32). It can reveal metallurgical weaknesses caused by improper sintering conditions or contamination with iron (section 3.4 in Chapter 3, "Manufacture and Characteristics of Stainless

Steel Powders"). Results for sintered 316L and several wrought stainless steels are shown in Table 9.7.

For visual inspection of the specimens, the following rating scale is used:

Category	Amount of attack	Visual inspection results
0	No attack	No blue spots
1	Light attack	Very weak blue spots without growth
2	Moderate attack	Blue spots with slow growth; no needle growth or large accumulation of blue dye
3	Severe attack	Blue spots with growth; needle growth or accumulation of the corrosion product on the surface

The base solution for this test is made by adding 0.99 g K_3 [$Fe(CN)_6$] and 1.27 g K_4 [$Fe(CN)_6$] $3H_2O$ to 1000 mL deionized water. From this solution, other solutions with chloride concentrations of 0.05, 0.1, and 0.5% Cl^- are made, giving redox potentials of 160, 165, and 180 mV

Table 9.6   Effect of density for 316L sintered at 1250 °C (2282 °F) for 120 min in pure hydrogen

Green density g/cm³	Density(a), g/cm³	Open pores(a), %	Average pore diameter(b), µm	Roundness(b)	NSS1, h	NSS2	Ferroxyl test, 0.5% Cl⁻	E_{stp}(c), mV SCE Cl⁻ 0.1%	0.5%
5.80	6.40	19.3	9.5	0.73	1336	9	0	350	150
5.91	6.51	17.8	8.8	0.74	>1500	10	0	250	150
6.05	6.65	15.9	8.0	0.72	>1500	10	0	275	100
6.19	6.75	14.4	8.7	0.79	>1500	10	0	250	125
6.25	6.83	13.4	8.0	0.74	>1500	10	0	300	100
6.38	6.93	11.8	7.3	0.75	1168	9	0	325	100
6.44	7.01	10.7	6.1	0.71	192	5	0	300	50

(a) Measured by oil impregnation technique. (b) Measured by image analysis. (c) Stepwise polarization. Source: Ref 31

Table 9.7 Ferroxyl test results for powder metallurgy (PM) and wrought stainless steels

AISI No.	Type, treatment	Visual rating(a) 0.05%
316L	PM, 1120 °C (2048 °F) 20 min, dissociated ammonia (–27 °C, or –17 °F)	3
316L	PM, 1250 °C (2282 °F) 30 min, H$_2$ (–35 °C, or –31 °F)	...
316L	PM, 1120 °C (2048 °F) 30 min, H$_2$ (–70 °C, or –94 °F)	...
316L	PM, 1120 °C (2282 °F) 120 min, H$_2$ (–70 °C, or –94 °F)	...
316L	PM, 1250 °C (2282 °F) 120 min, H$_2$ (–70 °C, or –94 °F)	...
431	Wrought	3
303	Wrought	3
304	Wrought	1
316	Wrought	...

(a) 0: No attack and no blue spots. 1: Light attack and very weak blue spots without growth. 2: Moderate attack. Blue spots with slow growth but no needle growth or large accumulation of blue dye. 3: Servere attack. Blue spots with needle growth or accumulation of corrosion product on the surface

SCE (±5 mV), respectively, at the test temperature of 25 °C (77 °F).

The test is fast and simple to perform and has been recommended for rapid testing of sintered stainless steel parts in a plant environment. However, agreement with salt spray testing and electrochemical measurements is only moderate.

9.1.5 Elevated-Temperature Oxidation Resistance

Only recently, with the development of sintered stainless steel automotive exhaust components, has the subject of elevated-temperature oxidation of sintered stainless steels assumed a more prominent role. Although complications arise from the presence of pores in sintered metals, attempts to improve their oxidation resistance are based on the principles of oxidation established for solid metals. Many high-temperature alloys rely on chromium to form a protective oxide scale. Solid metals that form protective oxide scales (for example, nickel, iron, and chromium) obey the parabolic equation for the time dependence of oxidation; that is, the diffusion of ions or the migration of electrons through the oxide scale control the rate of oxidation. Thus, in accordance with the Wagner theory of oxidation (Ref 33, 34), the concentration of ionic defects, and therefore the rate of oxidation, can be influenced by doping and by changing the phase structure through alloying.

Effect of Porosity. The presence of porosity in sintered metals causes the kinetics of oxidation, as measured by weight changes, to differ from those of solid metals and depends on both the size and porosity of a specimen. With high porosity, oxidation increases with increasing temperature, as is typical for solid materials. However, as porosity decreases and reaches a point where capillary inlets become blocked by oxidation products, a part of the internal surfaces ceases to participate in the oxidation process, and the oxidation rate then may even decrease with increasing temperature (Ref 35).

Kato and Kusaka (Ref 36) have determined the weight gains in air at 700 °C (1290 °F) for type 310L stainless steel parts that were vacuum sintered 1 h at 1250 °C (2280 °F), as a function of sintered density and mesh size of powder. The initial weight gain did not always show a parabolic course of oxidation. Within the density range studied, oxidation increased almost exponentially with decreasing density. Silicon-modified (4.06% Si) type 310L stainless steel showed weight gains that were less than 50% of those of regular type 310L. Higher sintering temperature and higher compacting pressure (higher densities) reduce surface porosity and specific pore surface area, thus lessening interior oxidation through pore closure. The lesser oxidation of the parts made from the finer powder fraction is probably due to the more difficult diffusion of oxygen through the finer pore structure. The maximum recommended operating temperature for these stainless steels is 700 °C (1290 °F).

Beneficial Effect of Pores. In the application for automotive exhaust flanges, one of the requirements includes alternate exposure to elevated temperature in air (677 °C, 1250 °F) and water quenching. For sintered stainless steels to pass this test, a minimum sintered density of 7.3 g/cm^3 was found to be critical (Ref 37). At this density, interconnected porosity is very low, and oxidation is restricted to the surface regions of a part. Aside from other benefits of the PM materials (Chapter 11, "Applications") in this application, the small degree of surface porosity is sufficient for anchoring the oxides that form during oxidation. As a result, the PM materials exhibit a small weight gain of less than 2% for various ferritic alloys, whereas wrought stainless steels exhibit weight losses of over 10% due to oxide spalling. Sintered parts possessing densities below approximately 7.2 g/cm³ had a high failure rate, because the oxidation extends into the interior of a part and causes thermal fatigue during the alternate heating and water quenching. Ferritic alloys such as 409LNb and 409Ni (Ref 38) appear to be superior to austenitic alloys because of the greater thermal conductivities and the lower coefficients of thermal expansion of the ferritics, which improve their thermal fatigue resistance.

Ishijima and Shikata (Ref 39) confirmed the rapid improvement in high-temperature oxidation resistance when the sintered density reached the regimen where interconnected porosity disappears (Fig. 9.13).

A further remarkable improvement in oxidation resistance of an Fe-17Cr alloy occurred with the addition as a fine dispersion.

Searson and Latanision (Ref 40) reported both improved room-temperature (aqueous environment) and elevated-temperature (oxidation) corrosion resistance of extruded type 303 stainless steel (18Cr-8Ni) made from rapidly solidified powder. The improvement over conventional wrought type 303 of the same composition was attributed to the presence of a uniform dispersion of fine MnS particles (because of the rapid quenching of the powder), which inhibit grain growth. The small distances between grain boundaries are believed to favor chromium diffusion and desirable chromium oxide formation as well as the formation of a fine network of the well-known SiO₂ intrusions. The

latter provide mechanical keying of the oxide film. The improvement in oxidation resistance was similar to that of a higher-alloyed (25Cr-20Ni) type 310 stainless steel.

The rapidly solidified material also exhibited a very low pit density after a 72 h immersion test in a 1% $FeCl_3$ solution, whereas the conventional alloy was highly susceptible to pitting, with some pits penetrating to over 1 mm (0.04 in.) in depth.

9.2 Corrosion Data of Sintered Stainless Steels

The corrosion data in this section are from published literature references of the past few years. Selection of corrosion data has been limited mainly to those references that include relevant processing information. This will permit the reader to compare the various sintering methods and to arrive at a solution that takes into account the entire PM process and represents the best compromise for an intended use.

Figure 9.14 shows corrosion data for the most common sintered austenitic stainless steels for immersion testing in a 5% aqueous solution of NaCl (Ref 41). Sintering conditions marked nonoptimized (N/O), designating processing that produces less than maximum corrosion resistance, are still widely used in the industry for sintering stainless steels. As previously mentioned, dewpoints of the sintering atmosphere, a critical variable, are not always monitored and are often omitted in published data on the corrosion resistances of sintered stainless steels. Accelerated cooling is typically practiced only in vacuum furnaces.

The suboptimal data of Fig. 9.14 are shown to highlight the strong effects of dewpoint and cooling rate on the corrosion resistances of the most widely used austenitic stainless steels. The individual data are averages of six specimens. The standard deviations of such averages are fairly large, sometimes exceeding 100%, for less-than-optimally sintered parts. They decrease to approximately 20 to 25% for optimally sintered parts. "Optimized" designates exclusion of all common corrosion defects (chromium carbides and nitrides, surface oxides formed during cooling, contamination with less noble metals, and crevice corrosion due to unfavorable pore sizes), except for the presence of residual oxides from incomplete reduction during sintering. The

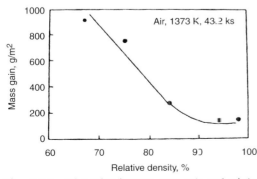

Fig. 9.13 Relationship between mass gain and relative density in Fe–17Cr alloy. Source: Ref 39. Reprinted with permission from MPIF, Metal Powder Industries Federation, Princeton, NJ

Sintering parameters

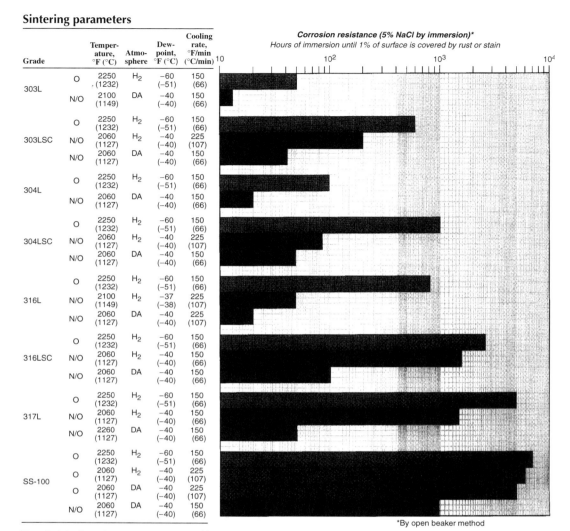

Fig. 9.14 Corrosion resistances (5% NaCl by immersion) obtainable for various grades of stainless steel, sintered under optimized (O) and nonoptimized (N/O) conditions. DA, dissociated ammonia. Sintered densities: 6.4 to 6.6 g/cm³. Sintering time: 45 min. Optimized designates exclusion of all common corrosion defects (chromium carbides and nitrides, surface oxides formed during cooling after sintering, contamination with less noble metals, and crevice corrosion due to unfavorable pore sizes), except the presence of residual oxides from incomplete reduction during sintering. Source: Ref 41

oxygen contents ranged from approximately 1400 to 2200 ppm.

The causes for less-than-optimal sintering conditions can be rationalized from the discussions in Chapter 5, "Sintering and Corrosion Resistance." In most cases, suboptimal properties are due to excessive dewpoints in the case of hydrogen sintering, and to inadequate cooling rates in the case of sintering in dissociated ammonia. Superior optimal values result from sintering above 1232 °C (2250 °F) and/or from longer reduction times, due to more intense oxide reduction under such conditions.

The bar graph shows clearly that for both optimal and suboptimal sintering conditions, corrosion resistance improves with increasing alloy content. Equally important, the greater the alloy content of a material, the less sensitive it is toward less favorable sintering conditions. Wrought 316L, tested under identical conditions, lasted approximately 5000 to 6000 h until 1% of the specimen surface was covered by stain or rust. This is approximately equal to the higher-alloyed sintered SS-100 material.

Table 9.8 (Ref 42) summarizes more neutral salt solution corrosion data, based on immersion

Table 9.8 Corrosion resistances of sintered and wrought stainless steels

Material	Corrosion Test(a)	Corrosion resistance rating(b)		Comments						
		A, h	B, h	Sintered density, g/cm³	Sintering atmosphere(c)	Sintering temperature °C	Sintering temperature °F	Sintering time, min	Type of furnace(d)	Reference(e)
Sintered stainless steels										
303L	I	...	5	6.7–6.8	DA	1150	2100	60	L	...
303LSC(f)	I	...	500	6.7–6.8	DA	1150	2100	60	L	...
304L	I	...	100	6.7–6.8	DA	1150	2100	60	L	...
316L	I	...	500	6.7–6.8	DA	1150	2100	60	L	...
	I	400	...	6.7–6.9	V	1205	2200	60	L	...
	NSS	600	...	6.7	H₂	1150	2100	30	I	18
	NSS	1110	...	6.3	H₂	1150	2100	30	I	18
	NSS	1056	...	6.6–6.7	V	1120	2050	30	L	18
316LSC(f)	I	...	1500	6.7–6.8	DA	1150	2100	60	L	...
	I	1000	1700	6.7–6.9	V	1205	2200	60	L	...
317	I	2400	4400	6.7–6.9	V	1205	2200	60	L	...
	NSS	>1500	...	6.7	H₂	1150	2100	30	I	...
SS100(g)	I	3400	5200	6.7–6.9	V	1205	2200	60	L	...
410L	I	...	200	7.0–7.1	V	1260	2300	60	L	...
434L	I	...	2200	7.0–7.1	V	1260	2300	60	L	...
Wrought stainless steels										
303	NSS	420	...	..	...	...	...	...	...	18
304	NSS	>1500	...	..	...	...	...	...	...	18
316	NSS	>1500	...	..	...	...	...	...	...	18
316L	I	5000	...	..	...	...	...	...	...	...
	NSS	1512	...	..	...	...	...	...	...	17
410	I	200	...	..	...	...	...	...	...	...
431	NSS	72	...	..	...	...	...	...	...	18
434	I	2200	...	..	...	...	...	...	...	...

(a) I, by immersion in 5% NaCl; NSS, neutral salt spray test ASTM B 117; ISO 4540–1980(E). (b) A, time in h until appearance of first stain or rust spot; B, time in h until 1% of surface of specimen is covered with stain or rust. (c) H₂, hydrogen; DA, dissociated ammonia; V, vacuum. (d) L, laboratory; I, industrial. (e) Data without reference numbers are author's data. (f) Proprietary grades of North American Hoganas. (g) 20 Cr17Ni5Mo. Source: Ref 42

Table 9.9 Properties of sintered 316L raw powder, and wrought 316L steel

Sintering	Unit	H₂(–30 °C, or –22 °F)				H₂ (–70 °C, or –94 °F)				Vacuum				Powder(a)	Wrought(b)
		1120 °C (2048 °F)/min		1250 °C (2282 °F)/min		1120 °C (2048 °F)/min		1250 °C (2282 °F)/min		1120 °C (2048 °F)/min		1250 °C (2282 °F)/min			
		30	120	30	120	30	120	30	120	30	120	30	120		
Density	g/m3	6.62	6.68	6.71	6.84	6.62	6.68	6.71	6.84	6.67	6.73	6.76	6.86	...	8.00
N	ppm	400	320	220	60	470	190	110	70	410	220	90	20	700	...
O	ppm	2400	2400	2200	1500	2300	2000	1900	1700	2200	2200	2100	1800	1900	...
C	ppm	230	220	190	130	240	250	170	110	60	60	20	10	180	300
I_{peak}	µA/cm²	150	90	87	83	10	10	7	9	4	7	8	9	...	0
I_{pass}	µA/cm²	29	21	28	19	14	10	12	11	9	13	12	7	...	0.5
E_{pit}	mV SCE	250	243	243	333	345	370	330	395	368	410	363	405	...	665(c)
E_{stp}	mV SCE	269	213	188	163	238	275	188	163	263	238	175	150	...	538(d)
NSS 1(e)	h	36	60	48	24	1392	1278	1260	1512	1056	1008	420	240	...	1512
NSS 2	h	13	24	13	2	1512	1140	1260	60	1512	1008	324	24	...	...

(a) Raw AISI 316L powder. (b) Wrought AISI 316L. (c) Measured with a crevice-free electrode. (d) Measured with a creviced electrode. (e) Time to corrosion in neutral salt spray test; 1, no pretreatment; 2, specimens filled with test solution. Source: Ref 17

and salt spray testing, for various sintered austenitic and ferritic stainless steels as well as some wrought stainless steels for comparison. Again, the broad sintering conditions employed are typical of those used in industry and will permit stainless steel parts producers to gage their progress regarding process optimization for optimal corrosion resistance.

Table 9.9 (Ref 17) summarizes electrochemical test data for 316L sintered in hydrogen and under vacuum. In these examples, inferior corrosion resistances, documented also by neutral salt spray data, are believed to be due to greater amounts of reoxidation of the surface of a specimen during cooling in a higher-dewpoint environment (section 5.2.3 in Chapter 5).

REFERENCES

1. *Wear and Erosion; Metal Corrosion*, Vol 3.02, *Annual Book of ASTM Standards*, ASTM International, 2002
2. L.M. Fedorchenko, A.P. Lyapunov, and V.V. Skorokhod, Phenomena Taking Place During Oxidation of Porous Metals at Elevated Temperatures, *Powder Metall.*, (No. 12), 1963, p 27–43
3. R.T. De Hoff and F.N. Rhines, "The Geometry and Mechanism of Sintering," Second European Symposium on Powder Metallurgy (Stuttgart, Germany), EPMA, 1968
4. E. Maahn and T. Mathiesen, "Corrosion Properties of Sintered Stainless Steel," U.K. Corrosion '91, Oct 1991 (Manchester, U.K.) NACE
5. L. Fedrizzi, J. Crousier, P.L. Bonora, and J.P. Crousier, Corrosion Mechanisms of an AISI Type 316L Sintered Stainless Steel in Sodium Chloride Solution, Werkst. Korros. Vol 42, 1991, p 403–409
6. G.E. Coates, Effect of Some Surface Treatments on Corrosion of Stainless Steel, Mater. Perform., Vol 29, (No. 8) 1990, p 61–65
7. P.K. Samal and J.B. Terrell, "On the Intergranular Corrosion of P/M 316L Stainless Steel," PM^2 TEC 2002 World Congress, June 16–21, 2002, (Orlando, FL), MPIF, Princeton, NJ
8. H.S. Nayar, R.M. German, and W.R. Johnson, The Effect of Sintering on the Corrosion Resistance of 316L Stainless Steel, *Modern Developments in Powder Metallurgy*, Vol 15, H. Hausner, H. Antes, and G. Smith, Ed., MPIF, Princeton, NJ, 1981
9. M.H. Tikkanen, Corrosion Resistance of Sintered P/M Stainless Steel and Possibilities for Increasing It, *Scand. J. Metall.*, Vol 11, 1982, p 211–215
10. T. Takeda and K. Tamura, Compacting and Sintering of Chrome-Nickel Austenitic Stainless Steel Powders, *Powder Powder Metall.* *(Japan)*, Vol 17, (No. 2), 1970, p 70–76
11. A. Kempster, J.R. Smith, and C.C. Hanson, Chromium Diffusion Coatings on Sintered Stainless Steel, *Metal Powder Report*, MPR Publishing Services Ltd., England, June 1986, p 455–460
12. T.J. Treharne, Corrosion Inhibition in Sintered Stainless Steel, U.S. Patent 4,536,228, Aug 20, 1985
13. P.K. Samal, E. Klar, and S.A. Nasser, On the Corrosion Resistance of Sintered Ferritic Stainless Steels, *Advances in Powder Metallurgy and Particulate Materials*, R. McKotch and R. Webb, Ed., Proc. of the 1997 Intl. Conf. on Powder Metallurgy and Particulate Materials, MPIF, Princeton, NJ, p 16–99 to 16–112
14. "Test Methods for Evaluating the Corrosion Resistance of Stainless Steel Powder Metallurgy (P/M) Parts/Specimens by Immersion in a Sodium Chloride Solution," B 895–05, ASTM International
15. D.W. Yuan, J.R. Spirko, and H.I. Sanderow, Colorimetric Corrosion Testing of P/M Stainless Steel, *Int. J. Powder Metall.*, Vol 33 (No. 2), 1977
16. "Metallic Coatings Cathodic to the Substrate," J. Porter and M. Phillips, Ed., ISO 4540–1980 (E), International Organization for Standardization.
17. T. Mathiesen and E. Maahn, Corrosion Behavior of Sintered Stainless Steels in Chloride-Containing Environments, *12th Scandinavian Corrosion Congress* (Helsinki, Finland), G.C. Sih, E. Sommer, and W. Dahl, 377, Martinus Nijhoff, 1992, p 1–9
18. T. Mathiesen, "Corrosion Behavior of Sintered Stainless Steel," Doctoral thesis, Technical University of Denmark, 1993 (in Danish)
19. C. Molins, J.A. Bas, J. Planas, and S.A. Ames, P/M Stainless Steel: Types and Their Characteristics and Applications, *Advances in Powder Metallurgy and Particulate Materials*, Vol 3, MPIF, Princeton, NJ, 1995, p 345–357
20. A.J. Sedriks, Effects of Alloy Composition and Microstructure on the Passivity of Stainless Steels, *Corrosion*, Vol 42 (No. 7), 1986, p 376–388
21. T. Mathiesen and E. Maahn, "Corrosion Behavior of Sintered Stainless Steel," *12th Scandinavian Corrosion Congress*, June 1992 (Helsinki, Finland), G.C. Sih, E. Sommer, and W. Dahl, 377, Martinus Nijhoff
22. T. Mathiesen and E. Maahn, "Alloying Elements in Sintered Stainless Steel," *U.K. Corrosion*, Oct 1992 (Manchester, U.K.) NACE.

23 . B. Shaw, Corrosion-Resistant Powder Metallurgy Alloys, *Powder Metal Technologies and Applications* Vol 7, *ASM Handbook*, ASM International, 1998, p 987

24. M.C. Baran, A.E. Segall, B.A. Shaw, H.M. Kopech, and T.E. Haberberger, Evaluation of P/M Ferritic Stainless Steel Alloys for Automotive Exhaust Applications, *Advances in Metallurgy and Particulate Materials*, R. McKotch and R. Webb, Eds., MPIF, Princeton, NJ, 1997, p 9–37

25. "Standard Reference Test Method for Making Potentiostatic and Potentiodynamic Anodic Polarization Measurements," G 5, ASTM International

26. "Standard Test Method for Conducting Potentiodynamic Polarization Resistance Measurements," G 59, ASTM International

27. "Test Method for Conducting Cyclic Potentiodynamic Polarization Measurements for Localized Corrosion Susceptibility of Iron-, Nickel-, or Cobalt-Based Alloys," G 61, ASTM Internatioinal

28. D. Ro and E. Klar, Corrosive Behavior of P/M Austenitic Stainless Steels, *Modern Developments in Powder Metallurgy*, Vol 13, H. Hausner, H. Antes, and G. Smith, Eds., MPIF, Princeton, NJ, 1980

29. A.V. Wartenberg, *Z. Anorg. Allg. Chem.*, Vol 79, 1913, p 71

30. T.L. Rosenfeld, *Localized Corrosion*, R.W. Staehle, B.F. Brown, J. Kruger, and A. Agrawal, Ed., NACE-3, Houston, TX, 1974

31. E. Maahn, S.K. Jensen, R.M. Larsen, and T. Mathiesen, Factors Affecting the Corrosion Resistance of Sintered Stainless Steel, *Advances in Powder Metallurgy and Particulate Materials*, Vol 7, MPIF, Princeton, NJ, 1994, p 253–271

32. T. Mathiesen and E. Maahn, Corrosion Testing of Stainless Steels, *Met. Powder Rep.* Vol. 49, (No. 4), 1994, p 42–46

33. O. Kubaschewski and B.E. Hopkins, *Oxidation of Metals and Alloys*, Butterworth & Co., London, 1962

34. K. Hauffe, *Oxidation of Metals*, Plenum Press, New York, 1965

35. I.M. Fedorchenko, A.P. Inapunov, and V.V. Skorokhod, Phenomena Taking Place During Oxidation of Porous Metals at Elevated Temperatures, *Powder Metall.*, (No. 12), 1963, pp 27–43

36. T. Kato and K. Kusaka, On Some Properties of Sintered Stainless Steels at Elevated Temperatures, *Powder Metall.*, Vol 27 (No. 5), July 1980, p 2–8

37. S.O. Shah, J.R. McMillen, P.K. Samal, and E. Klar, "Development of Powder Metal Stainless Steel Materials for Exhaust System Applications," Paper 980314, *SAE International Congress and Exposition*, Feb 1998 (Detroit, MI)

38. P.K. Samal and E. Klar, U.S. Patent 5,976,216, Nov 1999

39. Z. Ishijima and H. Shikata, Influence of the Dispersion of La_2O_3 on High-Temperature Oxidation of P/M Fe-Cr Alloys, *Advances in Powder Metallurgy and Particulate Materials*, V. Arnold, C. Chu, W. Jandesha, and H. Sanderow, Eds., Part 8, MPIF, Princeton, NJ, p 113–122

40. P.C. Searson and R.M. Latanision, The Corrosion and Oxidation Resistance of Iron- and Aluminum-Based Powder Metallurgical Alloys, *Corros. Sci.*, Vol 25 (No. 10), 1985, p 947–968

41. "Stainless Steel Powders," North American Hoganas Sales Brochure, 1998

42. E. Klar and P.K. Samal, Powder Metals, Chapter 59, *Corrosion Tests and Standards, ASTM Manual 20*, ASTM, 1995, p 551–557

CHAPTER 10

Secondary Operations

A GREAT MAJORITY of powder metallurgy (PM) parts are well suited for their intended applications in their as-sintered condition. However, additional processing is sometimes found to be desirable in order to enhance the performance and value of a part. In some situations, additional processing may be the most cost-effective option for enhancing dimensional accuracy, physical details, or the mechanical and physical properties of the sintered product, as well as for enabling fabrication of components with complex geometries by use of processes such as machining or joining. Also, additional processing may simply comprise a finishing step, such as sealing of pores, tumbling, or plating. The PM stainless steels are no exception to this rule. Secondary processes that are most commonly applied to PM stainless steels include machining (including drilling and tapping), pore sealing (by impregnation with organic fillers), coining or re-pressing, and joining (typically welding or brazing). Unlike many other PM steels, PM stainless steels are rarely subjected to heat treatment. Only the martensitic grades are provided any heat treatment, usually consisting of tempering for the purpose of improving the ductility and toughness of the material. Details on the heat treatment of martensitic stainless steels can be found in sections 2.4.3 in Chapter 2 and 7.4.1 in Chapter 7. Annealing is rarely carried out on PM stainless steels. One exception may be the annealing of heavily re-pressed/coined ferritic stainless steel components for the purpose of restoring their ductility, toughness, or magnetic performance (section 10.6).

10.1 Machining

Notwithstanding the ability of the PM process to produce near-net shape parts of complex geometry and fine details, machining of some form or other is deemed necessary for a significant number of PM parts. Features such as threads as well as grooves, holes, and undercuts that run in a direction perpendicular to that of compaction inevitably require machining. Approximately one-third of all PM parts produced for the automotive industry require machining, and over one-half of all PM fabricators carry out machining on their own premises.

The machining response of PM stainless steels can vary significantly, depending on the alloy composition and sintering parameters. Selection of optimal material and processing parameters can go a long way in controlling the cost of machining. In some situations, the choices may be limited, with the result that the cost of machining is a significant part of the overall cost of component manufacture.

10.1.1 Machinability of Wrought and PM Stainless Steels

Wrought and cast stainless steels, in particular austenitic alloys, pose a significantly greater challenge in machining when compared to carbon steels. Some of the positive attributes of stainless steels, such as high strength, toughness, and ductility, present themselves as problems when it comes to machining. Stainless steels tend to deform and work harden under the machining stresses, increasing both friction and cutting force. They produce long and stringy chips that seize or form built-up edges on the tool, leading to reduction of tool life and poor surface finish. Heat dissipation is slower with stainless steels, due to their low thermal conductivity, which causes the workpiece and cutting tool to operate at undesirably high temperatures. This accelerates wear in the cuttingtool and the tendency to develop cracks due to excessive thermal expansion (Ref 1). Efficiency of machining is also reduced by the need for frequent clearing of the long, stringy chips that adhere to the tooling and finished surfaces. In drilling, accumulation of chips in the hole and around the drill can contribute to loss of tool life and process efficiency. Several grades of wrought stainless steels are available in their free-machining versions. These typically contain a small amount of either sulfur or selenium. Improved machinability is attributed to two mechanisms: the additives coat and lubricate the tool tip, thus preventing built-up edges (Ref 2); and the additive particles promote formation of cracks in the primary shear zone, which assists chip breakage, thus reducing forces exerted on the tool (Ref 3). Free-machining grades cannot be specified for all applications, because they exhibit reduced corrosion resistance and may suffer cracking during hot forming.

The PM materials, including stainless steels, generally exhibit inferior machinability when compared with their wrought counterparts. Porosity decreases thermal conductivity, reducing the rate of heat transfer away from the cutting surface and the tool. The increased operating temperature is detrimental to the performance and life of the cutting tool. More importantly, porosity leads to interrupted cuts. Because the tool tip is repeatedly made to move from a pore to the solid material, it is subjected to cycles of impact loading and deflection, which leads to fatigue crack formation (Ref 4). Tool wear is actually fairly low for a low-density sintered material, and it deteriorates as the density increases into the intermediate range, due to the greater resistance of the material to deformation. As the density further increases to higher

levels, the trend reverses itself, due to increased thermal conductivity and reduction in porosity (Ref 3) (Fig. 10.1). Overall, tool tip degradation is more rapid when machining a porous material as compared with its pore-free counterpart. Machining of PM stainless steels is usually conducted without the aid of a liquid lubricant or coolant. This is because of the additional cost involved with the removal of such fluids from the pores, which can otherwise compromise its corrosion resistance.

Among the standard grades of PM stainless steels, 303L is formulated to offer enhanced machinability. A small addition of sulfur during melting of the steel (typically 0.2%) leads to the formation of fine precipitates of MnS in the solidified alloy powder. Figure 10.2 shows a typical microstructure of high-temperature, hydrogen-sintered 303L. Manganese sulfide precipitates are seen within the grains and along the grain boundaries as fine globules (comparison may be made with the microstructure of a similarly processed 304L shown in Fig. 9 in the Micrograph Atlas in the book). Corrosion resistance of 303L is significantly lower than that of 304L and 316L. An alternate means of enhancing machinability of a PM stainless steel is to admix a machinability-enhancing agent, such as MnS or MoS_2, with the powder prior to compacting and sintering (boron nitride is also a potential machinability–enhancing additive). However, the addition of MnS to a standard austenitic grade, such as

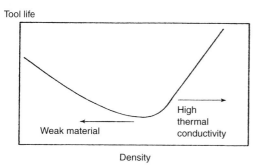

Tool life

Weak material

High thermal conductivity

Density

Fig. 10.1 Schematic depiction of porosity influence on tool life. Source: Ref 3. Reprinted with permission from MPIF, Metal Powder Industries Federation, Princeton, NJ

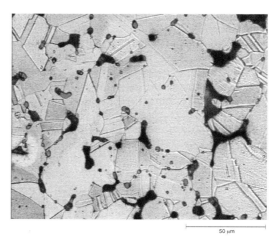

50 μm

Fig. 10.2 Micrograph of sintered 303L, showing the presence of fine globules of MnS within the grains and along the grain boundaries. Glyceregia etch. Reprinted with permission from MPIF, Metal Powder Industries Federation, Princeton, NJ

316L, does result in a significant reduction in the alloy's corrosion resistance, as is discussed later. Detrimental effects from MoS_2 addition are much smaller. Tin- and copper-modified austenitic PM stainless steels (LSCs) are also known to exhibit improved machinability in addition to their superior resistance to corrosion (discussed in detail in the following section) (Ref 5).

Compared to the austenitic grades, the ferritic grades of PM stainless steel are significantly easier to machine, especially when sintered in a 100% H_2 atmosphere or vacuum. Unless an application involves a very high volume of machining, it is not essential to use a machinability-enhancing additive in the case of a ferritic stainless steel. The PM martensitic stainless steels are difficult to machine due to their high hardness and low ductility. This also holds true for 400-series alloys that have been sintered in a nitrogen-bearing atmosphere and for those with high levels of residual carbon, because these materials inevitably consist of some amount of martensite, possibly combined with carbides and nitrides.

10.1.2 Factors Affecting Machinability of PM Stainless Steels

Several researchers have investigated the effects of sintering parameters, machinability additives, and the machining parameters on the efficiency and quality of machining PM stainless steels. A large majority of these studies have been based on drilling, which perhaps is quite appropriate because machining difficulties are greater with drilling, and also because drilling constitutes a large part of all PM stainless steel machining. Although the test parameters as well as the criteria used for evaluating machinability do differ from one study to another, their findings are in good agreement in a qualitative sense. Ambs (Ref 6) based his study on drilling time, keeping the drilling parameters fixed. Similarly, keeping the drilling parameters fixed, Samal and Terrell (Ref 7) compared the machining behavior of variously processed PM austenitic stainless steels in terms of drill life (number of holes drilled until drill failure) and the energy consumed in drilling. Kutsch and Beiss (Ref 8, 9) carried out an extensive investigation by varying drilling speed, feed rate, and lubrication, and they compared machining

performance in terms of its total cost and tool life. Total machining cost was computed using the equation:

$$C_{tot} = C_0 \cdot \{(t_s/m) + t_n + t_p + t_c \cdot (t_p/T)\} + C_t \cdot (t_p/T)$$

(Eq 10.1)

where

$$t_p = \frac{\pi \cdot d \cdot h}{1000 \cdot f \cdot v}$$

(Eq 10.2)

is the productive metal cutting time in drilling, and C_{tot} is total cost ($), T is tool life (min), v is surface cutting speed (m/min), f is feed (mm/rev), C_0 is operating cost ($/min), C_t is tool cost ($), m is lot size (1000), t_s is setup time (min), t_n is workpiece changing time (min), t_c is tool changing time (min), d is diameter of bore (mm), and h is depth of bore (mm).

Kutsch and Beiss (Ref 8) observed that ferritic 430L was significantly easier to drill in comparison to austenitic 316L under all conditions employed in their study.

Effect of Sintering Parameters. In the Ambs (Ref 6) study, sintering in 100% H_2 instead of dissociated ammonia at 1232 °C (2250 °F) reduced drilling time from 82 to 34 s/hole for PM 304L. When using dissociated ammonia as the sintering atmosphere, sintering at 1120 °C (2050 °F) resulted in increased drilling time as compared to sintering at 1232 °C (2250 °F), for 303L, 304L, and 316L. This is attributed to the higher equilibrium nitrogen contents of the alloys at the lower sintering temperature. In the Ambs study, none of the materials tested had any admixed machinability additive, and all drilling was carried out without any lubricant. Sanderow et al. (Ref 10) also noted that dissociated ammonia sintering resulted in much shorter tool lives when compared with vacuum sintering; the difficulty in machining was greater for the lower-temperature dissociated ammonia sintering in comparison to the higher-temperature dissociated ammonia sintering. Their study also shows a highly beneficial influence of resin impregnation on machinability over a wide range of sintering parameters (section 10.5).

In the Samal and Terrell study (Ref 7), additive-free 316L, sintered in 100% H_2 exhibited an average drill life of 6.9 holes compared to 2.2

holes for dissociated-ammonia-sintered material of the same composition. For 303L, 100% H_2-sintered material yielded a drill life of 41 holes compared to 7.5 holes for dissociated-ammonia-sintered material. Sintering temperature was 1232 °C (2250 °F) and the sintering time was 45 min for all tests. The nominal sintered density was 6.8 g/cm^3. Figures 10.3 and 10.4 show the effects of various levels of machinability additive on drilling efficiency and corrosion resistance for both 100% H_2-sintered and dissociated-ammonia-sintered 316L.

The results of the Kutsch and Beiss (Ref 8) study are shown in Fig. 10.5 and 10.6. In their study, the N_2-H_2 atmosphere comprised 70% N_2 and 30% H_2. In these figures, the horizontal lines placed just below the 0.1 \$/bore marker represent the fixed cost of machining. The variable cost of machining was roughly doubled when sintering was carried out in the nitrogen-rich sintering atmosphere as compared to 100% H_2. All samples were compacted at 600 MPa (87 ksi) and sintered at 1280 °C (2336 °F) for 30 min. Sintered densities were 6.87 and 6.57 g/cm^3 for 316L and 430L, respectively.

Effect of Machinability Additives. Historically, MnS is the most popular machinability additive for ferrous materials. Several proprietary modifications of MnS are currently available. The additive MnS+ is a chemically

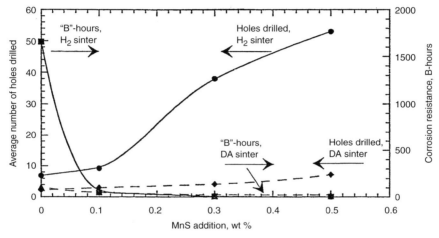

Fig. 10.3 Effects of sintering atmosphere and amount of MnS addition on the machinability and corrosion resistance of 316L. DA, dissociated ammonia. Source: Ref 7. Reprinted with permission from MPIF, Metal Powder Industries Federation, Princeton, NJ

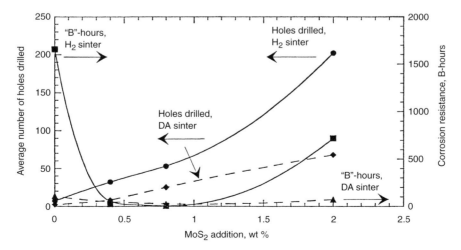

Fig. 10.4 Effects of sintering atmosphere and amount of MoS$_2$ addition on the machinability and corrosion resistance of 316L. DA, dissociated ammonia. Source: Ref 7. Reprinted with permission from MPIF, Metal Powder Industries Federation, Princeton, NJ

modified, low-carbon version of MnS that has superior resistance to oxidation. It contains nominally 5% Fe, which is chemically bonded with MnS. The additive MnSE is a high-purity version of MnS, containing 64.2% Mn, 35% S, and 0.8% O_2. In the Kutsch and Beiss study (Ref 8), a proprietary machinability-enhancing agent, called MnX, was used, and it also produced a small improvement in machinability over MnS in the case of N_2-H_2 atmosphere-sintered 316L. In the Samal and Terrell study (Ref 7), MnS+ addition showed a small improvement in machinability over MnS, but only for dissociated ammonia sintering. With hydrogen sintering, the improvements due to MnS+ or MnX addition over MnS were negligible. Typically, machinability improved as the amount of the additive (MnS, MnX, or MnS+) was increased, but mostly up to the 0.5% addition level; thereafter, further benefits were marginal. Hence, for these MnS-type additives, 0.5% addition level is considered optimum. For 100% H_2 sintering, molybdenum disulfide was found to be more effective than MnS or MnS+. Machinability continued to improve with MoS_2 addition up to 2.0%, the maximum level covered in the study. The MoS_2 addition also produced an appreciable improvement in machinability when sintering was carried out in dissociated ammonia. Dissociated-ammonia-sintered 316L containing 2% MoS_2 showed approximately the

same machinability as the hydrogen-sintered 316L containing 0.5% MnS.

Both MnS and MnS+ additions led to significant loss of corrosion resistance for PM 316L when tested by immersion in 5% aqueous sodium chloride solution (Ref 7), Wang (Ref 11) also observed a severe reduction in the corrosion resistance of PM 304L due to MnS addition in a mass-loss test involving immersion in 10% ferric chloride solution. Based on metallographic evaluation, he concluded that the interface between MnS and the matrix provided active sites for crevice corrosion.

In the case of wrought stainless steels, the loss of corrosion resistance is most often attributed to the anodic dissolution of MnS, leading to initiation of pitting in an otherwise fully dense matrix (Ref 12). Sulfur, with a solubility limit of 0.01% at room temperature in stainless steel, largely exists as the sulfides of manganese, chromium, and/or iron (the presence of FeS is highly undesirable due to its low melting point). Work by Kovach and Moskowitz (Ref 13) has shown that the type of sulfide that forms in a resulfurized stainless steel is determined by the manganese-to-sulfur ratio in the alloy. At low manganese-to-sulfur ratios of less than 0.4, only a chromium sulfide forms, which has a hexagonal structure and is brittle. At intermediate manganese-to-sulfur ratios, ranging from 0.4 to 1.8, a cubic, chromium-rich manganese sulfide

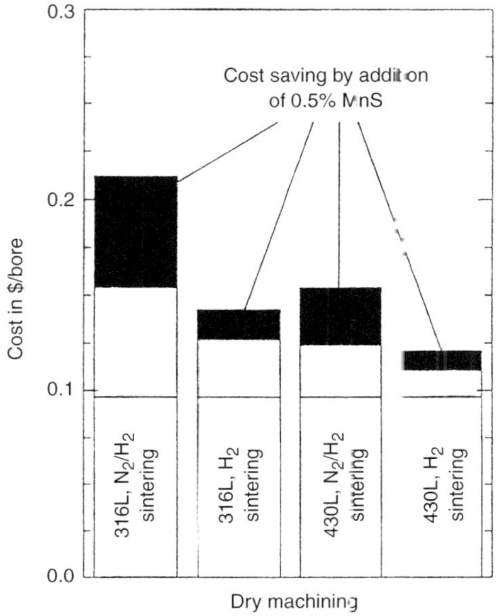

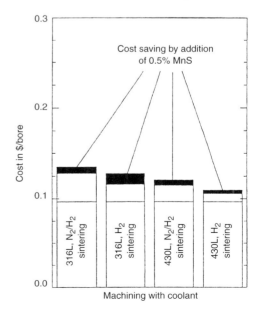

Fig. 10.5 Cost differential for machining in 100% H_2 versus a 70% N_2-30% H_2 atmosphere for various materials when no coolant was used. Source: Ref 8

Fig. 10.6 Cost differential for machining in 100% H_2 versus a 70% N_2-30% H_2 atmosphere for various materials when a coolant was used. Source: Ref 8

forms (some investigators have also reported the formation of a complex sulfide of chromium, iron, and manganese in low-manganese stainless steels). At manganese-to-sulfur ratios greater than 1.8, only pure MnS forms in the matrix. From the corrosion-resistance point of view, the pure form of MnS is most undesirable, while it has the highest beneficial effect from the machinability point of view. In PM 303L, the manganese-to-sulfur ratio is typically approximately 1.0, indicating that the sulfides in these alloys are chromium-rich manganese sulfides rather than pure MnS and hence are expected to be somewhat less detrimental than the admixed MnS. This is in agreement with the observation that MnS-added 316L exhibits significantly lower corrosion resistance compared to 303L (50 versus 7 B-hours) (Ref 5, 7).

Molybdenum disulfide additions led to significant loss in corrosion resistance at the 0.5 and 0.8% addition levels. However, at the 2.0% addition level, the loss of corrosion resistance is significantly smaller, and hence, if an application requires both good machinability and satisfactory corrosion resistance, the hydrogen-sintered 316L containing 2% MoS_2 can be the material of choice.

The MnS, MnS+, and MoS_2 additions, in the amounts used in these studies, had practically no adverse effects on the compacting properties of the powder blends or the room-temperature static mechanical properties of the sintered stainless steels.

Samal et al (Ref 14) studied the effects of MnS addition on the machinability of two PM 400-series stainless steels, using a drilling test. Alloys 409L and 409LNi, with and without MnSE (0.5% addition) were evaluated. All sintering was carried out in 100% H_2 at 1316 °C (2400 °F), and the sintered densities were typically 7.30 g/cm^3. The effect of MnSE addition on the corrosion resistance was marginal for both alloys. Table 10.1 lists the test parameters used, and Table 10.2 lists the results of the test,

along with machinability improvement of 434L due to MnSE and MnX addition (unpublished data). The 434L tests were conducted using the same drilling procedure as listed in Table 10.1. Sintering of the 434L sample sets was carried out at 1343 °C (2450 °F) in a 100% H_2 atmosphere, and the sintered densities were typically 7.25 g/cm^3.

The particle size distribution of the additive has a significant effect on the uniformity of distribution of the additive in the matrix of the sintered component, and, as a result, it can also influence the machinability. Although a finer particle size is usually more effective, very fine additives (typically smaller than approximately 2 μm in size) tend to interact substantially with the matrix and thus become less effective. The melting point of the pure form of MnS (or MnSE) is significantly higher than most temperatures employed for sintering stainless steel, and hence, the particle shape and size distribution of the admixed MnS (or MnSE) remain largely unchanged during sintering. In the case of the MnS that forms in situ in 303L, the microstructure of a high-temperature-sintered (>1316 °C, or 2400 °F) alloy often exhibits agglomerated and respheroidized inclusions of the sulfide, indicating that these chromium-(and possibly iron-) rich sulfides may have a lower melting point than the temperature employed for sintering. Hence, in a high-temperature-sintered 303L, if the green density is not high enough, the molten sulfide can migrate through the pores and form large agglomerates in the pores (Ref 5). Redistribution of sulfides in the sintered material will not only lead to erratic machining performance but also may leave the surface of the component devoid of the additive. This would be an undesirable situation if the machining of the surface layer is critical for the finished component.

Effect of Surface Modification. As discussed in Chapters 2 and 6, surface modification of

Table 10.1 Drilling test parameters (400-series)

Drill size: 3.5 mm (0.138 in.) diameter
Drill type: M-2-type high-speed steel, not coated (*DORMER made HSS jobber drill for stainless steel*, A 108, DIN 388, KG00)
Depth of holes: 7 mm (0.28 in.)
Spacing: 7 mm (0.28 in.), center-to-center
Speed: 2000 rpm
Feed rate: 0.06 mm (0.002 in.) per rotation
Lubricant: None
Test criterion: Number of holes drilled until drill failure
Number of tests on each material: 8

Table 10.2 Comparative machinability of powder metallurgy 400-series alloys

Alloy	Additive	Sintered density, g/cm^3	Hardness, HRB	Number of holes (average)	Standard deviation(a)
409L	None	7.32	57	45	1.9
410L	None	7.28	55	15	5.3
409LNi	None	7.32	88	20	1.2
434L	None	7.25	59	23	8.2
409L	MnSE	7.29	56	1644+	NA
409LNi	MnSE	7.32	86	1644+	NA
434L	MnSE	7.25	59	370	88.8
434L	MnX	7.25	59	361	92.6

(a) NA, not applicable

PM austenitic stainless steels by prealloying with tin and copper leads to significant improvement in their corrosion resistance. Studies by Kusaka et al. (Ref 15) and Samal et al. (Ref 5) determined that tin and/or copper modification improves machinability by a measurable amount. More importantly, it was determined that surface modification with tin and copper (and with copper alone) leads to a significant improvement in the corrosion resistance of 303L, thus offering materials that combine good corrosion resistance with good machinability. Surface modification of 304L led to a moderate improvement in machinability, while its corrosion resistance increased significantly—approaching that of optimally sintered 316L. Table 10.3 shows the effects of tin and/or copper modification on the machinability and corrosion resistance of 303L and 304L. In this study, the drilling test parameters were the same as listed in Table 10.1, except for a reduced drilling speed of 700 rpm and a greater drill depth of 12 mm (0.5 in). Corrosion tests were conducted by immersion in 5% NaCl (ASTM B 895).

Machinability improvements are attributed to the formation of a fine Ni-Cu-Sn eutectic precipitate in the matrix, which was first identified by Kusaka et al. and confirmed by Samal et al. (Ref 5) (Fig. 10.7).

Effect of Cutting Speed. The Beiss and Kutsch (Ref 9) study showed that even with the optimal feed rate, the tool life decreased rapidly when cutting speed was increased beyond a critical value. As a result, economical cutting speeds fell into a relatively narrow range, typically 5 to 15 m/min. (16 to 49 ft/min). Use of MnS (or MnX) as a machining additive not only increased the optimal cutting speed but also widened the range of economical cutting speed to 10 to 30 m/min (33 to 98 ft/min), typically. This makes the machining process more tolerant against missing the optimal cutting speed. Tool

feed rates employed in the study ranged from 0.10 to 0.25 mm/revolution. (0.004 to 0.01 in./revolution).

Effect of Machining Coolant. In the Beiss and Kutsch study, (Ref 9), the use of a machining coolant increased tool life significantly over dry machining, due to enhanced rates of heat transfer and chip removal. The benefits were greater for the materials that were more difficult to machine, such as additive-free and/or 70% N_2-30% H_2-sintered 316L (Fig. 10.5 and 10.6). However, benefits should be weighed against the cost of removal of coolant residues from the pores.

10.2 Welding

Welding of stainless steels, as that of many other steels, has been evolving for nearly a century and is currently a well-developed science. The metallurgical requirements governing welding practices vary from one family of stainless steel to another. A treatment of all underlying principles is beyond the scope of this book. An attempt is made to briefly cover some of the more critical issues in the welding of stainless steels. Readers interested in gaining an in-depth understanding of these principles are referred to the chapter on welding in *Stainless Steels*, ASM Specialty Handbook, 1994 (Ref 16).

Welding of PM stainless steels was not practiced to any great extent until the introduction of PM stainless steel exhaust flanges and hot exhaust gas outlet (HEGO) bosses in the early 1990s. Therefore, most publications addressing

Table 10.3 Effects of surface modification on the corrosion resistance and machinability of PM 303L and 304L

Alloy designation	Modified with Sn, wt%	Cu, wt%	Corrosion resistance A-hours	B-hours	Machinability(a) Average number of holes drilled
303L	0.0	0.0	24	54	22
303LSC	0.8	2.0	372	756	26
303LCu	0.0	2.1	210	452	32
304L	0.0	0.0	58	112	2
304LSC	0.8	2.0	620	1284	6

(a) Drilling speed used was 700 rpm. drilling depth was 12 mm (0.5 in.). All other parameters were the same as listed in Table 10.1

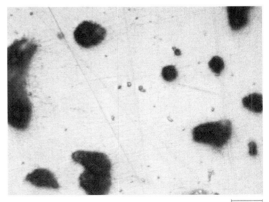

Fig. 10.7 As-polished cross section of sintered 304LSC, showing fine precipitates of the Cu-Ni-Sn eutectic phase, attributed to enhancement of machinability (unetched). Source: Ref 5. Reprinted with permission from MPIF, Metal Powder Industries Federation, Princeton, NJ

welding of PM stainless steels are of relatively recent origin. The metallurgical principles governing welding of wrought stainless steels are, to a large extent, applicable to PM stainless steels. Additional considerations must be given to potential difficulties arising from the porous nature of the PM materials. Unless the density of PM stainless steel is at least 6.8 g/cm^3, the presence of porosity can severely limit the choice of welding methods, as well as the parameters, that can be practically applied.

10.2.1 Basics of Welding Stainless Steel

Each of the five families of stainless steels has a different set of weldability considerations, stemming primarily from their different modes of phase transformation that occur during solidification from the welding temperature. Because the large majority of PM stainless steel components that do get welded belong to either the austenitic or the ferritic family, only these two families are considered here.

Filler metal selection is the most critical step in the welding of stainless steels. The final composition of the weld metal should be such that corrosion resistance of the assembly is not compromised. Most often, the filler metal selected is not of the same composition as that of the base alloy; frequently, it may be a "richer" alloy. The major challenge in filler metal selection is avoidance of weld metal cracking. Cracking can

occur just below the solidus temperature of the bulk alloy, taking the form of either a centerline crack or multiple transverse cracks. This type of cracking is called hot cracking or microfissuring, and it is a major concern with austenitic stainless steels. In these steels, this type of cracking is best avoided by promoting the formation of a small amount of ferrite in the weld. Determination of the optimal amount of ferrite needed to avoid hot cracking has been the subject of much research over the past four decades. In fact it was the main reason behind the development of the Schaeffler diagram in 1949 (section 2.3 in Chapter 2, "Manufacture and Characteristics of Stainless Steel Powders"). Since then, several modifications of the diagram have been developed that are capable of predicting the ferrite content of an austenitic weld metal with greater precision. One such modification is the DeLong diagram, which takes into account the effect of high nitrogen contents in the weld metal, while another, developed by the welding Research Council (and known as the WRC modification), takes into account the effect of manganese content in the weld metal. Figure 10.8 is a DeLong diagram showing various levels of ferrite in the alloy, represented as a ferrite number (FN) based on the chromium and nickel equivalents (Cr$_{eq}$ and Ni$_{eq}$). An FN number in the range of 4 to 8 is recommended for avoidance of hot cracking in austenitic stainless steels (up to an FN of 8, with

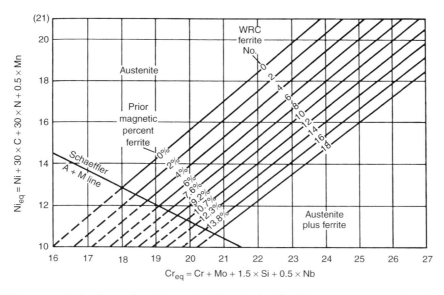

Fig. 10.8 DeLong constitution diagram for stainless steel weld metal. The Schaeffler austenitic-martensitic boundary is included for reference. Source: Ref 16. ASM Speciality Handbook Stainless Steels, p 340–341

ferrite numbers equaling weight percent ferrite). Magnetic test methods are suitable for determining the ferrite content of the weld metal. Table 10.4 lists a number of filler metals that are commonly used for welding wrought austenitic and ferritic stainless steels.

Other factors that contribute to hot cracking are the contour of the weld puddle and the presence of impurities, such as sulfur, selenium, boron, and phosphorus. A weld puddle that is tear drop shaped (or tailed) is more prone to cause hot cracking as opposed to one that is elliptical in shape. A tear-drop-shaped puddle allows the lower-melting components (those containing sulfur and/or phosphorus, typically) to concentrate along the centerline of the weld bead. Ferritic stainless steels are less affected, because they can tolerate higher concentrations of sulfur and phosphorus, due to the greater solubility of these elements in the ferrite matrix in comparison to the austenitic grades (Ref 18).

In the welding of an austenitic stainless steel, the control of the ferrite content of the weld metal is also important for avoidance of corrosion of the weldment. Slag formed during welding or the redeposited metallic vapors on the surface of the part can be the sites for initiation of corrosion. Acceptable welds will have no slag. Table 10.5 lists the maximum solubility of common slag-forming elements in a δ-iron (body-centered cubic, or bcc, ferrite) and γ-iron (face-centered cubic, or fcc, austenite). All these elements have much higher solubility in ferrite than in austenite. Hence, formation of some amount of ferrite in the austenitic weld metal is highly essential for elimination or reduction of slag formation. On the other hand, excessive ferrite content is also detrimental to corrosion resistance, because the δ-ferrite phase easily becomes the site for pitting corrosion. By studying a wide range of Cr_{eq} and Ni_{eq}, Collins and Williams (Ref 19) determined that the pitting tendency (as measured by ASTM G 150) increases precipitously when the ferrite content exceeds 3.5%. This limit coincided with a Cr_{eq}/Ni_{eq} ratio of 1.55 in their study. Their study was based on the Cr_{eq} and Ni_{eq} equations developed by Hammar and Svensson ($Cr_{eq} = Cr + 1.37$ Mo + 1.5 Si + 2 Nb + 3 Ti, and $Ni_{eq} = Ni + 0.31$ Mn + 22 C + 14.2 N + Cu) (Ref 20).

Compared to the low-carbon or alloy steels, welding of stainless steels (the austenitic grades in particular) requires more careful management of heat input due to their relatively poor thermal diffusivity (defined as thermal conductivity divided by specific heat and density) and their greater coefficient of thermal expansion. Steeper thermal gradients resulting from low thermal diffusivity can produce a high degree of thermal stress, which in turn can lead to weld cracking. It should also be noted that with good weld puddle contour control and heat management, it is possible to avoid hot cracking in a fully austenitic weld bead.

Cold cracking can occur at low temperatures, typically at 150 °C (302 °F), and is commonly attributed to excessive residual stresses present in the assembly. It commonly occurs in martensitic stainless steels or in ferritic stainless steel whose weld metal has transformed to martensitic due to sufficient carbon or nitrogen pickup. A deposit that is primarily austenitic will not cold crack.

The term *weldability* refers not only to the ease with which sound welds can be made but also to the satisfactory performance of these welds in service. Hence, it is essential that the welded structure exhibits satisfactory performance in terms of corrosion resistance, mechanical strength, ductility, and impact strength. In stainless steels, avoidance of intergranular corrosion requires that the carbon and nitrogen contents of the alloy are kept very low. Addition of a carbide stabilizer to the alloy, such as titanium and/or niobium (niobium is preferred for PM stainless steel made via water atomization; see section 5.2.4 in Chapter 5), ensures resistance to intergranular corrosion.

Table 10.4 Examples of filler metals

Alloy	Condition(a)	Filler metal
303	1 or 3	312
304	1	308
304L	1	347, 308L
316	1 or 2	316
316L	1 or 3	316Nb, 316L, 318
317L	1 or 3	317Nb
409	1 or 2	409, 409Nb, 308, 310
430	1	308, 309, 310
430	2	430

(a) Condition in which the weldment will be placed in service. 1, as-welded; 2, annealed; 3, stress relieved. Source: Ref 17

Table 10.5 Maximum solubility of slag-forming elements in δ-iron (ferrite) and γ-iron (austenite)

Element	Solubility in ferrite, wt%	Solubility in austenite, wt%
Ca	0.024	0.016
Si	10.9	1.9
Al	30	0.95
Ti	8.7	1.0
Zr	11.7	1.0

Embrittlement of the weld metal due to sigma-phase formation is frequently a concern with austenitic and high-alloy (greater than 16% Cr) ferritic alloys. Any delta ferrite formed in the weld bead can transform to sigma phase in service in the temperature range of 500 to 800 °C (932 to 1472 °F), leading to loss of both ductility and corrosion resistance. Sigma phase can be eliminated by heating to approximately 850 °C (1562 °F), then cooling the material rapidly in order to prevent 475 °C (885 °F) embrittlement (section 2.4.1 in Chapter 2, "Metallurgy and Alloy Compositions").

Both austenitic and ferritic stainless steels are prone to excessive grain growth, because these alloys solidify as a single phase. The problem is more severe with ferritic grades, because diffusion rates are greater in the bcc structure. Grain coarsening leads to a reduction in ductility and toughness of the heat-affected zone (HAZ). In ferritic stainless steels, martensite formation, due to pickup of carbon or nitrogen, can also lead to brittleness as well as loss of corrosion resistance.

The magnetic nature of ferritic stainless steels can divert the electric arc in the arc welding process, causing undesirable spatter. This problem may be corrected by demagnetizing the workpiece prior to arc welding.

The low electrical conductivity of austenitic stainless steels can pose a challenge with resistance (projection) welding (Ref 21).

Cleanliness of the weld zone is essential, not only to avoid pickup of deleterious elements, such as sulfur, boron, and phosphorus and any low-melting metals, such as copper, lead, tin, and zinc, but also to remove any oxides present. Oxides of nickel (NiO) and chromium (Cr_2O_3), due to their high melting points (2260 °C, or 4100 °F, and 1982 °C, or 3600 °F, respectively) will remain in the weld bead as stringers and will not be detected on x-ray. These may lead to early fatigue failure if used under cyclic load conditions.

In addition to taking steps for the avoidance of intergranular corrosion (i.e., reducing carbon and nitrogen contents, using a stabilizer, and rapid cooling), one must take precautions to avoid several other types of corrosion that may result from the welding process. Weld joints should be designed to avoid potential sites for crevice corrosion. Materials susceptible to stress corrosion cracking (austenitic stainless steels, in particular) should be given a stress-relief treatment. The filler metal selected may be the one that is more cathodic with respect to the base metal, in order to minimize the effects from galvanic corrosion. Corrosion fatigue may be avoided by minimizing stress concentration in the weld joint (e.g., blending welds to the base metal, using butt welds instead of fillet welds) and by blend grinding rough marks left behind from a machining or grinding operation.

10.2.2 Welding Methods Used with PM Stainless Steels

Welding methods that have been successfully employed for PM stainless steels include several fusion welding methods, such as gas tungsten arc welding (GTAW) and gas metal arc welding (GMAW), and a number of resistance welding methods, such as flash welding and high-energy impulse welding (HEIW). In the GTAW process, an arc is produced between a nonconsumable tungsten electrode and the base metal. In PM applications, this process may include the use of an auxiliary filler metal to compensate for any shrinkage occurring in the weld zone. An inert gas shield of argon, helium, or a mixture of the two is used to protect the weld pool from oxygen, nitrogen, and hydrogen contamination. A small amount of oxygen or carbon dioxide is often added to the inert gas mixture. For stainless steel welding, the amount of carbon dioxide, if used, is kept low (2 to 3%) in order to avoid carbon pickup. The GTAW process allows a greater degree of control over the welding process and, as such, provides satisfactory results in many situations. The GMAW process is similar to GTAW in terms of the protective shielding gas used; however, instead of the tungsten electrode, a filler metal is fed continuously to the weld. The GMAW process is relatively more expensive, but it is a faster process compared to GTAW. Both processes can produce high-quality, clean, slag-free welds, with a minimal number of defects. Resistance projection welding, such as HEIW, uses powerful capacitors to store energy so that currents as high as 400,000 A can be generated with a weld dwell time of 1/60 of a second. In this cases, welding takes place across a projection designed into the component, and because no filler metal is used, there is no concern for dilution effects. Microstructural changes are limited to the weld zone only. Laser welding is also a filler-metal-free welding process, employing relatively low heat inputs that result in fast cooling and steep temperature gradients.

10.2.3 Additional Considerations for PM Stainless Steels

Porosity plays a pivotal role in the welding of PM stainless steels. It can have significant influences in the ability of a material to withstand thermal stresses and to transfer heat and electricity efficiently to the weld zone. For any fusion welding that involves significant remelting of the base metal, such as arc welding, densities of 7.00 g/cm^3 or higher are found to be highly desirable. The PM materials with intermediate density levels (6.60 to 6.90 g/cm^3) only respond well to welding techniques that minimize the volume of fused metal, such as laser welding and resistance projection welding. The PM stainless steels with densities below 6.50 g/cm^3 may be totally impractical for welding. At these lower densities, particle remelting results in a greater degree of shrinkage in the weld zone. The high degree of shrinkage, combined with the lower interparticle bonding of the low-density base metal, can produce cold cracking (Ref 22). It should be noted that in the fusion welding of PM materials, a sufficient amount of filler metal feed must be provided in order to compensate for shrinkage in the weld zone. Porosity can harbor contaminants, such as moisture, machining coolants, and coining lubricants, that can interfere with the welding process. Any entrapped or filtering gases can also interfere with the inert shielding gas. Additives such as MnS, boron, molybdenum disulfide, and impregnating materials will not only interfere with the welding process but also may lead to hot shortness.

Using GMAW, Garver and Urffer (Ref 23) were successful in welding ferritic 409L stainless steel with densities as low as 6.80 g/cm^3 to wrought 409 tubing. More importantly at densities of approximately 7.20 g/cm^3 the response of PM stainless steel parts to GMAW welding became identical to that of wrought stainless steel. At a density of 7.20 g/cm^3 or higher, when practically all porosity is isolated, exposure of the molten weld metal to entrapped contaminants and gases is minimized. Figure 10.9 shows the microstructure of the weld zone of a PM 409L flange (7.20 g/cm^3 density) and wrought 409Nb tubing. Welding was carried out

with a 409Nb metal core wire and 98% argon, 2% oxygen shielding gas (Ref 24).

Hamill et al. (Ref 25) were successful in welding hydrogen-sintered PM 409L flange and bushings to wrought 409Nb tubing, using the GTAW process with auxiliary filler wire feed. Filler wire was 409Nb. They found that dissociated-ammonia-sintered 409L stainless steel was unsuitable for fusion welding due to its high interstitial content, and it led to large amounts of outgassing, excessive pore formation in the weld zone, and martensite formation in the HAZ. Both hydrogen- and vacuum-sintered 409L parts produced excellent welds. Dissociated ammonia sintering was carried out at 1232 °C (2250 °F) and the interstitial contents were 0.19% C 0.34% N$_2$, and 0.38% O$_2$. In this study, hydrogen sintering was carried out at 1260 °C (2300 °F), and the interstitial contents were 0.060% C, 0.042% N$_2$, and 0.14% O$_2$. Vaccum sinters were carried out at 1149, 1204, and 1260 °C (2100, 2200, and 2300 °F), and the interstitial levels for all three conditions were less than 0.017% C, less than 0.018% N$_2$, and less than 0.23% O$_2$.

Halldin et al. (Ref 26) were successful in welding PM 316L stainless steel using a number of different welding techniques. Their study included GTAW, GMAW, three types of resistance welding (namely, resistance spot welding, upset welding, and flash welding), friction welding (solid state), and arc stud welding. Except for friction welding, all other methods produced satisfactory welds when the density was 6.60 g/cm^3 or higher. The GMAW process tended to give the best results. Free-machining

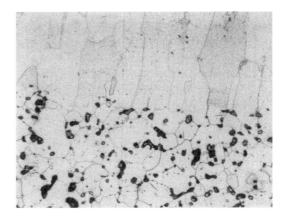

Fig. 10.9 Microstructure of the transition zone between PM 409L and the weld bead. Note the normal pore density at the interface as well as the grain continuity. Glyceregia etch

303L was not found to be as suitable for fusion welding.

Optimally sintered PM stainless steels can offer some advantages over their wrought counterparts. Pores present in PM stainless steels act as pinning sites for grain boundaries and restrict grain growth. Wrought materials are often associated with significant levels of residual stresses from a prior cold forming operation, and during welding, the relaxation of these stresses can lead to distortion of the assembly. The PM materials, on the other hand, are usually supplied in as-sintered condition, free of residual stresses.

In the welding of threaded HEGO bosses, Garver and Urffer (Ref 23) found that the lower rate of heat transfer of the PM parts provided an advantage over the cold-formed wrought HEGO bosses, in terms of reduced shrinkage and distortion of threads in the inside diameter of the parts.

10.3 Brazing

Brazing is primarily employed for joining stainless steels to dissimilar metals, including stainless steels of different composition, carbon and low-alloy steels, copper alloys, and nickel-base alloys. Unlike welding, brazing does not involve remelting of the base metal. In brazing, only the filler metal melts, and the joining of components is achieved via partial diffusion of filler metal into the base metal. Brazing filler metal, by definition, must melt above 450 °C (840 °F) and below the melting point of the base metal. It must be compatible with the base metal, wetting the base metal to form a strong bond. Preferably, the filler metal should have a narrow melting range, so that the heating rate of the brazing process becomes less critical. As a rule of thumb, brazing should take place at a temperature that is 55 to 110 °C (100 to 200 °F) higher than the liquidus of the filler alloy. In selecting the braze filler alloy, consideration must be given to any possibility of material degradation by mechanisms such as grain growth, sensitization in the case of stainless steels, oxidation, and the loss of mechanical strength due to annealing. The cost of the filler metal as well as the cost of operating the brazing furnace (dictated by temperature, time, and atmosphere) should also be taken into consideration. Eutectics and narrow-melting-range alloys can withstand superheating to increase fluidity, thus permitting the use of narrow joint gaps.

Based on the type of heating used, brazing processes are classified as furnace brazing, vacuum brazing, torch brazing, induction brazing, or electron beam brazing. Prevention of chromium oxide formation, either by keeping the brazing temperature below 593 °C (1100 °F) or by using a dry reducing atmosphere, is highly essential for brazing stainless steels. Alternately, a suitable flux can be used to prevent oxidation and improve wetting. Filler metals commonly used for stainless steel brazing fall under three families: silver base (silver, copper, zinc, etc.), nickel base (nickel, chromium, boron, silicon, iron, etc.), and precious metal base (gold, copper, etc.).

In the furnace brazing of austenitic stainless steels, consideration must be given to the high rate of thermal expansion of the alloy and to the possibility of stress-corrosion cracking from exposure to the molten filler metal while the component is under stress. Because the high-temperature strength of ferritic stainless steels declines significantly above 815 °C (1500 °F), fixturing may be required for these steels during furnace brazing. Sensitization is a concern unless the base material is an "L" grade of stainless steel and/or is stabilized with niobium or titanium. Oxides, nitrides, and sulfides are detrimental to sound brazing.

10.3.1 Basic Considerations in the Brazing of PM Stainless Steels (Ref 22, 23)

Sintered parts must have little or no interconnected porosity; the parts should have a sintered density of at least 7.20 g/cm^3. Stainless steels sintered at or above 1276 °C (2330 °F) respond well to brazing. Because the commonly used brazing temperatures are approximately 1121 °C (2050 °F), sinter-brazing at 1121 °C (2050 ° F) is not a viable alternative.

Flux must promote wetting and support heat transfer. It should tie up all surface oxides present, including any refractory oxides. AMS 3417 flux has an active range of 760 to 1204 °C (1400 to 2200 °F), which is suitable for fluxing refractory oxides. An ideal flux should have an active range' that is 111 °C (200 °F) higher than the liquidus and 55 °C (100 °F) lower than the solidus of the filler metal.

The joint gap must be designed to take into consideration the thermal expansion behavior of the mating parts. The optimal joint gap for PM stainless steel brazing is typically 0.10 to 0.15 mm (0.004 to 0.006 in.). Gap selection is critical, because it determines capillary forces, which in turn govern joint fill and joint strength. Mating components can be designed with stand-off projections to help maintain the gap and with channels formed on the joint surface to facilitate flow of the braze alloy around the joint.

Cleanliness of the joint surfaces is highly essential. Oil, dirt, cutting fluids, and soot are some of the possible contaminants in PM parts, and these must be completely removed.

Listed as follows are two brazing filler metals that have been successfully used for brazing PM stainless steels:

- Ancor Braze 72 (SKC 72 paste): 40 to 44% Ni, 38 to 42% Cu, 14 to 17% Mn, 1.6 to 2.0% Si, 1.3 to 1.7% B; solidus 890 °C (1635 °F), liquidus 982 °C (1800 °F)
- B Ni-1a (Wesgo foil): 13% Cr, 4% Fe, 4.5% Si, 2.9% B, balance nickel; solidus 966 °C (1770 °F), liquidus 1093 °C (2000 °F)

10.4 Sinter Bonding

Two concentric compacts of PM stainless steel can be joined together by sinter bonding, if they have suitable geometry and/or composition. Most conveniently an assembly for sinter bonding would consist of an austenitic core component and a ferritic outer component. Upon heating to the sintering temperature, the austenitic core, due to its greater coefficient of thermal expansion, will expand more than the ferritic component, ensuring good contact between the two components. Additionally, during sintering, the ferritic outer component will undergo a greater degree of shrinkage relative to the austenitic core, which will further enhance contact between the two surfaces. Intimate contact between the two surfaces will permit diffusion to take place across the interface, resulting in the sintering together of the two components.

This concept can also be used if only one of the components is a PM compact and the other component is a pore-free material. Overall, sinter bonding does not produce joint strengths as high as those obtained with welding or brazing, although it may be sufficient for some applications.

Okimoto et al. (Ref 27) demonstrated that PM 410L can be sinter bonded with PM permalloy (47% Ni, balance iron) via vacuum sintering at 1255 °C (2290 °F), in 1 h. Their investigation included the effects of compaction pressure and particle size distribution of the powder components on joint strength. Two types of joint designs were studied: in one, the permalloy was the core component, and in the other, 410L was the core component. When only coarse powders were used, joint strengths were fairly low; the design with permalloy as the core component had a higher joint strength compared to the design with 410L as the core component (50 versus 5 MPa, or 7000 versus 700 psi, typically). However, when only fine powders were used, both designs exhibited much higher joint strengths, of the order of 150 MPa (21.7 ksi). This process was aimed at substituting 410L/permalloy composites, which are conventionally made via roll cladding, for a magnetic shield application.

10.5 Resin Impregnation

Resin impregnation of sintered stainless steels eliminates or reduces interconnected porosity, and, as such, it offers a number of benefits. It is a low-cost way of producing a pore-free component, sufficiently suitable for meeting a good many application requirements. It can render the component leakproof under moderate fluid pressures, minimize entrapment of contaminants in pores, and also significantly improve machinability. All of these attributes can enhance the value of the PM stainless steel component in specific applications. For example, in food-processing applications, resin impregnation can eliminate the possibility of entrapping bacteria-causing food residues in the pores.

Resin-impregnated parts should not be used in service at temperatures above 232 °C (450 °F). The process does not coat the external surface of the part, and it also cannot be used to repair cracks. It results in a slight increase in the apparent hardness and wear resistance of the component (Ref 28).

10.5.1 Methods of Impregnation

Four methods are in common use for resin impregnation. These are known as wet vacuum, wet vacuum pressure, dry vacuum pressure, and pressure injection. In the wet vacuum method, parts are first immersed in the resin, and then the chamber is evacuated. As the chamber is backfilled with air to the atmospheric level, liquid resin fills the pores. Air from the pores is released as bubbles through the liquid resin. In the wet vacuum pressure method, instead of relying on the atmospheric pressure, a positive pressure is applied to force the resin into greater depth and finer pores. In both cases, penetration is a function of the properties of the resin, the structure of the porosity, and the hold time. Excess resin from the surface is removed by spinning the basket holding the parts, followed by washing in a suitable cleaning agent. Rinsing can be followed by immersion of the part in a catalytic activator solution in order to cure the resin at the surface of the pores. This creates a dense seal at the pore openings so that the resin in the pores is deprived of air, which assists in the anaerobic curing of the resin.

In the dry vacuum pressure method, the chamber containing the parts is first evacuated to remove most of the air from the pores, then the resin is poured over the parts. The chamber is brought to the atmospheric or a more positive pressure. This process is preferred for parts that have very small pores as well as for applications that are more demanding, such as hydraulic pump components. Washing and curing steps employed are similar to those used in the wet vacuum process.

In the pressure injection method, the porous part is fixtured in order to seal off passage ways to any internal cavities. The resin is then pumped into the part from a pressure chamber until it emerges on an exterior surface.

Resins commonly used are various combinations of methacrylate monomers, formulated to suit the parameters of the impregnation process. Formulation is aimed at lowering the viscosity and raising the boiling point. These resins cure into an inert, solid polymer in the pores by cross linking. Readers interested in further details of the method and materials are referred to Ref 29.

10.5.2 Benefits of Resin Impregnation

One of the most common reasons for resin impregnation of stainless steels is to improve machinability. In a study sponsored by the Center for Powder Metal Technology and carried out at Concurrent Technologies Corporation significant improvements were observed in tool life as a result of resin impregnation (Table 10.6) (Ref 10).

Enhanced machinability is also achieved in terms of improved surface finish, increased cutting speed, and ability to use lower-cost cutting tools. Resin impregnation also opens up the possibility of using a machining coolant, because the problem of coolant penetration into the pores is eliminated or minimized.

There are little or no published data indicating any enhancement in corrosion resistance arising from resin impregnation. This is partly due to the fact that resin impregnation only seals the pores in the material, and it has no influence on the chemistry and interstitial content of the as-sintered surface or the grain boundaries on the exterior surfaces. A part sintered under conditions that may have left its surface prone to corrosion will undergo corrosion with or without contribution from pores, in the form of crevice corrosion. In other words, resin impregnation is not a substitute for optimal sintering. In the authors' experience, resin-impregnated stainless steel parts are often found to have picked up iron powder contamination, because the same equipment is very often used to impregnate stainless steels and carbon steels. In one study (Ref 30), under optimal conditions of sintering, resin-impregnated 316L showed a corrosion life of only 20 h compared to 431 h for as-sintered samples (in terms of B-hours in a 5% NaCl immersion test). Upon close examination, resin-impregnated samples were found to have been contaminated by iron particles.

For food processing and drinking water (plumbing components) applications, resin impregnation offers significant benefits in terms of eliminating the potential for the entrapment of bacteria-causing residues. The National Sanitary Foundation has approved the use of resin-impregnated PM stainless steel components in equipment designed for drinking water supply.

Table 10.6 Effect of resin impregnation and sintering parameters on machinability

Grade	Sintering parameters	Resin impregnation	Drill life (number of holes)	Drill force	
				kN	lbf
303N1	100% DA;	No	37	2.5	558
304N1	1121 °C	No	2	3.5	777
316N1	(2050 °F),	No	1	3.4	773
316N1	32 min	Yes	192	0.5	121
303N2	90% DA;	No	45	2.3	514
303N2	10% N$_2$;	No	1	4.0	904
316N2	1316 °C	No	4	3.2	729
316N21	(2400 °F),	Yes	192	0.6	128
	48 min				
304L	Vacuum,	No	3	3.0	671
316L	500 µm H$_2$,	No	20	2.4	542
316L	1288 °C	Yes	192	0.5	121
	(2350 °F),				
	45 min				

Note: Sintered densities were 6.40 to 6.50 g/cm^3, in all cases. All data are for average of the three sets of tests. Maximum number of tests, permissible was 192. DA, dissociated ammonia. Source: Ref 10

As far as mechanical properties are concerned, resin impregnation has a small positive effect on tensile strength, hardness, and ductility. In the case of bushings and hollow cylinders, a significant enhancement of radial crush strength and ductility is achieved.

10.6 Re-pressing and Sizing

Sintered parts may be given an additional press-forming sequence, with a goal of increasing the final density and/or meeting dimensional requirements (including correction of any distortion that may have taken place during sintering). If enhancement of final density is the primary goal, then the operation is either called re-pressing or restrike densification. Alternately, if the primary goal is to meet dimensional tolerances and minimize the effects of distortion, the process is called sizing, coining, or restrike sizing.

These press-forming processes do require lubrication. A liquid lubricant is often used because of the ease of application. One must take into consideration any undesirable effects on the performance of the component arising from the entrapment of such lubricants in the pores. In comparison to powder compaction, these operations can be run at much higher speeds and with a high degree of automation. In some cases, depending the application, some additional processing may be required, either for stress relieving the component or for removal of a lubricant.

Re-pressing of high-density parts typically increases their final densities by up to 0.10 g/cm^3. Ferritic stainless steels are more suitable for repressing compared to the austenitic stainless steels, because they have a lower rate of work hardening. Because re-pressing entails some cold working, it leads to an increase in the yield and tensile strength and a decrease in ductility. Much of the surface porosity is sealed off, and the surface of the component becomes smoother. This condition can enhance the fatigue strength of the component. Enhancement of fatigue strength can also result from the increased density and strength achieved in re-pressing. In a study involving high-temperature (1288 °C, or 2350 °F), hydrogen-sintered 409LE, the 90% survival fatigue limit was improved from 262 to 310 Mpa (38 to 45 ksi) as a result of re-pressing. Re-pressing increased the density from 7.25 to 7.35 g/cm^3. In this study, the fatigue specimens (rotating-blending fatigue) were prepared by re-pressing sintered Charpy impact bars, followed by machining, and therefore, the improvement in the fatigue strength achieved was attributed only to work hardening and the increased density. No benefit was realized from surface modification in this case.

Re-pressing of lower-density or lower-temperature sintered materials may have an entirely different effect on mechanical properties, especially if the pores are angular in shape. The loss in ductility can be significant in these materials.

Sizing has recently become more of a necessity as high sintering temperatures become a common practice. The greater rate of shrinkage obtained with high-temperature sintering, especially with ferritic stainless steels, leads to both greater distortion and greater variations in dimensions, from part to part and lot to lot. Sizing operations usually require somewhat lower pressures compared to re-pressing, because only selected areas of the component are contacted by the die.

10.7 Other Surface Treatments

The PM stainless steel components are infrequently subjected to surface treatments such as tumbling, shot peening, grinding, polishing, buffing, or turning. In the event such treatments become necessary, extreme caution must be exercised to avoid smearing as well as contamination of surfaces with less noble metals. If carried out in a contamination-free environment, surface polishing can lead to enhanced corrosion resistance of PM stainless steels in the same manner as it influences the corrosion resistance of wrought stainless steels.

Wrought stainless steels are sometimes given a chemical passivation treatment that effectively removes light surface contamination of dirt, grease, oxides, and, in particular, smeared metal, much of which is usually transferred from rolling rolls, guide rolls, machining tools, slings, or other handling equipment. Typically, this is accomplished by treating with nitric or a mild organic acid. Nitric acid treatment ensures the required level of chromium in the protective film of the stainless steel. A clean surface naturally passivates on exposure to air or water, producing a thin, durable chromium oxide layer on the surface. No thermal treatment is necessary for passivation of stainless steels. The PM stainless steels, during cooling from the sintering temperature, inevitably develop some amount of silicon oxide on their surface in addition to the passive layer of chromium oxide. Any subsequent thermal treatment must avoid further promotion of silicon oxide formation in the interest of corrosion resistance.

REFERENCES

1. P. Belejchak, Machining Stainless Steel, *Adv. Mater. Process.,* Vol 12, 1997, p 23–25
2. R.M. German, *Powder Metallurgy Science,* 2nd ed., MPIF, Princeton, NJ, 1994, p 344–348
3. D.S. Madan, An Update on the Use of Manganese Sulfide (MnS) Powder in Powder Metallurgy Applications, *Advances in Powder Metallurgy and Particulate Materials,* ed. L. Pease III, R. Sansoucy, Vol 3, MPIF, Princeton, NJ, 1991, p 101
4. D. Graham, Machining PM Parts, *Manuf. Eng.,* Jan 1998, p 64–70.
5. P.K. Samal, O. Mars, and I. Hauer, Means to Improve Machinability of Sintered Stainless Steels, *Advances in Powder Metallurgy and Particulate Materials 2005,* Vol 7, C. Ruas and T.A. Tomlin, Compilers, MPIF, Princeton, NJ, 2005, p 66–78
6. H.D. Ambs, Machinability Studies on Sintered Stainless Steel, *Advances in Powder Metallurgy,* ed. L. Pease III, R. Sansoucy, Vol 3, MPIF, Princeton, NJ, 1991, p 89–100
7. P.K. Samal and J.B. Terrell, Effects of Various Machinability Additives on the Corrosion Resistance of P/M 316L Stainless Steel, *Advances in Powder Metallurgy and Particulate Materials,* ed. C. Rose, M. Thibodeau, Vol 9, MPIF, Princeton, NJ, 1999, p 9–15 to 9–28
8. U. Kutsch and P. Beiss, "Drilling of Stainless Sintered Steels," SAE Paper 980632, Proc. SAE International Congress and Exposition, Feb 23–26, 1998
9. P. Beiss and U. Kutsch, Machinability of Stainless Steel 430 LHC, *Proc. Euro PM 1995, Structural Parts,* Vol 1, (Birmingham, U.K.), EPMA, p 21–28
10. H. Sanderow, J. Spirko, and R. Corrente, The Machinability of P/M Materials as Determined by Drilling Tests, *Advances in Powder Metallurgy and Particulate*

Materials, ed. R. McKotch, R. Webb, Vol 2, Part 15, MPIF, Princeton, NJ, 1997, p 15–125 to 15–143
11. W.-F. Wang, Effect of MnS Powder Additions on Corrosion Resistance of Sintered 304L Stainless Steels, *Powder Metall.,* Vol 45, (No. 1), 2002, p 48–50
12. M. Henthrone, Corrosion of Re-Sulfurized Free-Machining Stainless Steels, *Corrosion,* Vol 26, (No. 12), Dec 1970, p 511–527
13. C.W. Kovach and A. Moskowitz, How to Upgrade Free-Machining Properties, *Met. Prog.,* Vol 91, Aug 1967, p 173–180
14. P.K. Samal, S.N. Thakur, M.T. Scott, and I. Hauer, "Exhaust Flanges and Oxygen Sensor Bosses—Machinability Enhancement of 400 Series Stainless Steels," presented at the SAE International Convention and Exposition, March 2005 (Detroit, MI)
15. K. Kusaka, T. Kato, and T. Hisada, Influence of Sulfur, Copper and Tin Additions on the Properties of AISI 304l Type Sintered Stainless Steel, *Modern Developments in Powder Metallurgy,* Vol 16, E.N. Aqua and C.I. Whitman, Ed., MPIF, Princeton, NJ, 1984, p 247–259
16. J.R. Davis, Ed., *Stainless Steels,* ASM Specialty Handbook, ASM International, 1994, p 340–401
17. Welding, Brazing, and Soldering, Vol 6, Metals Handbook, 9th ed., American Society for Metals, 1983, p 323
18. S. Lamb, Ed., *Practical Handbook of Stainless Steels and Nickel Alloys,* ASM International, 1999
19. S. Collins and P. Williams, Electropolished Tubing: Avoiding Corrosion in Welded Applications—Identifying Optimum AISI 316L Compositions, *Chem. Process.,* Dec 2000, p 33–36
20. O. Hammar and U. Svensson, *Solidification and Casting of Metals,* The Metals Society, London, 1979, p 401–410
21. J.A. Hamill, "Welding and Brazing of Stainless Steels," Short Course on PM Stainless Steels, March 1–2, 2000 (Durham, NC), MPIF, Princeton, NJ
22. J.A. Hamill, Welding and Joining Processes, *Powder Metal Technologies and Applications,* Vol 7, *ASM Handbook,* ASM International, 1998, p 656–662
23. F. Garver and J. Urffer, Welding of P/M Stainless Steel HEGO Fittings, *Advances in Powder Metallurgy and Particulate*

Materials, ed. R. McKotch, R. Webb, Vol 9, MPIF, Princeton, NJ, 1997, p 9–37 to 9–43

24. "Condensed Corrosion Testing of Welded Powder Metal Stainless Steel Exhaust Flanges," HazenTec l.c. Product Literature, Hazen, AR, 1997

25. J.A. Hamill, F.R. Manley, and D.E. Nelson, "Fusion Welding P/M Components for Automotive Applications," SAE Technical Paper 930490, Proc. SAE International Congress and Exposition, March 1993

26. G.W. Halldin, S.N. Patel, G.A. Duchon, Welding of 316L P/M Stainless Steel, *Prog. Powder Metall.*, Vol 39, 1984, p 267–280

27. K. Okimoto, K. Izumi, K. Iwamoto, T. Kuroda, S. Toyota, S. Hosakawa, and Y. Kato, Fabrication of Stainless Steel—Permalloy Composites by Sinter Joining,

Int. J. Powder Metall., Vol 37 (No. 8), 2001, p 55–62

28. R. Remler, "Resin Impregnation and Machinability of PM Stainless Steels," Short Course on PM Stainless Steels, March 1–2, 2000 (Durham, NC), MPIF, Princeton, NJ

29. C.M. Muisener, Resin Impregnation of Powder Metal Parts, *Powder Metal Technologies and Applications*, Vol 7, *ASM Handbook*, ASM International, p 688–692

30. P.K. Samal and J.B. Terrell, Corrosion Resistance of Boron Containing P/M 316L, *Advances in Powder Metallurgy and Particulate Materials*, ed. H. Ferguson, D. Whychell, Sr., Part 7, MPIF, Princeton, NJ, 2000, p 7–17 to 7–31

CHAPTER 11

Applications

INDUSTRIAL USE of sintered stainless steels began in the 1960s in North America, approximately a decade after the application of sintered carbon steels. The main factors responsible for the competitiveness of sintered stainless steels in the early years were similar to those that rendered sintered carbon steels competitive, mainly their low-cost net shape capability. However, the then-low level of corrosion resistance of sintered stainless steels was responsible for their low growth rate from the 1970s to approximately 1990. Initially, sintered stainless steels served many different market segments and some automotive applications (Fig. 11.1) with modest corrosion-resistance requirements. The first large automotive application was a rearview mirror bracket (section 11.1.1).

Gradually, and increasingly so now, corrosion resistance has become the primary consideration for using sintered stainless steels, as is the case for wrought and cast stainless steels. With the improvement of corrosion-resistance properties, and three major automotive uses—antilock brake system sensor rings, exhaust system flanges, and oxygen sensor bosses—the market distribution of sintered stainless steels has shifted to an automotive preponderance, as is typical of the powder metallurgy (PM) industry as a whole. Parallel with this shift, ferritic stainless steels now account for the major volume of sintered stainless steels.

Table 11.1 is an adaptation of the summary of applications of sintered stainless steels from *Powder Metal Technologies and Applications,* Volume 7, *ASM Handbook,* 1998 (Ref 2). Table 11.2 summarizes the characteristics and recommended uses of the more common grades of sintered stainless steels. Substantial quantities of sintered stainless steels are used in hardware, appliances, and electrical systems. Other applications include use in powder form as pigment flakes (Ref 3) and in thermal spraying (Ref 4). Substantial quantities are also used in parts with controlled interconnected porosity for filtration, metering of liquids and gases, and for sound attenuation in telephones, microphones, and hearing aids (Ref 5–7).

In the following, brief case histories for the major automotive applications of sintered stainless steels are presented, followed by a selection of stainless steel parts that received recognition in the annual Metal Powder Industries Federation (MPIF) "PM Part-of-the-Year Design Competition." It should be noted that award criteria included cost savings, tolerance control, part design uniqueness, and performance reliability. Corrosion resistance was not included as a direct criterion.

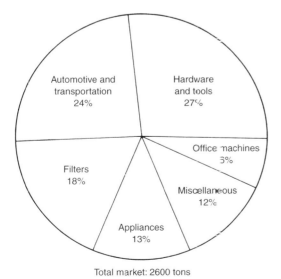

Total market: 2600 tons

Fig. 11.1 United States market distribution of powder metallurgy stainless steel products in 1979. Source: Ref 1

Table 11.1 Applications of sintered stainless steels

Application area	Alloy
Automotive	
Exhaust system flanges and sensor bosses	409L, 409LNi, 434L, 434LNb, 444L
Antilock brake system sensor rings	410L, 434L
Rearview mirror mounts	430L, 434L, 316L
Brake components	434L
Seat belt locks	304L
Windshield wiper pinions	410L
Windshield wiper arms	316L
Manifold heat control valves	304L
Hardware	
Lock components	304L, 316L
Threaded fasteners	303L
Fasteners	303L, 304L, 316L
Quick-disconnect levers	303L, 316L
Spacers and washers	316L
Electrical and electronic	
Limit switches	410L
G-frame motor sleeves	303L
Rotary switches	316L
Magnetic clutches	410L, 440A
Battery nuts	830
Electrical testing probe jaws	316L
Metal injection molding	
Medical apparatus	
Office equipment	
Automotive	Predominantly 316L
Telecommunications	and 17-4PH for most
Hardware	applications
Sporting goods	
Ornamental dental industry	
Industrial	
Water and gas meter parts	316L
Filters, liquid and gas	316L, 316L-Si
Recording fuel meters	303L
Fuel flow meter devices	410L
Pipe flange clamps	316L
Plumbing fixtures	303L
Sprinkler system nozzles	316L
Shower heads	316L
Window hardware	304L, 316L
Office equipment	
Nonmagnetic card stops	316L
Dictating machine switches	316L
Computer knobs	316L
Miscellaneous	
Coins, medallions	316L
Dental equipment	304L
Watch cases	316L
Fishing rod guides	304L, 316L
Photographic equipment	316L
Cam cleats	304L
Dishwasher components	304L
Can opener gears	410L

Source: Ref 2

11.1 Major Automotive Applications

11.1.1 Rearview Mirror Bracket

The rearview mirror bracket (Fig. 11.2) attaches the rearview mirror to the car windshield. It was developed in the 1960s and still accounts today for several hundred tons of the North American market for PM stainless steels.

These brackets are made mainly from 410L, 430L, and 434L by sintering in dissociated ammonia or H_2 at 1121 to 1149 °C (2050 to 2100 °F). Sintered density is ~7.1 g/cm^3. There are no demanding requirements regarding corrosion resistance or mechanical properties. However, the thermal expansion coefficient of the bracket must match that of glass, because the metal bracket is glued to the windshield window.

11.1.2 Antilock Brake System (ABS) Sensor Rings

The ABS sensor rings (Fig. 11.3) also known as tone wheels, were developed in the 1980s and presently account for over 20% of the total volume of sintered stainless steel parts.

As shown schematically in Fig. 11.4, a sensor ring constitutes an integral part of an antilock brake system (Ref 8). Rotating at the same speed as the car wheels, a magnetic sensor ring generates voltages in a stationary coil. The frequency of the induced voltage depends on the rotational speed of the ring or the wheel. The voltage signal is used to control brake and ignition functions via a computer to prevent locking of the car wheels. Thus, in addition to mechanical strength, ductility, and dimensional accuracy, the ring must possess adequate magnetic characteristics and sufficient corrosion resistance to survive its exposure to the elements of the road.

Various steel compositions and PM techniques have been developed to achieve the previously mentioned objectives. Sensor rings made from mild steel require a protective coating to resist abrasion and corrosion (Ref 9). The majority of ABS sensor rings are now made from 410L, 434L, and modified 434L (18Cr, 2Mo) stainless steel by hydrogen or vacuum sintering. Nitrogen contents are preferably <50 ppm to ensure good magnetic response (i.e., low coercive force, low remittance, high permeability, and high maximum induction) and adequate chloride corrosion resistance. High-temperature sintering produces lower levels of

Table 11.2   Characteristics of various grades of powder metallurgy (PM) stainless steels

Designation	Description	Characteristics
303L	Free-machining austenitic grade	Designed for parts that require extensive secondary machining operations. It has high strength and hardness. This alloy has marginal corrosion resistance. Sulfur added for machinability
303LSC and Ultra 303L	Enhanced corrosion-resistance version of 303L	These are cooper- and tin-modified versions of 303L alloy having all the characteristics of 303L, except for improved corrosion resistance.
304L	Basic austenitic grade	Most economical of austenitic grades. Used where material cost is large percentage of the total manufacturing cost. It has better corrosion resistance than 303L. Machinability is good. Copper- and tin-modified versions of 304L alloy (304LSC and Ultra 304L) are avilable for improved corrosion resistance.
316L	Standard austenitic grade	This alloy offers better corrosion resistance and machinability than 304L. With careful processing, it can meet the corrosion-resistance requirements of the more demanding applications. Copper- and tin-modified versions of 316L alloy offer even greater corrosion resistance than 316L alloy.
317L	Premium austenitic grade	It is a higher-molybdenum-content austenitic grade possessing excellent resistance to corrosion, especially to crevice corrosion (superior to 316LSC and Ultra 316L).
SS-100	Super premium austenitic grade	A highly alloyed austenitic grade superior to all other grades of PM stainless steel in corrosion resistance. Its corrosion resistance equals that of wrought 316L. In non-optimized sintering atmospheres it suffers a smaller loss of corrosion resistance compared to other grades of PM stainless steel.
409LNi	Weldable ferritic grade	A weldable grade of stainless steel containing niobium, which prevents sensitization. It is not recommended to make carbon additions to this grade. It is a magnetic alloy with good ductility and fair corrosion resistance.
410L	Standard ferritic/martensitic grade	This ferritic grade can be readily converted to a martensitic alloy by addition of small amounts of carbon prior to processing, which will also make it responsive to heat treatment. In the ferritic form, the alloy is ductile and machinable, whereas in the martensitic form, it is hard, with reduced ductility. In the martensitic form, it is used in wear-resistant applications. Both forms of the alloy are magnetic. The martensitic form has the lowest corrosion resistance of all PM stainless steel grades.
430L, 434L, 434LNb	Premium ferritic grades	Used for applications requiring some corrosion resistance but where economics (or magnetic requirement) preclude use of austenitic grade. Within the specified levels of carbon and nitrogen of standard compositions, these grades usually cannot be converted to a martensitic alloy. Color is compatible with chrome plate. Corrosion resistance is better than that of 410L. Machinability is slightly better than that of 410L.

Note: LSC and Ultra are proprietary grades of North American Hoganas and Ametek, respectively. Adapted from Ref 2

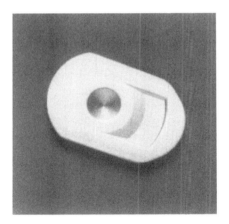

Fig. 11.2 Stainless steel bracket for automotive rearview mirror

interstitials, which is beneficial to both magnetic properties and corrosion resistance. The 410L stainless steel, sintered at 1260 °C (2300 °F) in hydrogen followed by re-pressing (690 Mpa, or 50 tsi) and annealing (900 °C, or 1652 °F, 45 min/H$_2$), resulted in a final density of 7.4 g/cm^3. Stainless steel sensor rings processed in this manner can develop various amounts of rust as well as a small number of pits when subjected to a 100 h 5% NaCl salt spray test in accordance with ASTM B 117. However, extensive field testing in various locations across the United States and Canada has shown that these rings are still functioning properly after 10 years and/or 100,000 miles, with only minor amounts of rust and no loss of mechanical strength (Ref 10). For aesthetic purposes, these rings can be nickel or chrome plated to give them a rust-free appearance.

11.1.3   Automotive Exhaust Systems

Because of stricter environmental regulations (no-leak exhaust systems) and consumer demands for extended service life, automotive exhaust

Fig. 11.3 Sensor rings for antilock brake systems

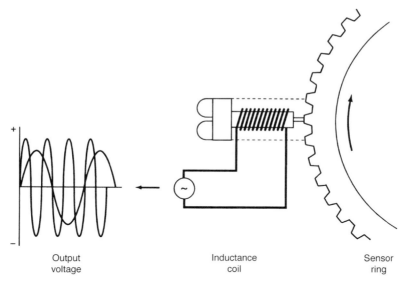

Output voltage   Inductance coil   Sensor ring

Fig. 11.4 Schematic of sensor system for antilock brakes. Source: Ref 8. Reprinted with permission from SAE paper 930449 ©1993 SAE International

systems are now increasingly made from stainless steel. For the connection of various components of an exhaust system (Fig. 11.5) as well as for measuring exhaust gas properties, flanges and hot exhaust gas outlet fittings, so-called sensor bosses (Fig. 11.6), are required. Both can be manufactured advantageously by PM technology.

Requirements for these parts include room-temperature and elevated-temperature mechanical integrity (hot tensile and compressive yield strength, creep resistance), gas-sealing quality, oxidation resistance, environmental corrosion resistance, and weldability (Ref 11, 12). Powder metallurgy offers the ability to produce parts with good surface finish, flatness, and dimensional accuracy—properties that are essential for achieving good gas-sealing capability. Low interstitial content of high-temperature hydrogen-sintered 409L and 434L satisfies the weldability requirement.

Flanges. The most demanding of the various specifications is a cyclic elevated-temperature oxidation/corrosion test in which a part is first

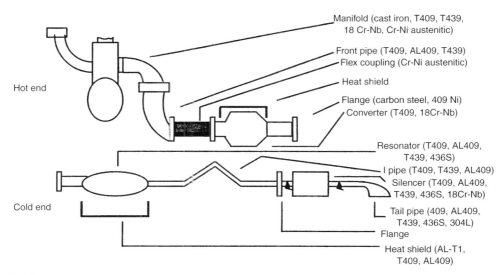

Fig. 11.5 Exhaust system components

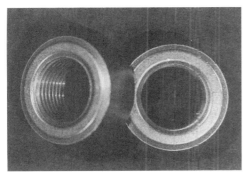

Fig. 11.6 Powder metallurgy flanges and sensor bosses for automotive exhaust

soaked for 14 min in 5% NaCl solution, then heated in air to 650 °C (1200 °F) and kept at this temperature for 90 min, then quenched in water for 1 min and exposed for 255 min in a humidity chamber (60 °C, or 140 °F, 85% relative humidity) for 200 cycles.

Another critical test is a condensed corrosion test for welded parts, in which a flange that is welded to the exhaust pipe (typically of 409Nb composition) is soaked for 15 min in a solution of 5000 ppm sulfate plus 100 ppm chloride at a pH of 2.5. After drying in air for 75 min, the part is exposed for 20 h to 85% relative humidity at 60 °C (140 °F). This is followed by soaking for 90 min in air at 427 °C (800 °F). After 25 cycles, the PM part is compared to a wrought welded 409Nb flange tested identically for signs of grain-boundary corrosion due to sensitization.

The third test, known as the hot vibration leak test, is designed to determine the leak tightness of the flange-manifold assembly, both in the as-assembled condition and after subjecting the assembly to heat, water quenching, and vibration.

Welded flanges made from various 400-series stainless steel compositions, at densities of 7.15 g/cm^3 and higher, were found to pass the condensed corrosion test. Both wrought and PM flanges developed light pitting; the depths of the pits were less than 1% of the thickness of the flange. Sensitization was absent in all PM flanges, including those made from niobium-free 410L due to its low carbon and nitrogen contents, less than 100 ppm of carbon plus nitrogen.

In order to pass the high-temperature oxidation/corrosion test, it was determined that the PM 400-series flanges must have a minimum sintered density of 7.20 g/cm^3. At lower sintered densities, the PM 400-series flanges failed due

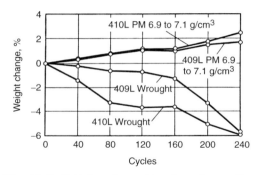

Fig. 11.7 Oxidation of wrought and powder metallurgy stainless steel flanges

to thermal fatigue and intergranular oxidation. The PM 304L flanges, processed to similar densities, did not pass the test. This was attributed to the lower thermal conductivity and higher coefficient of thermal expansion of austenitic stainless steels in comparison to ferritic grades, both of which lead to greater stress in a thermal fatigue test. The PM flanges exhibited a small amount of weight gain in the test, whereas the wrought stainless steel flanges suffered significant weight loss (Fig. 11.7).

The oxide scales formed on the wrought flanges spalled off periodically during water quenching, which resulted in weight loss. In the case of PM flanges, the oxide scale was adherent, being anchored in the surface pores (Ref 2).

In the hot vibration leak test, the objective is to determine the leak tightness of the flange/manifold assembly, as well as its serviceability. The latter refers to the ease of disassembly and reassembly of a flange at a service facility, along with its ability to provide leak tightness after reassembly. In this test, the performance of the PM 400-series stainless steel flanges was found to be superior to that of their wrought counterparts. This is attributed to the superior flatness and surface finish of PM flanges compared to those of stamped wrought flanges. The initial and post-test leak tightness of the PM flanges is attributed to their superior dimensional accuracy, including sphericity, compared to wrought stamped flanges.

Samal et al. (Ref 13) and Scott et al. (Ref 14) describe the superior performance of PM 409LNi flanges over wrought flat flanges made of heat treated 410Cb and wrought formed flanges made of wrought 410 in terms of gas-sealing ability, serviceability, and cost. The PM flange designs based on finite element analysis optimized weight reduction and stiffness.

Other properties, including elevated temperature, of these and similar stainless steels are described by Hubbard et al. (Ref 15) and Albee et al. (Ref 16).

Figure 11.8 illustrates the importance of density for several of the critical performance properties of exhaust flanges (Ref 17).

Oxygen Sensor Bosses. The function of hot exhaust gas oxygen sensors for automotive exhaust systems is to measure the oxygen content of the exhaust gas, which is related to the air/fuel ratio of the mixture undergoing combustion in the engine. The voltage output from the sensor is fed to an onboard computer that makes necessary adjustments to the air/fuel ratio. This optimizes fuel consumption and minimizes the generation of undesirable by-product gases in the exhaust stream.

The high-temperature oxidation and corrosion requirements are similar to those of the exhaust flanges described previously. In addition, however, all sensor bosses must be weldable, machinable for thread tapping, and resistant to galling after exposure to temperatures as high as 1000 °C (1832 °F). Galling is a major concern in this application, because damages caused to a sensor boss during sensor replacement can lead to replacement of a major portion of the exhaust system. Typically, a PM sensor boss is pitched against a wrought 304L stainless steel sensor boss made via cold heading or screw machining. Powder metallurgy ferritic materials are found to minimize galling with the outer cases of oxygen sensors that are made of an austenitic stainless steel (Ref 18).

Optimally processed 400-series stainless steels can meet these requirements at sintered densities of at least 7.20 to 7.25 g/cm^3 (Ref 19, 20). The preferred composition, however, appears to be 409L, which, because of its stabilization with niobium, can tolerate higher carbon levels.

11.2 Stainless Steel Filters and Other Porous Stainless Steels

Uses of PM stainless steels where porosity is functional include filters, self-lubricating bearings, parts for metering and distributing of liquids or gases (spargers, aerators), flame arrestors, and parts for sound attenuation in telephones, microphones, and hearing aids. More recently, highly porous stainless steels have also

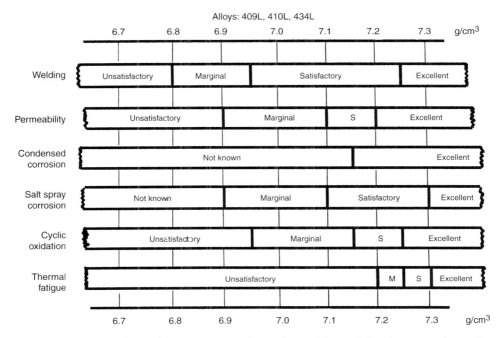

Fig. 11.8 Performance versus density of three 400-series powder metallurgy stainless steels for exhaust system flanges. Source: Ref 17. Reprinted with permission from MPIF, Metal Powder Industries Federation, Princeton, NJ

been made in the form of cellular structures and metal foams.

Filters and other applications of similar porosities are usually made from water-atomized stainless steel powders by pressing and sintering, similar to the manufacture of structural parts. However, loose powder sintering in trays or molds, cold isostatic compaction, and binder-assisted extrusion are also used to produce large, flat sheets and seamless cylinders (Ref 21). Powders are also mixed with binders and pore-forming agents to increase porosity.

The advantages of stainless steel metallic filters over nonmetal filters, based on organic materials, include use at elevated temperatures, higher strength and ductility, higher thermal conductivity and resistance to thermal shock, and good corrosion resistance. Metallic filters can also be cleaned and recycled.

Porosities for stainless steel filters range from approximately 20 to 70%, with mean pore sizes from 1 to 165 μm. For good efficiency, filters are made from narrow screen fractions. For a given density, the quality of a filter is determined by its permeability (Ref 22, 23). Permeability, which is dependent on porosity, pore size and shape, surface area, and tortuosity of the pore space, determines a filter pressure

drop, as illustrated in Fig. 11.9 for 1.6 mm (1/16 in.) thick 316L filters of various mean pore sizes (Ref 24). Tensile and yield strengths for 316L stainless steel, as a function of the porosity/density ratio, are shown in Fig. 11.10.

Spherical gas-atomized powders—because of their more uniform porosities, topologies, and smoother pore surfaces—provide more controlled and superior permeabilities and filtration efficiencies. Filters made from irregular particle-shaped powders, however, significantly widen the range of critical filter properties, for example, particle capture. Hoffman and Kapoor (Ref 6) describe the characteristics of filters manufactured from three 316L powders of varying particle shape. Such powders can be produced by controlling the water-atomization process (section 3.1.2 in Chapter 3, "Manufacture and Characteristics of Stainless Steel Powders") as well as the chemical composition (section 3.1.3 in Chapter 3) of the stainless steel powder. Figure 11.11 illustrates the relationship of filter characteristics with permeability for four sieve fractions of 316 E-F, a powder consisting of 10% spheroidal and 90% high length-to-diameter ratio particles.

German has shown that for a more complete description of flow rate in stainless steel filters, particularly at high velocities or high pressures,

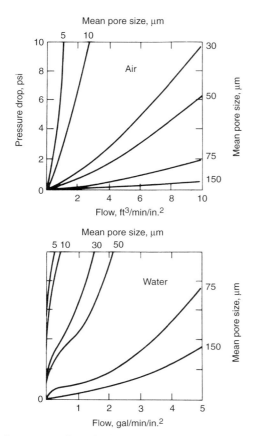

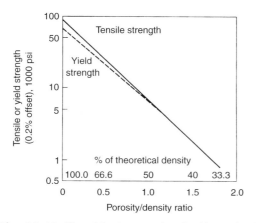

Fig. 11.9 Relationship between flow and pressure drop for 1.59 mm (1/16 in.) thick 316L porous stainless steel. Mean pore sizes range from 5 to 150 μm. Source: Ref 24

Fig. 11.10 Effect of density on tensile and yield strengths of 316L stainless steel

it is necessary to take into account inertial energy losses of the flow medium arising from changes in the direction the medium takes when passing through the tortuous pore structure. To include this effect, the well-known Darcy equation for flow is expanded to include an inertia term in addition to its viscosity term, as follows:

$$P \times \Delta P/P_m = t\,\mu\,Q/\alpha_o + t\,\rho\,Q^2/\beta$$

where P is the average pressure, ΔP is the pressure drop within the filter, P_m is the pressure at which Q is measured, t is the thickness of the filter, μ is the dynamic viscosity of the gas, Q is the flow rate, α_o is the viscous permeability coefficient, ρ is the density of the gas, and β is the inertia permeability coefficient.

The equation is very accurate for gas velocities in the range of 0.1 to 20 m/s. ISO standard 4022 (Ref 25) provides a detailed procedure for the experimental determination of the viscous and inertial permeability coefficients of a filter material through measurements of pressure drop and volumetric flow rates.

Figure 11.12 shows the shear strength, the relevant strength property for filters, of the three powders as a function of porosity. Particle size fraction had no effect on shear strength. Hoffman and Kapoor also cite explicit relationships between permeability and filter grade, permeability and maximum pore size (as determined by the bubble point test), maximum and minimum particle size of the powder used, and porosity.

With optimal sintering, the corrosion resistances of stainless steel filters will follow those of optimally sintered structural stainless steel parts, taking into account the lower densities of the former; that is, in an acidic environment, corrosion resistance declines with decreasing density (increasing surface area), as illustrated in Fig. 5.4 in Chapter 5, "Sintering and Corrosion Resistance." In a neutral chloride environment, however, typical stainless steel filters are not subject to crevice corrosion, due to their low densities (Chapter 5).

Stainless steel filters with superior flow efficiencies are made from stainless steel fibers with porosities of up to 90% (Ref 26). Stainless steel filters are used in the chemical, food, pharmaceutical, and cryogenic industries, among others. Figure 11.13 shows examples of 316L filters.

Figure 11.14 illustrates the effect of porosity on acoustic response. Attenuation increases with increasing density, and the high-frequency peak is eliminated in all sample materials (Ref 24).

Figure 11.15 (Ref 7) shows a porous sparger element for the introduction of fine gas bubbles

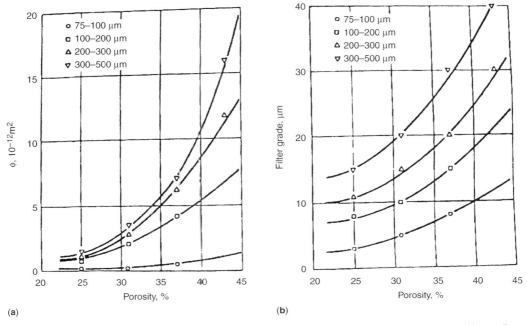

(a)   (b)

Fig. 11.11 Filter characteristics of 316 B-F powder as a function of porosity and sieve fraction. (a) Viscous permeability coefficient. (b) Filter grade by glass bead test. Source: Ref 6. Reprinted with permission from MPIF, Metal Powder Industries Federation, Princeton, NJ

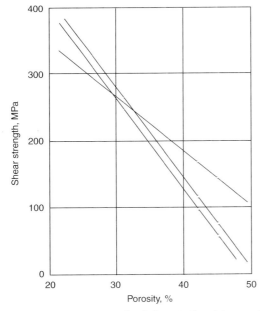

Fig. 11.12 Shear strength of three 316L stainless steel powders as a function of porosity. Source: Ref 6. Reprinted with permission from MPIF, Metal Powder Industries Federation, Princeton, NJ

Fig. 11.13 Typical porous powder metallurgy stainless steel filter parts (type 316L plate and three-element filter)

Metal Foams and Cellular Structures. In recent years, highly porous materials, including stainless steels, have also been made in the form of cellular metals and metal foams. Both open-cell and closed-cell structures can be produced. They are used in different applications: open cell for heat exchangers, filters, and batteries; closed cell for lightweight structures, energy absorption, vibration damping, and thermal insulation (Ref 27). For low-melting metals, foaming agents may be directly added to the liquid metal. For high-melting metals, various approaches have been shown to be technically feasible. One of the methods uses a water-based metal powder slurry (i.e., metal injection

below the surface of a liquid. The porous tubes were made by cold isostatic pressing and sintering, then tungsten inert gas welded onto the stainless steel plenum structure.

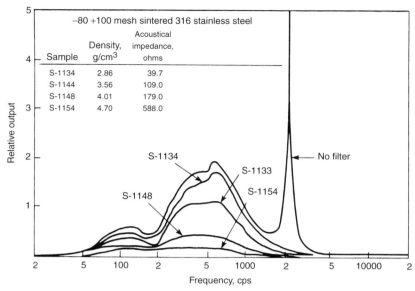

Fig. 11.14 Effect of porosity on acoustic response of sintered type 316L stainless steel. Source: Ref 24

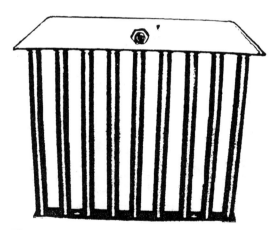

Fig. 11.15 Porous sparger element. Source: Ref 7. Reprinted with permission from MPIF, Metal Powder Industries Federation, Princeton, NJ

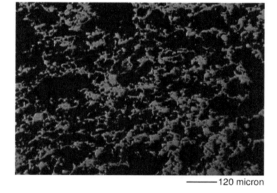

Fig. 11.16 SUS 304L stainless steel foam surface sintered at 1200 °C (2192 °F) Source: Ref 28. Reprinted with permission from MPIF, Metal Powder Industries Federation, Princeton, NJ

molding-type stainless steel powder in an aqueous solution of polyvinyl alcohol) that is mixed with paraffin wax and a dispersing agent (i.e., a neutral detergent) (Ref 28). The resulting emulsion is frozen while the water-based binder is gelling. After extraction of the paraffin wax with supercritical CO_2, the formed body is sintered. Figure 11.16 shows an SEM of the SUS 304L stainless steel foam surface (80% porosity, pore sizes from 20 to several 100 μm). Figure 11.17 shows a cross section of this foam.

Another manufacturing method is based on hollow metal spheres that are made by a fluid

bed spray coating process, where a polymer substrate material (i.e., foamed polystyrene) is coated with a metal powder slurry to form individual spheres that, after removal of the binder and expanded rigid polystyrene plastic by a heat treatment, can be formed and sintered into structures (Ref 27). Hollow particle sizes range from 0.5 to 10 mm (0.02 to 0.4 in.), and the wall thicknesses of the hollow particles are adjustable from 20 to 1000 μm.

Reviews of manufacturing processes and properties of porous metals made by these

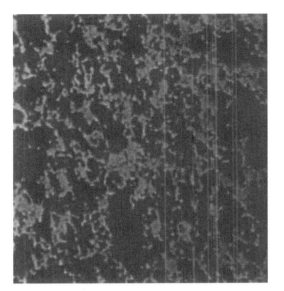

Fig. 11.17 Cross section of SUS 304L stainless steel foam. Source: Ref 28. Reprinted with permission from MPIF, Metal Powder Industries Federation, Princeton, NJ

processes were conducted by Barhart (Ref 29), Stephani and Andersen (Ref 30), and Claar et al. (Ref 31).

11.3 Metal Injection Molding

Metal injection molding (MIM), also referred to as powder injection molding, is a process used to produce quantities of precision parts with complex shapes not conducive to production by the traditional (press-and-sinter) PM method. Successful applications of MIM production has been an area of growth, because the process allows low production costs, shape complexity, tight tolerances, applicability to several materials, and high final properties. Precision shapes can be achieved at a significant cost reduction compared to other manufacturing processes, and the use of fine powders promotes densification and the development of high properties. Mold design and part handling are critical aspects of MIM processing.

Stainless steels are the most common material used in MIM components, followed closely by iron-nickel steels. Stainless steel 316L is the most commonly used alloy. Ferritic and duplex stainless steels are also used. Figure 11.18 (Ref 32) shows an assortment of stainless steel MIM parts. One application area is the production of orthodontic components, representing a worldwide industry on the order of $600 million

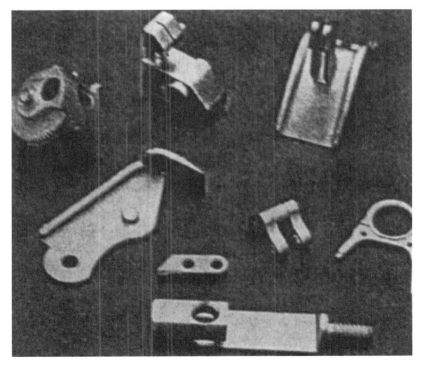

Fig. 11.18 Assortment of stainless steel metal injection molded parts. Source: Ref 32. Reprinted with permission from MPIF, Metal Powder Industries Federation, Princeton, NJ

(U.S. dollars) annually. The orthodontic bracket in Fig. 11.19 (Ref 33) was originally produced by investment casting but switched to MIM production in 2001.

Watchcases and Watch Components (ETA SA Fabriques d'Ebauches, Switzerland). Sonderegger (Ref 34) et al. describe the competitive development of metal powder injection molding technology for the production of shock- and water-resistant watchcases and other watch parts from 316L (Fe-17Cr-12Ni-2Mo-2Mn). In addition to high tolerance requirements, strict corrosion requirements had to be met. The authors describe several problems and their solutions, including powder-binder separation at the surface and macropores, leading to increased cycle times for polishing; unexpected dimensional scattering due to feedstock inconsistencies, creating computer numerical controlled machining problems; an ovalization of the case during sintering due to the eccentrically located battery hole, leading to parts out of tolerance; and longer-than-expected sintering times due to the lack of a rapid cooling option on the batch furnace.

11.4 Stainless Steel Award-Winning Parts

The MPIF sponsors an annual "Part-of-the-Year Design Competition" among PM parts manufacturers to recognize outstanding applications of PM technology. Submitted PM parts are judged by a panel of PM experts on the basis of innovative design, design complexity, precision, reliability, tooling requirements, and cost-effectiveness. Winners are awarded prizes at the MPIF annual meeting. The following examples are a selection of stainless steel parts that received awards in recent years. Part descriptions are from the year the award was received.

Stainless Steel Rotor Hub. The SSI Sintered Specialties Division of SSI Technologies Inc. was recognized with an MPIF parts award in 2002 for production of 316L stainless steel rotor hubs (Fig. 11.20) that operate in positive displacement food-processing pumps. The net-shape parts are sinter bonded together to create a structurally sound interface that prevents food particles from becoming trapped. The parts were vacuum sintered at 1343 to 1399 °C (2450 to 2550 °F), with a partial pressure of nitrogen. Sintered density was 7.7 g/cm^3, and the part weights ranged from 0.1 to 2.9 kg (0.22 to 6.4 lb). The parts have an ultimate tensile strength of 586 MPa (85,000 psi), a yield strength of 310 MPa (45,000 psi) and 45% elongation. These properties are comparable to those of annealed wrought metal and exceed cast metal. Secondary operations are limited to broaching the center holes. Cost reductions ranged up to 60% compared with castings or machining the hubs from bar stock. The pumps process viscous materials such as cheese, peanut butter, and ham pieces.

Stainless Steel Mortise Deadbolt. ASCO Sintering received an MPIF parts award in 2001 for production of a 316L stainless steel mortise deadbolt (Fig. 11.21). The part is made from MPIF material SS-316N1-25, sintered at 1129 °C (2065 °F) for 25 min in an atmosphere of 45% H$_2$-55% N$_2$. Sintered density is 6.6 g/cm^3. The typical tensile strength is 276 MPa (40,000 psi),

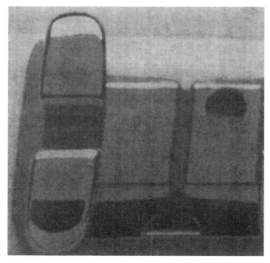

Fig. 11.19 Orthodontic bracket (17-4PH). Source: Ref 33. Reprinted with permission from MPIF, Metal Powder Industries Federation, Princeton, NJ

Fig. 11.20 Stainless steel rotor hubs (2002 MPIF parts award recipient). Source: Ref 35. Reprinted with permission from MPIF, Metal Powder Industries Federation, Princeton, NJ

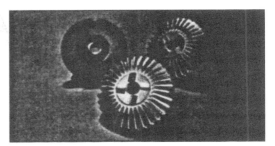

Fig. 11.22 Stainless steel bevel gear/indexing ratchet (2001 MPIF parts award recipient). Source: Ref 36. Reprinted with permission from MPIF, Metal Powder Industries Federation, Princeton, NJ

Fig. 11.21 Stainless steel mortise deadbolt (2001 MPIF parts award recipient). Source: Ref 36. Reprinted with permission from MPIF, Metal Powder Industries Federation, Princeton, NJ

Fig. 11.23 Stainless steel ball guide (2001 MPIF parts award recipient). Source: Ref 36. Reprinted with permission from MPIF, Metal Powder Industries Federation, Princeton, NJ

and the minimum yield strength is 170 MPa (25,000 psi). The part passed severe break tests and a 96 h salt spray test; the PM deadbolt withstood 200 h.

Powder metallurgy was significantly less expensive than competing processes, such as machining from bar and investment casting. Secondary operations include deburring and sinter bonding a stainless steel pin into the deadbolt.

Stainless Steel Bevel Gear/Indexing Ratchet. In 2001, Allied Sintering also received an MPIF part award in the stainless steel category for a bevel gear/indexing ratchet (Fig. 11.22). The 304L bevel gear/indexing ratchet is used in the drive mechanism of a surgical stapler. The part is sintered at 1177 °C (2150 °F) in dissociated ammonia to a typical density of 6.6 g/cm^3. There are no secondary operations. Yield strength is 207 MPa (30,000 psi). Typical hardness is 63 HRB. Powder metallurgy replaced a two-piece machined and welded assembly, offering a 70% cost savings.

Stainless Steel Ball Guide. A stainless steel ball guide (Fig. 11.23) produced by FMS Corporation was awarded with an MPIF parts award in 2001. The one-way complex ball guide

is made from MPIF material SS-316N1-25. The part is sintered at 1188 °C (2170 °F) in dissociated ammonia to a density of 6.6 g/cm^3. It has a minimum yield strength of 172 MPa (25,000 psi) and a typical tensile strength of 282 MPa (41,000 psi).

The part functions in a one-way ball valve in a high-pressure stainless steel pump for spraying paints and solvents. Its complex inside diameter (ID) configuration of three ribs, not connected in the center, allows fluid to flow more efficiently through the ball valve than a standard design. The ball guide retains a ball in a one-way valve. The valve allows liquid to flow in one direction as three ID tangs retain a ball. As liquid flows in the opposite direction, the ball seats against an orifice, preventing liquid from flowing.

Actuator Output Gear. An AGMA class 7 output gear (Fig. 11.24) used as an actuator in an automobile engine manifold is a complex part made from SS-304N1-30 to a minimum sintered density of 6.4 g/cm^3. This 1999 MPIF award-winning part was produced by Keystone Powdered Metal Company. The part is formed on a multilevel compaction press, allowing independently controlled tool members to achieve correct density distribution. Sintering is performed at 1166 to 1182 °C (2130 to 2160 °F) in a hydrogen-base atmosphere. The net-shape part meets critical tolerances, with an inside diameter of 4.80 to 4.85 mm (0.188 to 0.190 in.) and a measurement under wire of

Fig. 11.24 Actuator output gear; foreground, matching gear (1999 MPIF parts award recipient). Source: Ref 37. Reprinted with permission from MPIF, Metal Powder Industries Federation, Princeton, NJ

15.44 mm/15.31 mm (0.6080/0.6030 in.). It has a typical ultimate tensile strength of 296 MPa (43,000 psi), minimum yield strength of 207 MPa (30,000 psi), a typical transverse rupture strength of 772 MPa (112,000 psi), and a 61 HRB apparent hardness.

The part functions with a mating 304 stainless PM gear. More than one million output gears have been produced. Powder metallurgy replaced a hobbed steel part, offering significant savings.

Valve Handle Assembly. This three-piece 316 valve handle assembly (Fig. 11.25) is used in a chemical process application recognized in 1995 and was produced by Intech Metals, Inc. The PM assembly replaced a handle formed by plastic injection molding. The parts are sintered at 1150 °C (2100 °F) in hydrogen to a density of 6.3 g/cm^3. The parts must pass a 5000 h saltwater test and meet a critical wall-thickness specification. The parts have an average ultimate tensile strength of 262 MPa (38,000 psi) and a yield strength of 186 MPa (27,000 psi). Powder metallurgy offered a more than 50% cost savings over other manufacturing techniques. Secondary operations include drilling, tapping, and inserting the pin in the handle.

Valve Handle Insert Lockout Assembly. This three-piece assembly (handle base, plunger, and trigger) (Fig. 11.26) is made from a proprietary 316 stainless steel alloy and represents a replacement product in high-pressure valve systems. The product was produced by Intech Metals Inc. and was recognized in 1994.

Sintering is performed at 1150 °C (2100 °F) in hydrogen. Sintered densities are 6.2, 6.4, and 6.5 g/cm^3 for the handle base, plunger, and trigger, respectively. The parts have a minimum

Fig. 11.25 Valve handle assembly. From left, base with tapped hole, spring guide, and handle with inserted pin (1995 MPIF parts award recipient). Source: Ref 38. Reprinted with permission from MPIF, Metal Powder Industries Federation, Princeton, NJ

Fig. 11.26 Valve handle insert lockout assembly (1994 MPIF parts award recipient). Source: Ref 39. Reprinted with permission from MPIF, Metal Powder Industries Federation, Princeton, NJ

ultimate tensile strength of 303 MPa (44,000 psi), a yield strength of 206 MPa (30,000 psi), an 8% elongation, and a 30 HRB hardness.

The fabricator meets a 0.152/0.165 mm (0.0060/0.0065 in.) boss tolerance on the handle base, in addition to a high corrosion-resistance requirement. The parts must pass a 5000 h 6% salt solution test. Secondary operations are limited to minor machining of the handle base.

Stainless Steel Spring Seals. This miniature pressure-limit valve (Fig. 11.27), produced by ASCO Sintering Company and recognized by MPIF in 1993, is used in a variety of products, including adjusting flow in a kidney dialysis machine. Air or fluids flow through the openings in the spring seat. The valve opens when the desired pressure is reached. The pressure is preset by adjusting the seat against the spring on the threaded shaft.

The part is made to MPIF specification SS-316L-15. It is sintered for 45 min at 1260 °C (2300 °F) in hydrogen to a density of 6.6 g/cm^3. Minimum yield strength is 227 MPa (33,000 psi).

The PM part replaced a two-piece assembly and a screw-machined part. Powder metallurgy offered significant cost savings, and the one-piece design eliminated serious production problems and excessive rejects related to the two-piece assembly.

11.5 Stainless Steel Flake Pigments

Stainless steel flake pigments, mainly 316L, are used in corrosion-resistant industrial coatings and as additives to plastics. Flakes are produced from water-atomized stainless steel powder by dry or wet ball milling or in high-energy mills. The addition of a lubricant (i.e., stearic acid) facilitates milling and prevents welding of the powder particles.

Stainless steel flake coatings are typically formulated in a variety of organic vehicles selected for their use. Weathering is said to actually improve the appearance of the stainless steel because of its polishing action.

Figure 11.28 shows an SEM and a particle size distribution of stainless steel flake pigment (Ref 41).

REFERENCES

1. D.L. Dyke and H.D. Ambs, Stainless Steel Powder Metallurgy, *Powder Metallurgy—Applications, Advantages, and Limitations*, E. Klar, Ed., American Society For Metals, 1983

(a)   10 μm x 1000

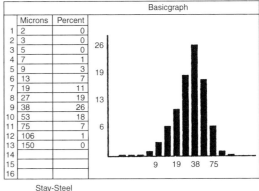

Basicgraph

	Microns	Percent
1	2	0
2	3	0
3	5	0
4	7	1
5	9	3
6	13	7
7	19	11
8	27	19
9	38	26
10	53	18
11	75	7
12	106	1
13	150	0
14		
15		
16		

Stay-Steel

(b)

Fig. 11.28 (a) SEM and (b) particle size distribution of stainless steel flake pigment. Source: Ref 41. Reprinted with permission of John Wiley & Sons, Inc.

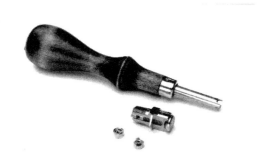

Fig. 11.27 Spring seats and pressure-limit valve (1993 MPIF parts award recipient). Source: Ref 40. Reprinted with permission from MPIF, Metal Powder Industries Federation, Princeton, NJ

2. E. Klar and P.K. Samal, Powder Metallurgy Stainless Steels, *Powder Metal Technologies and Applications*, Vol 7, *ASM Handbook*, ASM International, 1998, p 774–785

3. S.A. Humphrey and P.J. Laden, Stainless Steel Flake, *Properties and Economics*, Vol 1, *Pigment Handbook*, 2nd ed., P.A. Lewis, Ed., John Wiley & Sons, 1988

4. K.M. Kulkarni, Metal Powders Used for Hardfacing, *Powder Metallurgy*, Vol 7, *Metals Handbook*, 9th ed., American Society For Metals, 1984

5. N. Nickolas and R. Ray, Porous Stainless Steel—The Unique Filter Medium, *Modern Developments in Powder Metallurgy*, Vol 5, MPIF, Princeton, NJ, 1971, p 187–199

6. G. Hoffman and D. Kapoor, Properties of Stainless Steel P/M Filters, *Int. J. Powder Metall. Powder Technol.*, Vol 12 (No. 4), 1976, p 281–296

7. W.R. Johnson and R.M. German, Gas Flow Controlled by Porous P/M Media, *Modern Developments in Powder Metallurgy*, Vol 12, H. Hausner, H. Antes, and G. Smith, Ed., MPIF, Princeton, NJ, 1981, p 821–833

8. S. Shah, P.K. Samal, and E. Klar, "Properties of 410-L P/M Stainless Steel Antilock Brake Sensor Rings", Paper 930449, Society of Automotive Engineers International Congress and Exposition, March 1–5, 1993, (Detroit, MI)

9. M. Hanada, N. Amano, Y. Takeda, Y. Saegusa, and T. Koiso, "Development of Powder Metallurgy Sensor Ring for Use in Antilock Brake System", Paper 890407, SAE Technical Paper Series, Int. Congress and Exhibition, Feb 27 to March 3, 1989 (Detroit, MI)

10. S. Shah, J.R. McMillen, P.K. Samal, S.A. Nasser, and E. Klar, "On the Real Life Performance of Sintered Stainless Steel ABS Sensor Rings", Paper 970423, SAE Technical Paper Series, Int. Congress and Exhibition, Feb 24–27, 1997 (Detroit, MI)

11. P.F. Lee, S. Saxion, G. Regan, and P. dePoutiloff, "Requirements for Stainless Steel P/M Materials in Automotive Exhaust System Applications", Advances in P/M Technology Seminar (Dearborn, MI); PM²TEC '97 Conference (Chicago, IL), MPIF, Princeton, NJ

12. P.F. Lee, "Requirements for P/M Stainless Steel Materials in Order to Meet Future Exhaust System Performance Criteria", Paper 980311, Society of Automotive Engineers International Congress and Exposition, Feb 23–26, 1998, (Detroit, MI)

13. P.K. Samal, J.B. Terrell, S.O. Shah, and M.T. Scott, Material and Design Optimization for Improved Performance of PM Stainless Steel Exhaust Flanges, *Proceedings of Euro PM 2000 Conference* (Munich, Germany), EPMA

14. M.T. Scott, S.O. Shah, J.R. McMillen, N.W. Elsenety, and P.K. Samal, "Improved P/M Stainless Steel Exhaust Flanges Based on Innovative Design Concepts", Paper 2000-01-0336, SAE 2000 World Congress, March 6–9, 2000 (Detroit, MI)

15. T. Hubbard, K. Couchman, and C. Lall, "Performance of Stainless Steel P/M Materials in Elevated Temperature Applications", Paper 980326, presented at SAE International Congress and Exposition, Feb 1998, (Detroit, MI)

16. T.R. Albee, P. dePoutiloff, G.L. Ramsey, and G.E. Regan, "Enhanced Powder Metal Materials for Exhaust Systems Applications", Paper 970281, presented at SAE International Congress and Exhibition, Feb 1997 (Detroit, MI)

17. P.K. Samal, "Development of P/M Stainless Steel for Automotive Exhaust Flange Applications", New Developments and Applications in P/M, Stainless Steel Seminar, April 2–3, 1998 (Cleveland, OH), sponsored by Metal Powder Industries Federation, Princeton, NJ

18. G.L. Ramsey, G.E. Regan, P.A. dePoutiloff, and P.K. Samal, "P/M Stainless Steel Flanges and Sensor Bosses Meet Critical Qualification Requirements for Exhaust Applications", Paper 2000-01-1002, SAE International Congress and Exposition March, 2000, (Detroit, MI)

19. S.O. Shah, J.R. McMillen, P.K. Samal, and J.B. Terrell, Requirements for Powder Metal Stainless Steel Materials in the Oxygen Sensor (HEGO) Boss Application, *P/M Applications*, SAE SP-1447, Society of Automotive Engineers, 1999, p 105–113

20. F. Garver and J. Urffer, Welding P/M Ferritic Stainless Steel HEGO Fittings, *Advances in Powder Metallurgy and Particulate Materials*, Vol 9, R. McKotch and R. Webb, Ed., MPIF, Princeton, NJ, 1997, p 9-37 to 9-43

21. N. Nicholaus and R. Ray, Porous Stainless Steel—The Unique Filter Medium, *Modern Developments Powder Metallurgy*, Vol 5, MPIF, Princeton, NJ, 1970, p 187–199

22. R.M. German, Gas Flow Physics in Porous Metals, *Int. J. Powder Metall. Powder Technol.*, Vol 15, 1979, p 23–30

23. W. Schatt and K.P. Wieters, *Powder Metallurgy*, European Powder Metallurgy Association, 1997, p 346–352

24. D.L. Dyke and H.D. Ambs, Stainless Steel Powder Metallurgy, *Powder Metallurgy—Applications, Advantages, and Limitations*, E. Klar, Ed., American Society for Metals, 1983, p 123–144

25. "Permeable Sintered Metal Materials—Determination of Fluid Permeability", 4022, International Organization for Standardization, 1987

26. W. Schatt and K.-P. Wieters, *Powder Metallurgy Processing and Materials*, EPMA, 1997, p 357

27. U. Waag, G. Stephani, F. Bretschneider, and H. Venghaus, Mechanical and Acoustical Properties of Highly Porous Materials Based on Metal Hollow Spheres, *Advances in Powder Metallurgy and Particulate Materials*, Part 7, MPIF, Princeton, NJ, 2002, p 207–212

28. T. Shimizu and A. Kitajima, New Methods to Produce Foam Metals Using Hydro-Gel Binder, *Advances in Powder Metallurgy and Particulate Materials*, Part 7, V. Arnold, C. Chu, W. Jandeska, and H. Sanderow, Ed., MPIF, Princeton, NJ, 2002, p 207–212

29. J. Banhart, Manufacture, Characterization and Application of Cellular Metals and Metal Foams, *Prog. Mater. Sci.*, Vol 46, 2000, p 559–632

30. G. Stephani and O. Andersen, Solid State and Deposition Methods, *Handbook of Cellular Metals*, H.-P. Degischer and B. Kriszt, Ed., VCH-Wiley, 2002, p 56–70

31. T.D. Clarr, C.J. Yu, D. Kupp, and H. Eifert, "Production of Ultra-Lightweight Components Using Powder Metallurgy Processes," in *Advances in Powder Metallurgy and Particulate Materials*, Vol 12, p 12-71 to 12-86, 2000, H. Ferguson and D. Whychell, Sr, Ed., MPIF, Princeton, NJ

32. G. Fridman, Metalor 2000 on Target for MIM Success, *Met. Powder Rep.*, Vol 50 (No. 5), 1995, p 28–31

33. R. Cornwall, PIM 2001 Airs Industry's Successes and Challenges, *Met. Powder Rep.*, Vol 56 (No. 6), 2001, p 10–13

34. M. Sonderegger, B. Unternaehrer, and A. Oberli, Application of the MIM-Technology for Swatch-Irony Watchcases and Watch Components, *Second European Symposium on Powder Injection Moulding*, Oct 18–20, 2000 (Munich, Germany), EPMA, p 235–242

35. *Int. J. Powder Metall.*, Vol 38 (No. 5), 2002, p 27.

36. *Int. J. Powder Metall.*, Vol 37 (No. 4), 2001, p 49, 50

37. *Int. J. Powder Metall.*, Vol 35 1999

38. *Int. J. Powder Metall.*, Vol 31 (No. 3), 1995, p 219

39. *Int. J. Powder Metall.*, Vol 30 (No. 3), 1994, p 270

40. *Int. J. Powder Metall.*, Vol 29 (No. 3), 1993, p 281

41. S.A. Humphrey and P.J. Laden, Stainless Steel Flake, *Properties and Economics*, Vol 1, *Pigment Handbook*, 2nd ed., P.A. Lewis, Ed., John Wiley & Sons, 1988, p 819–821

Atlas of Microstructures

Powder Morphologies

Fig. 1 SEM image of a water atomized stainless steel powder (316L) having a moderately irregular particle shape, leading to a good combination of apparent density, green strength, compressibility, and flow rate

Fig. 2 SEM image of a stainless steel powder (409L) having a highly irregular particle shape, leading to low apparent density, high green strength, low compressibility, and marginal flow rate

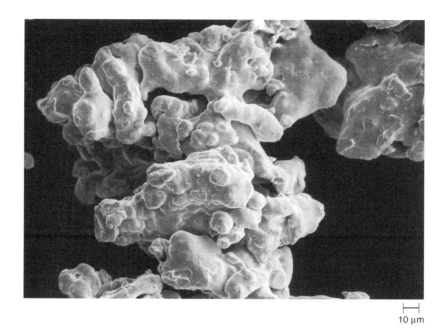

10 µm

Fig. 3   A high magnification SEM image of a particle shown in Figure 2

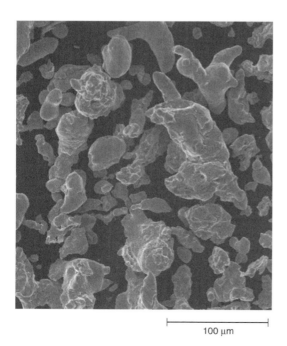

100 µm

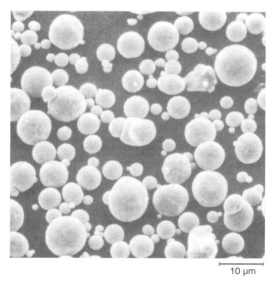

10 µm

Fig. 4   SEM image of a stainless steel powder having a marginally irregular particle shape, leading to high apparent density, low green strength, high compressibility, and a high flow rate

Fig. 5   SEM of a gas atomized 316L powder having the typical spherical particle shape. Such powders are used in MIM and in hot isostatic compaction. Source: Courtesy of Roberto Garcia, N.C. State University

Effect of Compaction Pressure on Porosity

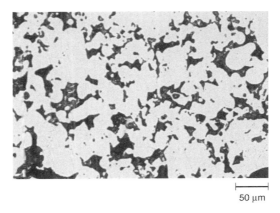

50 μm

Fig. 6   Pore structure of sintered 316L produced by compacting to 6.25 g/cm³ green density (compaction pressure 30 TSI), and then sintering in 100% hydrogen at 1205 °C (2200 °F) for 30 minutes (sintered density of 6.43 g/cm³). As-polished

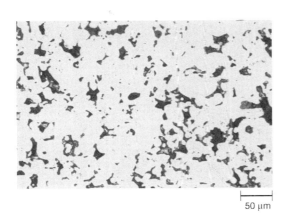

50 μm

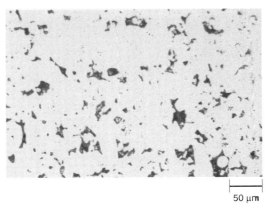

50 μm

Fig. 7   Pore structure of sintered 316L produced by compacting to 6.60 g/cm³ green density (compaction pressure 45 TSI), and then sintering in 100% hydrogen at 1205 °C (2200 °F) for 30 minutes (sintered density of 6.75 g/cm³). As-polished

Fig. 8   Pore structure of sintered 316L produced by compacting to 6.84 g/cm³ green density (compaction pressure 55 TSI), and then sintering in 100% hydrogen at 1205 °C (2200 °F) for 30 minutes (sintered density of 6.95 g/cm³). As-polished

Austenitic Stainless Steels

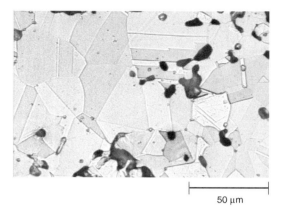

50 μm

Fig. 9   Photomicrograph of high temperature 1315 °C (2400 °F), hydrogen sintered 304L showing well-rounded porosity, precipitate-free grain boundaries, and abundant twin boundaries. Sintering time 30 minutes and dew point of sintering atmosphere –46 °C (–55 °F). Glyceregia etch

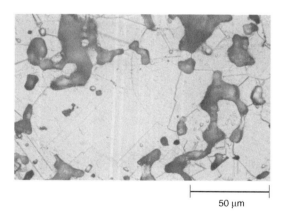

50 μm

Fig. 10   Photomicrograph of high temperature 1288 °C (2350 °F), 30 minutes, hydrogen sintered 316L showing well-rounded porosity, precipitate-free grain boundaries, and twin boundaries. Glyceregia etch

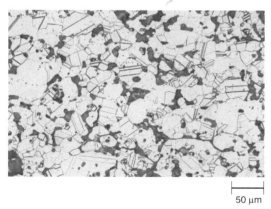

50 μm

Fig. 11   A low magnification photomicrograph of the sample shown in Fig. 10 showing well-rounded porosity, large grains, precipitate-free grain boundaries, and twin boundaries. Glyceregia etch

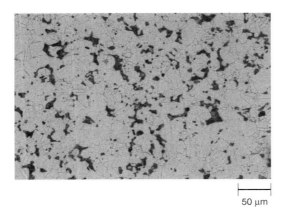

50 µm

Fig. 12 Photomicrograph of 1205 °C (2200 °F) hydrogen sintered (30 minutes) 316L, showing well-developed twin boundaries, medium grain size, and precipitate-free grain boundaries. Glyceregia etch

50 µm

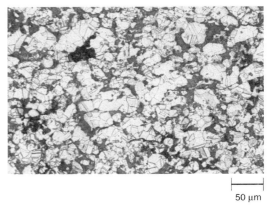

50 µm

Fig. 13 Photomicrograph of 1121 °C (2050 °F) hydrogen sintered (30 minutes) 316L showing fine grain structure, presence of oxides on some prior particle boundaries, and relatively fewer twin boundaries. Glyceregia etch

Fig. 14 A low magnification photomicrograph of sample shown in Fig. 13, showing fine grain structure, presence of oxides on some prior particle boundaries, angular porosity, and relatively fewer twin boundaries. Glyceregia etch

Ferritic Stainless Steels

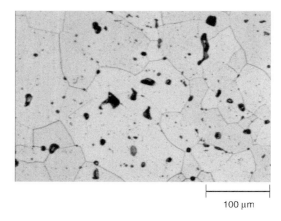

100 µm

Fig. 15 Photomicrograph of 1366 °C (2490 °F), hydrogen sintered 409L showing a predominantly coarse grain structure and fine precipitates of columbium compounds in the grain boundaries and within the grains. Glyceregia etch

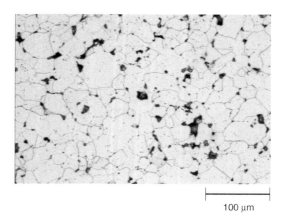

100 µm

Fig. 16 Photomicrograph of 1238 °C (2260 °F), hydrogen sintered 409L showing a relatively fine grain structure, and columbium compounds in the matrix as fine precipitates. Glyceregia etch

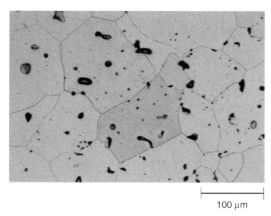

100 µm

Fig. 17 Photomicrograph of 1360 °C (2480 °F), hydrogen sintered 434L showing a coarse grain structure. Note pinning of grain boundaries at pores. Glyceregia etch

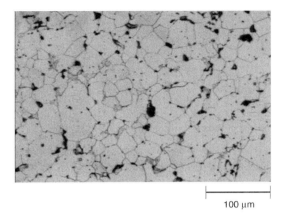

100 μm

Fig. 18 Photomicrograph of 1227 °C (2240 °F), hydrogen sintered 434L showing a fine grain structure and presence of some martensite in the microstructure. A residual carbon content of 0.052% was responsible for formation of martensite. Glyceregia etch

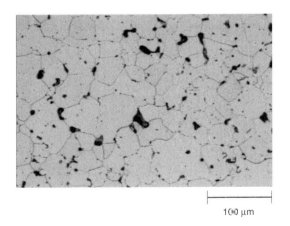

100 μm

Fig. 19 Photomicrograph of 1260 °C (2300 °F), hydrogen sintered 434L showing a mixed microstructure of coarse and fine grains. Glyceregia etch

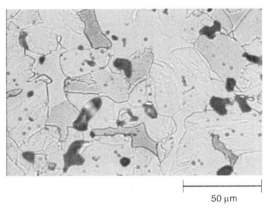

50 μm

Fig. 20 Photomicrograph of 1316 °C (2400 °F), hydrogen sintered 409LNi showing martensitic (light colored) and ferritic grains (dark colored). Columbium compounds are seen as fine precipitates within the grains. Glyceregia etch

Oxides in Sintered Stainless Steel

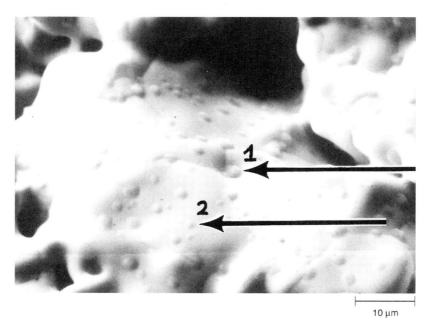

10 μm

Fig. 21 SEM image of as-sintered surface of a 316L part showing spherical oxides formed during cooling from the sintering temperature. These are oxides of silicon, and their formation is promoted by a high dew point of the sintering atmosphere and slow rate of cooling from sintering temperature

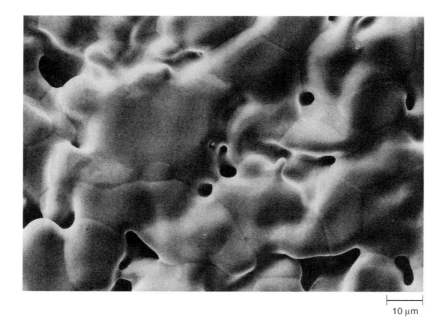

10 μm

Fig. 22 SEM image of as-sintered surface of a 434L part showing an oxide free surface, achieved by sintering in a low dew point sintering atmosphere, followed by rapid cooling

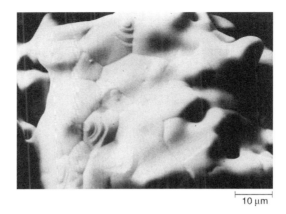

Fig. 23   SEM of oxide-free surface of a sintered 316L

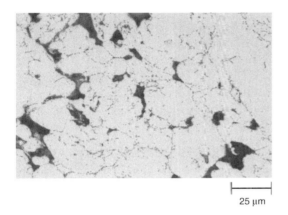

25 µm

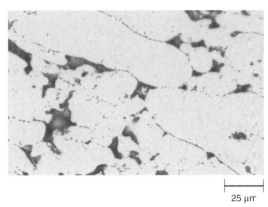

25 µm

Fig. 24   Photomicrograph of as-polished cross-section of a low temperature-sintered 316L part, showing presence of oxides along prior particle boundaries and within the particles. Most of these oxides were formed during air delubrication. carried out at at 649 °C (1200 °F). No reduction of oxides occurred during sintering, at 1121 °C (2050 °F) in 100% hydrogen for 30 minutes

Fig. 25   Photomicrograph of as-polished cross-section of a moderate temperature-sintered 316L part, showing presence of oxides mainly along prior particle boundaries. Most of these oxides were formed during air delubrication, at 649 °C (1200 °F). Partial reduction of oxides took place during sintering, at 1205 °C (2200 °F) in 100% hydrogen for 30 minutes

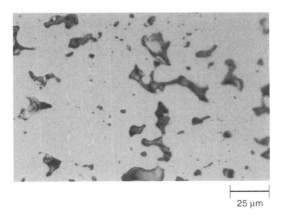

25 µm

Fig. 26   Photomicrograph of as-polished cross-section of a high temperature-sintered 316L part, showing nearly complete reduction of oxides in the microstructure during sintering. Delubrication was carried out in air at 649 °C (1200 °F) and sintering at 1288 °C (2350 °F) in 100% hydrogen for 30 minutes

Carbides in Sintered Stainless Steel

20 μm

Fig. 27 Photomicrograph of low temperature-sintered 316L showing necklace type carbides along grain boundaries. Carbon content was 0.12%. Sintering was carried out at 1150 °C (2100 °F) in 100% hydrogen. Glyceregia etch

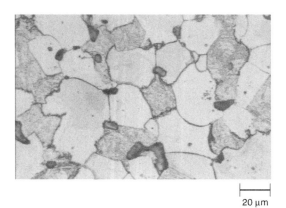

20 μm

Fig. 28 Photomicrograph of high carbon containing 410L, showing carbide network along the grain boundaries. The dark colored grains are martensitic. Carbon content was 0.10%. Sintering was carried out at 1315 °C (2400 °F) in 100% hydrogen. Glyceregia etch

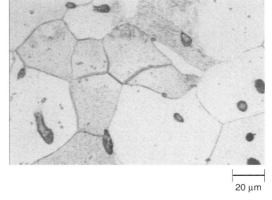

20 μm

Fig. 29 Photomicrograph of high carbon containing 434L, showing carbide network along the grain boundaries. Carbon content was 0.10%. Sintering was carried out at 1315 °C (2400 °F) in 100% hydrogen. Glyceregia etch

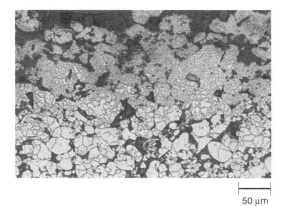

50 μm

Fig. 30 Photomicrograph of a sintered 316L part, taken near the exterior surface, showing a heavy network of chromium carbide precipitates in the grain boundaries. as well as within the grains, caused by contamination with carbonaceous deposits (lubricant residue) in the sintering furnace. Glyceregia etch

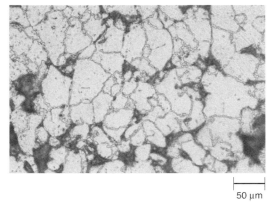

50 μm

Fig. 31 Photomicrograph of as-sintered 316L part, showing continuous network of chromium carbide precipitates along the grain boundaries. Glyceregia etch

Nitrides in Sintered Stainless Steel

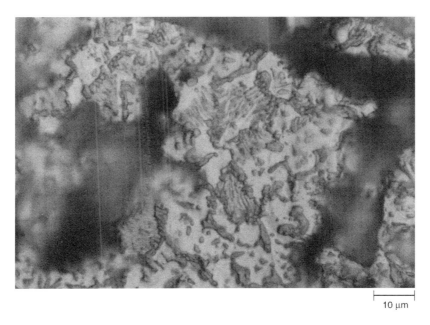

10 µm

Fig. 32 Optical photograph of the surface of an as-sintered 316L, showing chromium nitride precipitates along grain boundaries and within grains. Not etched

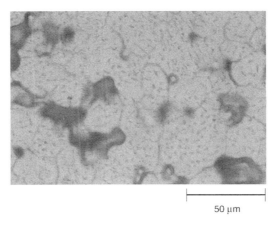

50 µm

Fig. 33 Optical photograph of the surface of an as-sintered 304L showing chromium nitride precipitates along grain boundaries and within grains. Not etched

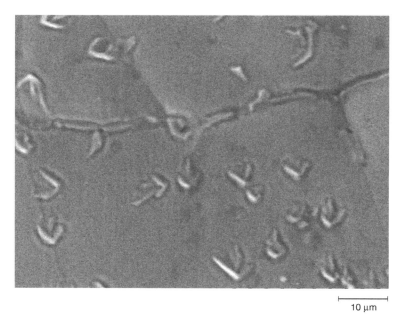

10 µm

Fig. 34 SEM image of the surface of as-sintered 316L part, showing chromium nitride precipitates along the grain boundaries and in the grains

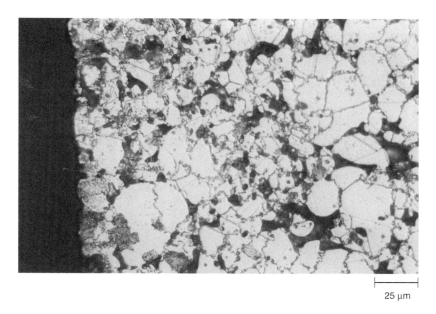

25 µm

Fig. 35 Photomicrograph of cross-section of as-sintered 316L, showing lamellar precipitates of chromium nitrides within grains and chromium nitride precipitated along grain boundaries. A heavier concentration of nitrides is seen at the surface of the part (left side) which indicates nitride formation occurred due to slow cooling in a nitrogen bearing atmosphere

Corrosion of PM Stainless Steel

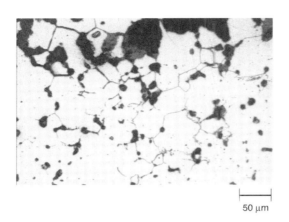

50 µm

Fig. 36 As-polished microstructure of corrosion tested 434L, showing intergranular type of corrosion attack in a salt spray test, caused by depletion of chromium along grain boundaries. Carbon content was 0.07%. Sintering was carried out at 1315 °C (2400 °F) in 100% hydrogen

50 µm

Fig. 37 As-polished microstructure of corrosion tested 410L, showing intergranular type of corrosion attack in a salt spray test, caused by depletion of chromium along grain boundaries. Carbon content was 0.06%. Sintering was carried out at 1315 °C (2400 °F) in 100% hydrogen

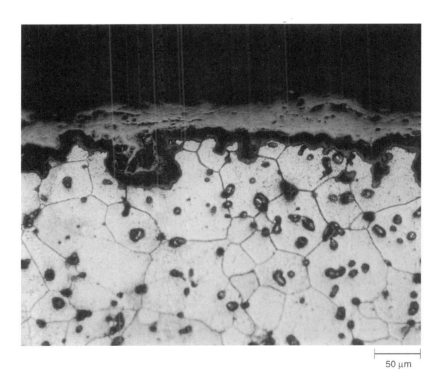

50 µm

Fig. 38 Pitting type corrosion in PM 409L (welded) subjected to condensed corrosion test for 25 cycles (25 weeks). An adherent layer of corrosion products is seen on the sample surface. Sample as-made by hydrogen sintering to 7.20 g/cc density. (See section 11.1,3 for test procedure and significance.) The response of PM Stainless is similar to that of wrought stainless steel (Fig. 39)

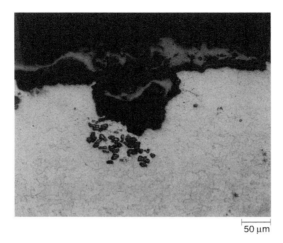

50 µm

Fig. 39 Pitting type corrosion in wrought 409L (welded) subjected to condensed corrosion test for 25 cycles (25 weeks). An adherent layer of corrosion products is seen on the sample surface. Some degree of intergranular corrosion is also noted near the large pit

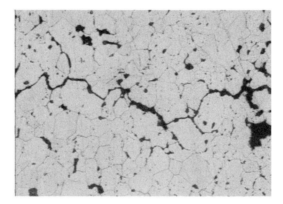

Fig. 40 Photomicrograph of the cross section of a PM 434L part showing intergranular path of corrosion damage caused by stress corrosion cracking

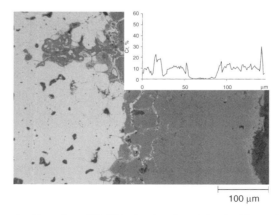

100 µm

Fig. 41 Chromium line scan of oxide scale on PM 409L, showing alternate bands of iron oxide and spinel. Anchoring of the scale is also seen at a surface pore. Good bonding between the scale formed and the PM stainless steel leads to minimal loss of mass thickness in cyclic oxidation test

Fractographs of PM Stainless Steel

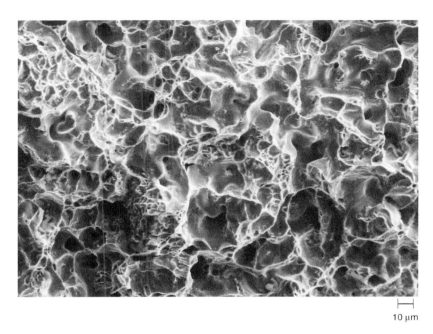

10 μm

Fig. 42 SEM image of dimpled fracture surface of a well-sintered, ductile PM 410L

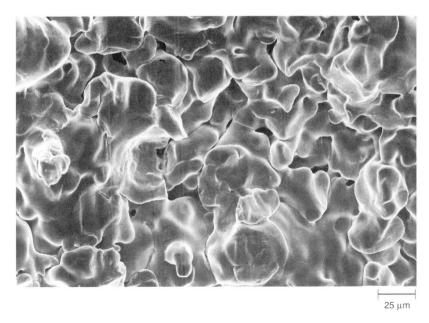

25 μm

Fig. 43 SEM image of the fracture surface of a PM 410L, showing original particle surfaces, which most likely originated as a green crack

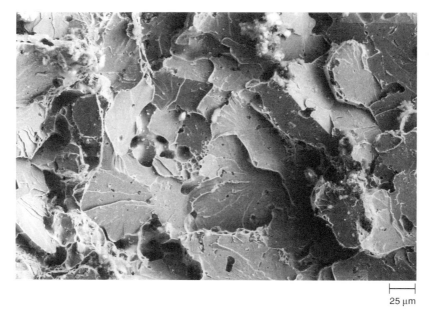

25 µm

Fig. 44   SEM image of PM martensitic 410 surface with cleavage type fracture associated with a high degree of brittleness

25 µm

Fig. 45   SEM image of the fracture surface of a PM 434L sintered part which failed due to intergranular stress corrosion cracking. Fracture progressed along the grain boundaries of the well-sintered sample

Appendix 1: 300-Series PM Stainless Steels

Powder metallurgy material properties

| Material designation code(a) | Minimum values | | | Tensile properties | | | Typical values(c) | | | | | Hardness, Rockwell HRB | | | |
| | Strength(b), 10³psi | | Elongation (in 1 in.), % | Ultimate strength, 10³psi | Yield strength (0.2%), 10³psi | Elongation (in 1 in.), % | Elastic constants | | Unnotched Charpy impact energy, ft-lbf | Transverse rupture strength, 10³psi | Compressive yield strength (0.1%), 10³psi | Macro (apparent) | Micro (converted) | Fatigue strength 10⁷ cycle, 10³psi | Density, g/cm³ |
	Yield	Ultimate					Young's modulus, 10⁶psi	Poisson's ratio							
SS-303N1-25	25	···	0.0	39	32	0.5	15.5	0.25	3.5	86	38	62	ND	13	6.4
SS-303N2-35	35	···	3.0	55	44	3.0	16.5	0.25	19.0	98	46	63	ND	16	6.5
SS-303N2-30	38	···	6.0	68	45	10.0	20.0	0.27	35.0	ND	46	70	ND	21	6.9
SS-303L-12	12	···	12.0	39	17	17.5	17.0	0.25	40.0	82	21	21	ND	15	6.6
SS-303L-15	15	···	15.0	48	24	20.0	20.0	0.27	55.0	ND	29	35	ND	19	6.9
SS-304N1-30	30	···	0.0	43	38	0.5	15.5	0.25	4.0	112	38	61	ND	15	6.4
SS-304N2-33	33	···	5.0	57	40	10.0	16.5	0.25	25.0	127	47	62	ND	18	6.5
SS-304N2-38	38	···	8.0	70	45	13.0	20.0	0.27	55.0	ND	47	68	ND	23	6.9
SS-304H-20	20	···	7.0	40	25	10.0	17.0	0.25	20.0	85	25	35	ND	(c)	6.6
SS-304L-13	13	···	15.0	43	18	23.0	17.0	0.25	45.0	ND	22	30	ND	17	6.6
SS-304L-18	18	···	18.0	57	26	26.0	20.0	0.27	80.0	ND	28	45	ND	21	6.9
SS-316N1-25	25	···	0.0	41	34	0.5	15.5	0.25	5.0	108	36	59	ND	11	6.4
SS-316N2-33	33	···	5.0	60	39	10.0	16.5	0.25	28.0	125	44	62	ND	14	6.5
SS-316N2-38	38	···	8.0	70	45	13.0	20.0	0.27	48.0	ND	46	65	ND	19	6.9
SS-316H-20	20	···	5.0	35	25	7.0	17.0	0.25	20.0	85	25	33	ND	(c)	6.6
SS-316L-15	15	···	12.0	41	20	18.5	17.0	0.25	35.0	80	22	20	ND	13	6.6
SS-316L-22	22	···	15.0	57	30	21.0	20.0	0.27	65.0	ND	29	45	ND	17	6.9

(a) N1: Nitrogen alloyed. Good strength, low elongation. Sintered at 1149 °C (2100 °F) in dissociated ammonia. N2: Nitrogen alloyed. High strength, medium elongation. Sintered at 1288 °C (2350 °F) in dissociated ammonia. L: Low carbon. Lower strength, highest elongation. Sintered at 1288 °C (2350 °F) in partial vacuum. Cooled to avoid nitrogen absorption. Processing parameters used to generate these data; other conditions could be used. (b) Suffix numbers represent minimum strength values in 10³ psi yield in the as-sintered condition and ultimate in the heat treated condition. (c) Mechanical property data derived from laboratory-prepared test specimens sintered under commercial manufacturing conditions. ND, not determined. Source: MPIF Std 35, 2007 Edition

Appendix 2: 400-Series PM Stainless Steels

Powder metallurgy material properties

Material designation code(a)	Minimum values Strength(b)(c), 10³psi Yield	Minimum values Strength(b)(c), 10³psi Ultimate	Minimum values Elongation (in 1 in.), %	Typical values(d) Tensile properties Strength, 10³psi Ultimate	Typical values(d) Tensile properties Strength, 10³psi Yield (0.2%)	Typical values(d) Tensile properties Elongation (in 1 in.), %	Typical values(d) Elastic constants Young's modulus, 10⁶psi	Typical values(d) Elastic constants Poisson's ratio	Typical values(d) Unnotched Charpy impact energy, ft-lbf	Typical values(d) Strength, 10³psi Transverse rupture	Typical values(d) Strength, 10³psi Compressive yield (0.1%)	Typical values(d) Hardness, Rockwell Macro (apparent)	Typical values(d) Hardness, Rockwell Micro (converted)	Typical values(d) Fatigue strength 90% survival, 10³psi	Typical values(d) Density, g/cm³
SS-410-90HT	...	90	0.0	105	(e)	< 0.5	18.0	0.25	2.5	113	93	23 HRC	55 HRC	35	6.5
SS-410L-20	20	...	10.0	48	26	16.0	24.0	0.27	50	ND	28	45 HRB	ND	18	6.9
SS-430N2-28	28	...	3.0	60	35	5.0	25.0	0.27	25	ND	33	70 HRB	ND	25	7.1
SS-430L-24	24	...	14.0	50	30	20.0	25.0	0.27	80	ND	33	45 HRB	ND	25	7.1
SS-434N2-28	28	...	4.0	60	35	8.0	24.0	0.27	15	ND	33	65 HRB	ND	22	7.0
SS-434L-24	24	...	10.0	50	30	15.0	24.0	0.27	65	ND	33	50 HRB	ND	22	7.0

(a) N2: Nitrogen alloyed. High strength, medium elongation. Sintered at 1288 °C (2350 °F) in dissociated ammonia. L: Low carbon. Lower strength, highest elongation. Sintered at 1288 °C (2350 °F) in partial vacuum. Cooled to avoid nitrogen absorption. HT: Martensitic grade, heat treated. Highest strength. Sinter hardened at 1149 °C (2100 °F) in dissociated ammonia. Processing parameters used to generate these data; other conditions could be used. (b) Suffix numbers represent *minimum* strength values in 10³ psi yield in the as-sintered condition and ultimate in the heat treated condition. (c) Tempering temperature for heat treated (HT) materials: 177 °C (350 °F). (d) Mechanical property data derived from laboratory-prepared test specimens sintered under commercial manufacturing conditions. (e) Yield and ultimate tensile strength are approximately the same for heat treated materials. ND, not determined. Source: MPIF Std 35, 2007 Edition

Appendix 3: Brief Glossary of Terms*

A

absolute pore size. The maximum pore opening of a porous material, such as a filter, through which no large particle will pass.

accelerated corrosion test. Method designed to approximate, in a short time, the deteriorating effect under normal long-term service conditions.

acicular powder. A powder composed of needle or sliverlike particles.

activated sintering. A sintering process during which the rate of sintering is increased, for example, by addition of a substance to the powder or by changing sintering conditions.

activation. The enlargement of the surface free energy or lattice binding energy of a solid. Also, the changing of a chemically passive surface of a metal to a chemically active state. Contrast with *passivation.*

activation energy. Generally, the energy required for initiating a chemical reaction or physical process such as diffusion or plastic flow.

activator. The additive used in activated sintering, also called a dopant.

active potential. The potential of a corroding material.

adhesion. The force of attraction between the atoms or molecules of two different phases.

aeration. (1) Exposing to the action of air. (2) Causing air to bubble through. (3) Introducing air into a solution by spraying, stirring, or a similar method. (4) Supplying or infusing with air, as in sand or soil.

agitator. A device to intensify mixing. Example: a high-speed stirrer or paddle in a blender or drum of a mill.

agglomerate (noun). Several particles adhering together.

agglomerate (verb). To develop an adherent cluster of particles.

aggregate (noun). A mass of particles.

aggregate (verb). To create a mass of particles. See *agglomerate.*

air classification. The separation of a powder into particle size ranges by means of an air stream of controlled velocity.

alloy powder, alloyed powder. A metal powder consisting of at least two constituents that are partially or completely alloyed with each other.

amorphous powder. A powder that consists of particles that are substantially noncrystalline in character.

angle of repose. The angular contour that a powder pile assumes.

annealed powder. A powder that is heat treated to render it soft and compactible.

anodic polarization. The change of the electrode potential in the noble (positive) direction due to current flow. See also *polarization.*

antechamber. The entrance vestibule of a continuously operating sintering furnace.

aperture size. The opening of a mesh, as in a sieve.

apparent density. The weight of a unit volume of powder, usually expressed as grams per cubic centimeter, determined by a specified method.

apparent hardness. The value obtained by testing a sintered object with standard indentation hardness equipment. Because the reading is a composite of pores and solid material, it is usually lower than that of a wrought or cast material of the same composition and condition. Not to be confused with particle hardness.

apparent pore volume. The total pore volume of a loose powder mass or a green compact. It may be calculated by subtracting the apparent density from the theoretical density of the substance.

atomization. The dispersion of a molten metal into particles by a rapidly moving gas or liquid stream or by other means.

atomized metal powder. Metal powder produced by the dispersion of a molten metal by a rapidly moving gas or liquid stream, or by mechanical dispersion.

---

* Adapted from "Terms and Definitions," *Powder Metallurgy*, Vol 7, *Metals Handbook*, 9th ed., American Society for Metals, 1984

automatic press. A self-acting machine for repeated compacting, sizing, or coining. See *press.*

average density. The density measured on an entire body or on a major number of its parts whose measurements are then averaged.

axial loading. The application of pressure on a powder or compact in the direction of the press axis.

B

bake (verb). (1) To remove gases from a powder at low temperatures. (2) To heat treat a compacted powder mixture of a metal and polymer at the curing temperature.

ball mill. A machine in which powders are blended or mixed by ball milling.

ball milling. Grinding, blending, or mixing in a receptacle of rotational symmetry that contains balls of a metal or nonmetal harder than the material being milled.

batch. (1) The total output of one mixing, some times called a lot. (2) The tray or basket of compacts placed in a sintering furnace.

batch sintering. Presintering or sintering in such a manner that compacts are sintered and removed from the furnace before additional unsintered compacts are placed in the furnace.

billet. A compact, green or sintered, that will be further worked by forging, rolling, or extrusion; sometimes called an ingot.

binder (noun). A cementing medium; either a material added to the powder to increase the green strength of the compact, and which is expelled during sintering, or a material (usually of lower melting point) added to a powder mixture for the specific purpose of cementing together powder particles that alone would not sinter into a strong body.

binder metal. A metal used as a binder.

binder phase. The soft metallic phase that cements the carbide particles in cemented carbides. More generally, a phase in a heterogeneous sintered material that gives solid coherence to the other phase(s) present.

binder removal. The chemical or thermal extraction of the binder from a compact.

blank. A pressed, presintered, or fully sintered compact, usually in the unfinished condition, to be machined or otherwise processed to final shape or condition.

blend (noun). Thoroughly intermingled powders of the same nominal composition.

blend, blending (verb). The thorough intermingling of powder fractions of the same nominal composition to adjust physical characteristics.

blistering. The formation of surface bubbles on the compact during sintering, caused by dynamic evolution of air or gases hold the green compacts during passage through a continuous sintering furnace.

bonding. The joining of compacted or loose particles into a continuous mass under the influence of heat.

breakdown potential. The last noble potential where *pitting* or *crevice corrosion,* or both, will initiate and propagate.

bridging. The formation of arched cavities or pores in a loose or compacted powder mass.

briquet(te). A self-sustaining mass of powder of defined shape. See preferred term *compact* (noun).

buffer gas. A protective gas curtain at the charge or discharge end of a continuously operating sintering furnace.

bulk density. Powder in a container or bin expressed in mass per unit volume.

bulk volume. The volume of the powder fill in the die cavity.

burnoff. The removal of additives (binder or lubricant) by heating.

burr. An edge protrusion on a pressed compact or a coined part caused by plastic flow of metal into the clearance space between a punch and a die cavity. Synonymous with *flash.*

C

cake. A coalesced mass of unpressed metal powder.

capillary attraction. The driving force for the infiltration of the pores of a sintered compact by a liquid.

carbonyl powder. Powders prepared by the thermal decomposition of a metal carbonyl compound such as nickel tetracarbonyl $Ni(CO)_4$ or iron pentacarbonyl $Fe(CO)_5$.

charge. The powder fed into a die for compacting.

chemical decomposition. The separating of a compound into its constituents.

chemical deposition. The precipitation of a metal from a solution of its salt by the addition to the solution of another metal or a reagent.

chemically precipitated powder. A metal powder that is produced as a fine precipitate by chemical displacement.

chemical vapor deposition. The precipitation of a metal from a gaseous compound onto a solid or particulate substrate. Also known as CVD.

CIP. The acronym representing the words cold isostatic pressing.

classification, classifying, classify. Separation of a powder into fractions according to particle size.

closed pore. A pore completely surrounded by solid material and inaccessible from the surface of the body.

cloth. Metallic or nonmetallic screen or fabric used for screening or classifying powders.

coarse fraction. The large particles in a powder spectrum.

cold compacting. See preferred term *cold pressing.*

cold isostatic pressing. The pressing of a powder, compact, or sintered object at ambient temperature by nominally equal pressure from every direction.

cold pressing. The forming of a compact at or below room temperature.

cold welding. Cohesion between two surfaces of metal, generally under the influence of externally applied pressure at room temperature.

compact (noun). The object produced by compression of metal powder, generally while confined in a die.

compact, compacting, compaction (verb). The operation or process of producing a compact; sometimes called pressing.

compactibility. The ability of powder to be consolidated into a usable green compact. A conceptual term related to the powder characteristics of compressibility, green strength, and edge retention.

compacting crack. A crack in a compact that is generated during the major phases of the pressing cycle, such as load application, load release, and ejection.

compressibility. (1) The ability of a powder to be formed into a compact having well-defined contours and structural stability at a given temperature and pressure; a measure of the plasticity of powder particles. (2) A density ratio determined under definite testing conditions. Also referred to as compactibility.

compressibility curve. A plot of the green density of a compact with increasing pressure.

compressibility test. A test to determine the behavior of a powder under applied pressure. It tells of the degree of densification and cohesiveness of a compact as a function of the magnitude of the pressure.

compression crack. See *compacting crack.*

compression ratio. The ratio of the volume of the loose powder in a die to the volume of the compact made from it.

concentration cell. An electrolytic cell, the electromotive force of which is caused by a difference in concentration of some component in the electrolyte. This difference leads to the formation of discrete cathode and anode regions.

concentration polarization. That portion of the polarization of a cell produced by concentration changes resulting from passage of current through the electrolyte.

corrosion fatigue. The process in which a metal fractures prematurely under conditions of simultaneous corrosion and repeated cyclic loading at lower stress levels or fewer cycles than would be required in the absence of the corrosive environment.

corrosion potential (E_{corr}). The potential of a corroding surface in an electrolyte, relative to a reference electrode. Also called rest potential, open-circuit potential, or freely corroding potential.

corrosion rate. Corrosion effect on a metal per unit of time. The type of corrosion rate used depends on the technical system and on the type of corrosion effect. Thus, corrosion rate may be expressed as an increase in corrosion depth per unit of time (penetration rate, for example, mils/yr) or the mass of metal turned into corrosion products per unit area of surface per unit of time (weight loss, for example, $g/m^2/yr$). The corrosion effect may vary with time and may not be the same at all points of the corroding surface. Therefore, reports of corrosion rates should be accompanied by information on the type, time dependency, and location of the corrosion effect.

crevice corrosion. Localized corrosion of a metal surface at, or immediately adjacent to, an area that is shielded from full exposure to the environment because of close proximity between the metal and the surface of another material.

cross-product contamination. The unintentional mixing of powders with distinct differences in either physical characteristics or chemical composition.

D

decomposition. Separation of a compound into its chemical elements or components.

degassing. Specifically, the removal of gases from a powder by a vacuum treatment at ambient or at elevated temperature.

delube. The removal of a lubricant from a powder compact, usually by burnout, or alternatively by treatment with a chemical solvent.

demixing. (1) The undesirable separation of one or more constituents of a powder mixture. (2) Segregation due to overmixing.

dendritic powder. Particles, usually of electrolytic origin, typically having the appearance of a pine tree.

density, absolute. (1) The ratio of the mass of a volume of solid material to the same volume of water. (2) The mass per unit volume of a solid material expressed in grams per cubic centimeter.

density, dry. The mass per unit volume of an unimpregnated sintered part.

density ratio. The ratio of the determined density of a compact to the absolute density of metal of the same composition, usually expressed as a percentage. Also referred to as percent theoretical density.

density, wet. The mass per unit volume of a sintered part impregnated with oil or other nonmetallic material.

dewaxing. The removal of wax from a powder compact by treatment with a chemical solvent or by burnout.

dewpoint. The temperature at which water vapor begins to condense. An index of water vapor content in a gas. Example: –40 °C (–40 °F) dewpoint contains 0.02% water vapor by volume.

diametrical strength. A property that is calculated from the load required to crush a cylindrical sintered test specimen in the direction perpendicular to the axis.

die. The part or parts making up the confining form in which a powder is pressed or a sintered compact is re-pressed or coined. The term is often used to mean a die assembly.

die barrel. A tubular liner for a die cavity.

die body. The stationary or fixed part of a die assembly.

die bolster. The external steel ring that is shrunk fit around the hard parts comprising the die barrel.

die breakthrough. The bursting of the die.

die cavity. That portion of the die body in which the powder is compacted or the sintered compact is re-pressed or coined.

die fill. A die cavity filled with powder.

die insert. A removable liner or part of a die body or punch.

die liner. A thin, usually hard and wear-resistant lining of the die cavity, such as produced by hard chromium plating. It is usually thinner than a die insert.

die lubricant. A lubricant applied to a punch or the walls of a die cavity to minimize die-wall friction and to facilitate pressing and ejection of the compact.

die opening. Entrance to the die cavity.

die plate. The base plate of a press into which the die is sunk.

die set. (1) The aligned mountings onto which punch and die assemblies are secured. (2) The die system ready to install in the press.

die volume. See preferred term *fill volume.*

die-wall lubricant. Synonomous with *die lubricant.*

diffusion. The movement of atoms within a substances, usually from an area of high constituent concentration to an area of low constituent composition, in order to achieve uniformity.

diffusion-alloyed powder. Partially alloyed powder produced by means of a diffusion anneal.

diffusion porosity. The porosity that is caused by the diffusion of one metal into another during sintering of an alloy. Also known as Kirkendall porosity.

dimensional change. Object shrinkage or growth resulting from sintering.

disintegration. Reduction of massive material to powder.

dispersing agent. A substance that increases the stability of a suspension of particles in a liquid medium by deflocculation of the primary particles.

dispersion strengthening. The strengthening of a metal or alloy by incorporating chemically stable submicron-sized particles of a non-metallic phase that impede dislocation movement at elevated temperature.

dissociated ammonia. A frequently used sintering atmosphere. Also reffered to as cracked gas.

distribution contour. The shape of the particle size distribution curve.

double-action press. A press that provides pressure from two sides, usually opposite each other, such as from top and bottom.

double pressing. A method whereby compaction is carried out in two steps. It may involve removal of the compact from the die after the first pressing for the purpose of storage, drying, baking, presintering, sintering, or other treatment, before reinserting into a die for the second pressing.

double sintering. A method consisting of two separate sintering operations with a shape change by machining or coining performed in between.

drum test. A test of the green strength of compacts by tumbling them in a drum and examining the sharpness of the edges and corners.

E

edge stability. An indicator of strength in a green compact, as may be determined by tumbling in a drum (Referred to as drum test).

edge strength. The resistance of the sharp edges of a compact against abrasion, as may be determined by tumbling in a drum. See *drum test.*

ejection. Removal of the compact after completion of pressing, whereby the compact is pushed through the die cavity by one of the punches. Also called *knockout.*

electrolytic powder. Powder produced by electrolytic deposition or by pulverizing of an electrodeposit.

elutriation. A test for particle size in which the speed of a liquid or gas is used to suspend particles of a desired size, with larger sizes settling for removal and weighing, while smaller sizes are removed, collected, and weighed at certain time intervals.

endothermic atmosphere. A gas mixture produced by the partial combustion of a hydrocarbon gas with air in an endothermic reaction. Also known as endogas.

exfoliation. The spallation of a face layer of a compact, usually the result of air entrapment or faulty pressing technique.

exothermic atmosphere. A gas mixture produced by the partial combustion of a hydrocarbon gas with air in an exothermic reaction. Also known as exogas.

exudation. The action by which all or a portion of the low-melting constituent of a compact is forced to the surface during sintering; sometimes referred to as bleedout or sweating.

F

feedstock. A moldable mixture of powder and binder (for metal injection molding)

fill density. See preferred term *apparent density.*

fill depth. Synonymous with *fill height.*

fill factor. The quotient of the fill volume of a powder over the volume of the green compact after ejection from the die. It is the same as the quotient of the powder fill height over the height of the compact. Inverse parameter of *compression ratio.*

fill height. The distance between the lower punch face and the top plane of the die body in the fill position of the press tool.

fill position. The position of the press tool that enables the filling of the desired amount of powder into the die cavity.

fill ratio. See *compression ratio.*

fill volume. The volume that a powder fills after flowing loosely into a space that is open at the top, such as a die cavity or a measuring receptacle.

filter. Metal filters are porous products made from either wires and fibers or sintered powders.

final density. The density of a sintered or repressed product.

fines. The portion of a powder composed of particles smaller than a specified size, usually 44 μm (−325 mesh).

flake powder. Flat or scalelike particles whose thickness is small compared to the other dimensions.

flash. Excess metal forced out between the punches and die cavity wall during compacting or coining. See also *burr.*

floating die. A die body that is suspended on springs or an air cushion, which causes the die to move together with the upper punch over a stationary lower punch; the rate of die movement is lower than that of the upper punch and is a function of the friction coefficient of the powder in relation to the wall of the die cavity.

floating die pressing. The compaction of a powder in a floating die, resulting in densification at opposite ends of the compact. Analogous to double-action pressing.

flow factor. See preferred term *flow rate.*

flow meter. A metal cylinder whose interior is funnel shaped and whose bottom has a calibrated orifice of standard dimensions to permit passage of a powder and the determination of the flow rate.

flow rate. The time required for a powder sample of standard weight to flow through an orifice in a standard instrument according to a specified procedure.

fraction. That portion of a powder sample which lies between two stated particle sizes.

fragmentation. The process of breaking a solid into finely divided pieces.

fugitive binder. An organic substance added to a metal powder to enhance the bond between the particles during compaction and thereby increase the green strength of the compact, and which decomposes during the early stages of the sintering cycle.

fully dense material. A material completely free of porosity and voids.

G

galvanic corrosion. Accelerated corrosion of a metal because of an electrical contact with a more noble metal or nonmetallic conductor in a corrosive electrolyte.

gas classification. The separation of a powder into its particle size fractions by means of a gas stream of controlled velocity flowing counterstream to the gravity-induced fall of the particles. The method is used to classify submesh-sized particles.

granular powder. A powder having equidimensional but nonspherical particles.

granulation, granulating. The production of coarse metal particles by pouring molten metal through a screen into water or by agitating the molten metal violently while it is solidifying.

green. Unsintered (not sintered).

green compact. An unsintered compact.

green density. The density of a green compact.

green strength. The ability of a green compact to maintain size and shape during handling and storage prior to sintering.

grit, grit size. The particle size of an abrasive powder, such as carborundum, corundum, silicon carbide, or diamond, used in cutting and machining operations.

growth. An increase in compact or part size as a result of excessive pore formation during sintering.

H

HIP. The acronym representing the words hot isostatic pressing.

hot densification. Rapid deformation of a heated powder preform in a die assembly for the purpose of reducing porosity. Metal is usually deformed in the direction of the punch travel. See *hot pressing*.

hot isostatic pressing. A process for simultaneously heating and forming a compact in which the powder is contained in a sealed flexible sheet metal or glass enclosure and the so-contained powder is subjected to equal pressure from all directions at a temperature high enough to permit plastic deformation and sintering to take place.

hot pressing. Simultaneous heating and forming of a compact.

hydride powder. A powder produced by removal of the hydrogen from a metal hydride.

hydrogen loss. The loss in weight of metal powder or a compact caused by heating a representative sample according to a specified procedure in a purified hydrogen atmosphere. Broadly, a measure of the oxygen content of the sample when applied to materials containing only such oxides as are reducible with hydrogen and no hydride-forming element.

hydrostatic compacting. See *hydrostatic pressing*.

hydrostatic mold. A sealed flexible mold made of rubber, a polymer, or pliable sheet made from a low-melting metal such as aluminum.

hydrostatic pressing. A special case of isostatic pressing that uses a liquid such as water or oil as a pressure-transducing medium and is therefore limited to near-room-temperature operation.

I

impact sintering. An instantaneous sintering process during high-energy-rate compacting that causes localized heating, welding, or fusion at the particle contacts.

impregnation. The process of filling the pores of a sintered compact with a nonmetallic material such as oil, wax, or resin.

infiltrant. Material used to infiltrate a porous sinter. The infiltrant as positioned on the compact is called a slug.

infiltration. The process of filling the pores of a sintered or unsintered compact with a metal or alloy of lower melting temperature.

injection molding. A process similar to plastic injection molding using a plastic-coated metal powder.

intercommunicating porosity. See preferred term *interconnected porosity*.

interconnected pore volume. The volume fraction of pores that are interconnected within the entire pore system of a compact or sintered product.

interconnected porosity. A network of connecting pores in a sintered object that permits a fluid or gas to pass through the object. Also referred to as interlocking or open porosity.

interface. A surface that forms the boundary between phases in a sintered compact.

interface activity. A measure of the chemical poential between the contacting surfaces of two particles in a compact or two grains in a sintered body.

intergranular corrosion. Corrosion occurring preferentially at grain boundaries, usually with slight or negligible attack on the adjacent grains. Also called intercrystalline corrosion.

internal oxidation. The preferential in situ oxidation of certain constituents or phases within the bulk of a solid alloy, accomplished by diffusion of oxygen into the body. The process is suitable for the production of dispersion-strengthened alloys, if the constituent to be oxidized forms a stable oxide and the major alloy component permits a high rate of oxygen diffusion.

irregular powder. Particles lacking symmetry.

isostatic mold. A sealed container of glass or sheet of carbon steel, stainless steel, or a nickel-base alloy. See *isostatic pressing*.

isostatic pressing. Cold or hot pressing of a powder using equal pressure from all directions.

K

keying. The deformation of metal particles during compacting to increase interlocking and bonding.

knockout (verb). Ejecting of a compact from a die cavity.

knockout punch. A punch used for ejecting compacts.

L

lamination. (1) A discontinuity, crack, or separation in a plane perpendicular to the axis of applied pressure that may be the result of air entrapment or misalignment of the pressing tools during compacting. (2) A thin compressed or rolled powder product with two or more layers.

linear shrinkage. The shrinkage in one dimension of a compact during sintering. Contrast with *volume shrinkage*.

liquid disintegration. The process of producing powders by pouring molten metal on a rotating surface.

liquid-phase sintering. Sintering of a compact or loose powder aggregate under conditions where a liquid phase is present during part of the sintering cycle.

loose powder. Uncompacted powder.

loose powder sintering. Sintering of uncompacted powder using no external pressure.

lubricant. A substance mixed with a powder to facilitate compacting and subsequent mold ejection of compact; often a stearate or a proprietary wax. It may also be applied as a film to the surfaces of the punches or the die cavity wall, such as by spray coating.

lubricating. Mixing or incorporating a lubricant with a powder to facilitate compacting and ejecting of the compact from the die cavity; also, applying a lubricant to die walls and/or punch surfaces.

M

macropore. Pores in pressed or sintered compacts that are visible with the naked eye.

master alloy powder. A prealloyed powder of high concentration of alloy content, designed to be diluted when mixed with a base powder to produce the desired composition.

matrix metal. The continuous phase of a polyphase alloy or mechanical mixture: the physically continuous metallic constituent in which separate particles of another constituent are embedded.

mesh size. The width of the aperture in a cloth or wire screen.

metal injection molding (MIM). A process in which feedstock (of powder in a binder) is forced under pressure into a die.

metal powder. Elemental metals or alloy particles, usually in the size range of 0.1 to 1000 μm.

micromesh. A sieve with precisely square openings in the range of 10 to 120 μm produced by electroforming.

micromesh sizing. The process of sizing micromesh particles using an air or liquid suspension process.

micropore. The pores in a sintered product that can only be detected under a microscope.

milling. The mechanical comminution of a metal powder or a metal powder mixture, usually in a ball mill, to alter the size or shape of the individual particles, to coat one component of a mixture with another, or to create uniform distribution of components.

milling fluid, milling liquid. An organic liquid, such as hexane, in which ball milling is carried out. The liquid serves to reduce the heat of friction and resulting surface oxidation of the particles during grinding, and to provide protection from other surface contamination.

mill scale powder. Pulverized iron oxide scale that is a by-product of hot rolling of steel. The material is readily reduced to a soft, spongy iron powder free of mineral inclusions and other solid impurities.

mixed powder. Powder made by mixing two or more powders as uniformly as possible.

molding. See *compact, compacting, compaction.*

multiple-die pressing (verb). The simultaneous compaction of powder into several identical parts with a press tool consisting of a number of components.

multiple-punch press. A mechanical or hydraulic press that actuates several punches individually and independently of each other.

N

nitrogen alloying. Alloying by transfer of nitrogen from a furnace atmosphere to a powder or powder metallurgy part.

nodular powder. Irregular powder having knotted, rounded, or similar shapes.

O

open-circuit potential. The potential of an electrode measured with respect to a reference electrode or another electrode when no current flows to or from it.

open pore. A pore open to the surface of a compact. See *intercommunicating porosity.*

open porosity. See *interconnected porosity.*

overfill. The fill of a die cavity with an amount of powder in excess of specification.

overmix (verb). Mixing of a powder longer than necessary to produce adequate distribution of powder particles. Overmixing may cause particle size segregation.

oversinter (verb). The sintering of a compact at higher temperature or for longer time periods than necessary to obtain the desired microstructure or physical properties.

oversize powder. Powder particles larger than the maximum permitted by a particle size specification.

overvoltage. The difference between the actual electrode potential when appreciable electrolysis begins and the reversible electrode potential.

oxide network. Continuous or discontinuous oxides that follow prior-particle boundaries.

P

partially alloyed powder. Powder in which the alloy additions are metallurgically bonded to elemental or prealloyed powders.

particle morphology. The form and structure of an individual particle.

particle shape. The appearance of a metal particle, such as spherical, rounded, angular, acicular, dendritic, irregular, porous, fragmented, blocky, rod, flake, nodular, or plate.

particle size. The controlling linear dimension of an individual particle as determined by analysis with screens or other suitable instruments.

particle size distribution. The percentage by mass, by numbers, or by volume of each fraction into which a powder sample has been classified with respect to size.

passivation. (1) A reduction of the anodic reaction rate of an electrode involved in corrosion. (2) The process in metal corrosion by which metals become passive. (3) The changing of a chemically active surface of a metal to a much less reactive state. Contrast with *activation.*

PIM. The acronym representing the words powder injection molding. See also *metal injection molding.*

pitting. Localized corrosion of a metal surface, confined to a point or small area, that takes the form of cavities.

plasticizer. A substance added to a powder or powder mixture to render it more formable during cold pressing or extrusion.

polarization. (1) The change from the open-circuit electrode potential as the result of the passage of current. (2) A change in the potential of an electrode during electrolysis, such that the potential of an anode becomes more noble, and that of a cathode more active than their respective reversible potentials. Often accomplished by formation of a film on the electrode surface.

polarization curve. A plot of current density versus electrode potential for a specific electrode-electrolyte combination.

pore. An inherent or induced cavity (void) within a particle or within an object.

porosity. The amount of pores (void) expressed as a percentage of the total volume of the powder metallurgy part.

powder. An aggregate of discrete particles that are usually in the size range of 1 to 1000 µm.

powder designation. A code number identifying a specific powder.

powder fill. The filling of a die cavity with powder.

powder flow meter. An instrument for measuring the rate of powder flow.

powder forging. Not densification by forging of an unsintered, presintered, or sintered preform.

powder injection molding (PIM). See *metal injection molding.*

powder lubricant. An agent or component incorporated into a mixture to facilitate compacting and ejecting of the compact from its mold.

powder metallurgy. The technology and art of producing metal powders and of the utilization of metal powders for the production of massive materials and shaped objects.

powder metallurgy part. A shaped object that has been formed from metal powders and sintered by heating below the melting point of the major constituent. A structural or mechanical component made by the powder metallurgy process.

prealloyed powder. A metallic powder of two or more elements that are alloyed in the powder manufacturing process and in which the particles are of the same nominal composition throughout.

preform. The initially pressed compact to be subjected to re-pressing or forging.

premix (noun). A uniform mixture of components prepared by a powder producer for direct use in compacting.

premix (verb). A term sometimes applied to the preparation of a premix.

presintering. Heating a compact to a temperature below the final sintering temperature, usually to increase the ease of handling or shaping of a compact or to remove a lubricant or binder (burnoff) prior to sintering.

press (noun). The machine used for compacting, sizing, or coining. Presses may be mechanical:

eccentric, crank, cam, toggle, knuckle joint, rotary (table); or hydraulic: single action, multiple action, double action; or combination mechanical-hydraulic.

pressed density. The weight per unit volume of an unsintered compact. Same as green density.

protective atmosphere, protective gas. The atmosphere in the sintering furnace designed to protect the compacts from oxidation, nitridation, or other contamination from the environment.

protective potential. The threshold value of the *corrosion potential* that has to be reached to enter a protective potential range.

R

reaction sintering. The sintering of a powder mixture consisting of at least two components that chemically react during the treatment.

reduced powder. Generic term for any metal or nonmetal powder produced by the reduction of an oxide, hydroxide, carbonate, oxalate, or other compound without melting.

reduction of oxide. The process of converting a metal oxide to metal by applying sufficient heat in the presence of a solid or gaseous material, such as hydrogen, having a greater attraction for the oxygen than does the metal.

reduction ratio. (1) The quotient of the reduced oxygen content into the total initial oxygen content of a powder. (2) The quotient of the reduced cross section into the original cross section in metal working such as extrusion; an indication of the degree of plastic deformation.

refractory metal. A metal characterized by its high melting temperature, generally above 2000 °C (3600 °F).

re-pressing. The application of pressure to a sintered compact, usually for the purpose of improving a physical or a mechanical property or for dimensional accuracy.

resintering. (1) A second sintering operation. (2) Sintering a re-pressed compact.

restrike. Additional compacting of a sintered compact.

S

screen. The woven wire or fabric cloth, having uniformly sized openings, used in a sieve for retaining particles greater than the particular mesh size. The U.S. standard, ISO, or Tyler screen sizes are commonly used.

screen analysis. See *sieve analysis.*

screen classification. See *sieve classification.*

screening. Separation of a powder according to particle size by passing it through a screen having the desired mesh size.

secondary operation. Any operation performed a sintered compact, such as sizing, coining, re-pressing, impregnation, infiltration, heat or steam treatment, machining, joining, plating, or other surface treatment.

shrinkage. The decrease in dimensions of a compact occurring during sintering.

sieve. A powder separator using a set of graduated mesh size screens.

sieve analysis. Particle size distribution, usually expressed as the weight percentage retained on each of a series of standard screens of decreasing mesh size and the percentage passed by the screen of finest size; also called screen classification.

sieve classification. The separation of powder into particle size ranges by the use of a series of graded sieves. Also called screen analysis.

sieve fraction. That portion of a powder sample that passes through a sieve of specified number and is retained by some finer mesh sieve of specified number.

sinter, sintering (verb). The bonding of adjacent particles in a powder mass or compact by heating to a temperature below the melting point of the main constituent.

sinter, sinterings (noun). See preferred term *powder metallurgy part.*

sintered density. The quotient of the mass (weight) over the volume of the sintered body expressed in grams per cubic centimeter.

sintered density ratio. The ratio of the density of the sintered body to the solid, pore-free body of the same composition or theoretical density.

sintering atmosphere. See *protective atmosphere.*

sintering cycle. A predetermined and closely controlled time-temperature regime for sintering compacts, including the heating and cooling phases.

sintering temperature. The maximum temperature at which the compact is sintered. The temperature is either measured directly on the surface of the body by optical pyrometer, or indirectly by thermocouples installed in the furnace chamber.

sintering time. The time period during which the compact is at sintering temperature.

size fraction. A separated fraction of a powder whose particles lie between specified upper and lower size limits.

sizing. The pressing of a sintered compact to secure a desired dimension.

sizing die. A die used for the sizing of a sintered compact.

solid-state sintering. A sintering procedure for compacts or loose powder aggregates during which no component melts. Contrast with *liquid-phase sintering* hot pressing method that provides for the surface activation of the powder particles by electric discharges generated by a high alternating current applied during the early stage of the consolidation process.

specific gravity. The ratio of the density of a material to the density of some standard material, such as water at a specified temperature, or (for gases) air at standard conditions of pressure and temperature. Also referred to as relative density.

specific surface area. The surface area of a powder expressed in square centimeters per gram of powder or square meters per kilogram of powder.

spherical powder. A powder consisting of ball-shaped particles.

spheroidal powder. A powder consisting of oval or rounded particles.

subsieve analysis. Size distribution of particles that will pass through a standard 325-mesh sieve having 44 μm openings.

subsieve fraction. Particles that will pass through a 44 μm (325-mesh) screen.

subsieve size. See preferred term *subsieve fraction.*

superfines. The portion of a powder composed of particles smaller than a specified size, currently less than 10 μm.

T

tap density. The density of a powder when the volume receptacle is tapped or vibrated under specified conditions while being loaded.

theoretical density. The density of the same material in the wrought condition. See *density, absolute.*

transpassive region. The region of an *anodic polarization* curve, noble to and above the passive potential range, in which there is a significant increase in current density (increased metal dissolution) as the potential becomes more positive (noble).

transverse-rupture strength. The stress, as calculated from the flexure formula, required to break a specimen as a simple beam supported near the ends while applying the load midway between the centerlines of the supports.

triple-action press. A press that provides pressure from three sides, such as from top and bottom and from one side, either to impress a side indentation or recess, or to clamp a segmented die on top of a press table.

U

undersized powder. Powder particles smaller than the minimum permitted by a particle size specification.

uniaxial compacting. Compacting of powder along one axis, either in one direction or in two opposing directions. Contrast with *isostatic pressing.*

unidirectional compacting. Compacting of powder in one direction.

uniform corrosion. (1) A type of corrosion attack (deterioration) uniformly distributed over a metal surface. (2) Corrosion that proceeds at approximately the same rate over a metal surface. Also called general corrosion.

upset pressing. The pressing of a powder compact in several stages, which results in an increase in the cross-sectional area of the part prior to its ejection.

V

vacuum sintering. Sintering at subatmospheric pressure, such as in a technical vacuum or in a high vacuum.

void. See preferred term *pore.*

volume filling. Filling the volume of a die cavity or receptacle with loose powder, and striking off any excess amount.

volume fraction. The volume percentage of a constituent or of porosity in a sintered body. Example: the amount of a refractory oxide phase in a dispersion-strengthened alloy.

volume ratio. The volume percentage of solid in the total volume of the sintered body.

volume shrinkage. The volumetric size reduction a compact undergoes during sintering. Contrast with *linear shrinkage.*

W

warpage. The distortion that occurs in a compact during sintering.

SOURCES

- *ASM Materials Engineering Dictionary,* ASM International, 1992
- *"Definitions and Terms,"* MPIF 9-71, Metal Powder Industries Federation, Princeton, NJ
- *"Glossary of Terms Relating to Powders,"* British Standard 2955, British Standards Institution
- *International Powder Metallurgy Glossary, Metal Powder Report,* MPR Publishing Services, Shrewsbury, England
- *"Powder Metallurgical Materials and Products—Vocabulary,"* ISO/DIS 3252, International Organization for Standardization
- *"Standard Definitions of Terms Used in Powder Metallurgy,"* B 243, ASTM International
- Terms and Definitions, *Powder Metallurgy,* Vol 7, *Metals Handbook,* 9th ed., American Society for Metals, 1984

Index